Introduction to Continuum Mechanics for Engineers

Introduction to Continuum Mechanics for Engineers

REVISED EDITION

RAY M. BOWEN
Professor of Mechanical Engineering
President Emeritus
Texas A&M University

DOVER PUBLICATIONS, INC.
Mineola, New York

Bibliographical Note

This Dover edition, first published in 2009, is an unabridged republication of the 2007 revised edition of the work originally published in 1989 by Plenum Press, New York, as Volume 39 of *Mathematical Concepts and Methods in Science and Engineering*.

Library of Congress Cataloging-in-Publication Data

Bowen, Ray M., 1936–
Introduction to continuum mechanics for engineers / Ray M Bowen. — Rev. ed.
p. cm.
Includes bibliographical references and index.
ISBN-13: 978-0-486-47460-1
ISBN-10: 0-486-47460-7
1. Continuum mechanics. I. Title.

QA808.2.B68 2009
531—dc22

2009024685

Manufactured in the United States by Courier Corporation
47460701
www.doverpublications.com

Preface (First Edition)

This textbook is intended to introduce engineering graduate students to the essentials of modern Continuum Mechanics. The objective of an introductory course is to establish certain classical continuum models within a modern framework. Engineering students need a firm understanding of classical models such as the linear viscous fluids (Navier-Stokes theory) and infinitesimal elasticity. This understanding should include an appreciation for the status of the classical theories as special cases of general nonlinear continuum models. The relationship of the classical theories to nonlinear models is essential in light of the increasing reliance, by engineering designers and researchers, on prepackaged computer codes. These codes are based upon models which have a specific and limited range of validity. Given the danger associated with the use of these computer codes in circumstances where the model is not valid, engineers have a need for an in depth understanding of continuum mechanics and the continuum models which can be formulated by use of continuum mechanics techniques.

Classical continuum models and others involve a utilization of the balance equations of continuum mechanics, the second law of thermodynamics, the principles of material frame-indifference and material symmetry. In addition, they involve linearizations of various types. In this text, an effort is made to explain carefully how the governing principles, linearizations and other approximations combine to yield classical continuum models. A fundamental understanding of these models evolve is most helpful when one attempts to study models which account for a wider array of physical phenomena.

This book is organized in five chapters and two appendices. The first appendix contains virtually all of the mathematical background necessary to understand the text. The second appendix contains specialized results concerning representation theorems. Because many new engineering graduate students experience difficulties with the mathematical level of a modern continuum mechanics course, this text begins with a one dimensional overview. Classroom experience with this material has shown that such an overview is helpful to many students. Of course, more advanced students can proceed directly to the Chapter II. Chapter II is concerned with the kinematics of motion of a general continuum. Chapter III contains a discussion of the governing equations of balance and the entropy inequality for a continuum. The main portion of the text is contained in Chapter IV. This long chapter contains the complete formulation of various general continuum models. These formulations begin with general statements of constitutive equations followed by a systematic examination of these constitutive equations in light of the restrictions implied by the second law of thermodynamics, material frame-indifference and material symmetry. Chapter IV ends with an examination of the formal approximations necessary to specialize to the classical models mentioned above. So as to illustrate further applications of continuum mechanics, the final chapter contains an introductory discussion of materials with internal state variables.

The book is essentially self contained and should be suitable for self study. It contains approximately two hundred and eighty exercises and one hundred and seventy references. The references at the end of each chapter are divided into References and General References. The

References are citations which relate directly to the material covered in the proceeding chapter. The General References represent additional reading material which relate in a general way to the material in the chapter.

This text book evolved over an extended period of time. For a number of years, early versions of the manuscript were used at Rice University. I am indebted for the assistance my many students gave me as the lecture notes evolved into a draft manuscript. The final manuscript has been utilized at the University of Kentucky by my colleague, Professor Donald C. Leigh, in an introductory graduate course. I am indebted to him for his many comments and suggestions.

Ray M. Bowen
Lexington, Kentucky

Preface (2004 Revised Edition)

This electronic textbook is a revision to the textbook, Introduction to Continuum Mechanics which was published by Plenum Press in 1989. A small amount of new material has been added in Chapters 1, 3 and 4. In addition, an effort has been made to correct numerous typographical errors that appeared in the first edition. It is inevitable that other typographical errors creep into the manuscript when it is retyped. I hope there has been a net reduction in these kinds of errors from the first edition to this revised edition

I remain indebted to my colleagues that have pointed out errors over the years. A special mention needs to be made to my good friends Dr. C.-C. Wang of Rice University and Dr. Donald C. Leigh of the University of Kentucky. Not only were they kind enough to adopt the first edition as a textbook, they informed me of many corrections and improvements that could be made.

I am also indebted to my students at Texas A&M University that endured my teaching from the revised edition after being out of the classroom for many years.

It is my desire and intention that this revised textbook be made available for free to anyone that wishes to have a copy. For the immediate future, the access will be provided by posting it on the website of the Mechanical Engineering Department of Texas A&M University. I would appreciate being informed of any typographical and other errors that remain in the posted version.

Preface (2007 Revised Edition)

The 2007 revisions mainly involve the correction of typographical errors that have entered the text as it has been retyped and revised over the years. A small amount of new material has been added in Chapter 1. The index and Table of Contents have been revised to reflect the addition of this material. For those that acquire the text as a pdf file, the search utility within Adobe Acrobat provides an excellent alternative to the index.

Ray M. Bowen
rbowen@tamu.edu
College Station, Texas

Contents

1

One-Dimensional Continuum Mechanics

It is often not clear to engineering students that there is a common basis for their courses in thermodynamics, fluid mechanics and elasticity. The pace of most undergraduate curriculums is such that there is no opportunity to stress the common features of these courses. In addition, many undergraduate engineering students have limited skill with vector analysis and Cartesian tensor analysis. These problems make it awkward to teach a modern introductory course in Continuum Mechanics to first year engineering graduate students. Experience has shown that an elementary preview of the modern course can be a great asset to the student. This chapter contains such a preview. It is a brief survey of the elements of continuum mechanics presented for one dimensional continuous bodies. This survey allows the student to encounter a new notation and several new concepts without the problem of learning three dimensional vector and tensor analysis.

This chapter contains a development of the one dimensional forms of the equations of balance of mass, momentum and energy. The entropy inequality is presented, and it is utilized to derive the thermodynamic restrictions for a particular material model.

1.1. Kinematics of Motion and Strain

We shall denote by $\mathscr{B}$ the one dimensional body. The symbol X denotes an element or *particle* of the body $\mathscr{B}$. It is useful at this point not to distinguish between the body $\mathscr{B}$ and the portion of one dimensional space it occupies. Thus, X is a real number. It is customary to refer to X as the *material* or *Lagrangian* coordinate of the particle. The set of material coordinates is a subset of the real numbers called the *reference configuration*. If t denotes the time and $\mathscr{R}$ is the set of real numbers, then the *deformation function* is a function $\chi(\cdot,t):\mathscr{B}\rightarrow\mathscr{R}$ which, for each t, maps $\mathscr{B}$ into its *present configuration*. We write

$$x=\chi(X,t) \tag{1.1.1}$$

where x is the *spatial* position or coordinate of the particle X at the time t. The spatial coordinates are also called *Eulerian* coordinates. We shall assume that for each t, χ has an inverse χ^{-1} such that

$$X=\chi^{-1}(x,t) \tag{1.1.2}$$

Theses assumptions insure that X and x are in one to one correspondence for each t and are, in effect, a statement of permanence of matter. The particle X cannot break into two particles as a result of the deformation, and two particles X_1 and X_2 cannot occupy the same spatial position x at the same instant of time.

The *velocity* of X at time t is

$$\dot{x} = \frac{\partial \chi(X,t)}{\partial t} \tag{1.1.3}$$

The *acceleration* of X at the time t is

$$\ddot{x} = \frac{\partial^2 \chi(X,t)}{\partial t^2} \tag{1.1.4}$$

The *displacement* of X at the time t is

$$w = \chi(X,t) - X \tag{1.1.5}$$

Because of (1.1.2), we can regard $\dot{x}$, $\ddot{x}$ and w to be functions of (x,t) or (X,t). The pair (X,t) are called *material variables*, and the pair (x,t) are called *spatial variables*. Clearly, by use of (1.1.1) and (1.1.2) any function of one set of variables can be converted to a function of the other set.

If ψ is a function of (X,t), then its *material derivative*, written $\dot{\psi}$, is defined by

$$\dot{\psi} = \frac{\partial \psi(X,t)}{\partial t} \tag{1.1.6}$$

If the function $\widehat{\psi}$ of (x,t) is defined by

$$\widehat{\psi}(x,t) = \psi(\chi^{-1}(x,t),t) \tag{1.1.7}$$

an elementary application of the chain rule yields

$$\dot{\psi} = \frac{\partial \psi(X,t)}{\partial t} = \frac{\partial \widehat{\psi}(x,t)}{\partial t} + \frac{\partial \widehat{\psi}(x,t)}{\partial x}\dot{x} \tag{1.1.8}$$

Equation (1.1.8) gives the material derivative in terms of spatial derivatives. For notational simplicity, we shall write (1.1.8) as

$$\dot{\psi} = \frac{\partial \psi}{\partial t} + \frac{\partial \psi}{\partial x}\dot{x} \tag{1.1.9}$$

where it is understood that $\partial \psi / \partial t$ is computed at fixed x, and $\partial \psi / \partial x$ is computed at fixed t. As an illustration of (1.1.9), we can take $\psi = \dot{x}$ and obtain

$$\ddot{x} = \frac{\partial \dot{x}}{\partial t} + \frac{\partial \dot{x}}{\partial x}\dot{x} \tag{1.1.10}$$

The *deformation gradient* is defined by

$$F = \frac{\partial \chi(X,t)}{\partial X} \tag{1.1.11}$$

Since χ has an inverse, it is trivially true that $\chi(\chi^{-1}(x,t),t) = x$ and, thus,

$$\frac{\partial \chi(X,t)}{\partial X}\frac{\partial \chi^{-1}(x,t)}{\partial x} = 1 \tag{1.1.12}$$

Equation (1.1.12) shows that $F \neq 0$ and

$$F^{-1} = \frac{1}{F} = \frac{\partial \chi^{-1}(x,t)}{\partial x} \tag{1.1.13}$$

The *displacement gradient* is defined by

$$H = \frac{\partial w(X,t)}{\partial X} \tag{1.1.14}$$

It follows from (1.1.11), (1.1.14), and (1.1.5) that H and F are related by

$$H = F - 1 \tag{1.1.15}$$

If dX denotes a differential element of the body at X, then it follows from (1.1.1) and (1.1.11) that

$$dx = FdX \tag{1.1.16}$$

Equation (1.1.16) shows that if dX is a material element at X then $FdX = dx$ is the deformed element at x. Therefore, F measures the *local deformation* of material in the neighborhood of X.

If $\dot{x}$ is expressed as a function of (x,t), then

$$L = \frac{\partial \dot{x}}{\partial x} \tag{1.1.17}$$

is the *velocity gradient*. Because

$$\frac{\partial}{\partial X}\frac{\partial \chi(X,t)}{\partial t} = \frac{\partial \dot{x}}{\partial X} = \frac{\partial \dot{x}}{\partial x}\frac{\partial \chi(X,t)}{\partial X} \tag{1.1.18}$$

it follows from (1.1.11), (1.1.17) and (1.1.6) that

$$\dot{F} = LF \tag{1.1.19}$$

Exercise 1.1.1

Show that

$$\overline{F^{-1}}^{\cdot} = -F^{-1}L \tag{1.1.20}$$

Next we shall state and prove an important result known as *Reynold's theorem* or the *transport theorem*. Consider a fixed portion of the body in $X_1 \le X \le X_2$. The deformation function, for each t, deforms this portion into a region in $\mathscr{R}$. Without loss of generality, we can take this region to be $x_1 \le x \le x_2$, where $x_1 = \chi(X_1,t)$, $x_2 = \chi(X_2,t)$ and $x = \chi(X,t)$. Reynold's theorem states that if Ψ is a sufficiently smooth function of x and t, then

$$\overline{\int_{x_1}^{x_2} \Psi(x,t)dx}^{\cdot} = \int_{x_1}^{x_2} \frac{\partial \Psi(x,t)}{\partial t} dx + \Psi(x_2,t)\dot{x}(x_2,t) - \Psi(x_1,t)\dot{x}(x_1,t) \tag{1.1.21}$$

The derivation of (1.1.21) is best approached by utilizing a formalism which does not depend upon the use of a deformation function, and its associated motion. We first let $\Phi(x,t)$ be the indefinite integral, with respect to x, of a function $\Psi(x,t)$. In other words, let

$$\Phi(x,t) = \int \Psi(x,t)dx + \text{ constant} \tag{1.1.22}$$

It follows then that

$$\frac{\partial \Phi(x,t)}{\partial x} = \Psi(x,t) \tag{1.1.23}$$

and, for two points x_1 and x_2,

$$\int_{x_1}^{x_2} \Psi(x,t)dx = \Phi(x_2,t) - \Phi(x_1,t) \tag{1.1.24}$$

If we allow the limits in (1.1.24) to be functions of time, i.e.,

$$\int_{x_1(t)}^{x_2(t)} \Psi(x,t)dx = \Phi(x_2(t),t) - \Phi(x_1(t),t) \tag{1.1.25}$$

The total derivative of $\int_{x_1(t)}^{x_2(t)} \Psi(x,t)dx$ is

$$\begin{aligned}\frac{d}{dt}\int_{x_1(t)}^{x_2(t)}\Psi(x,t)dx &= \frac{d\Phi(x_2(t),t)}{dt}-\frac{d\Phi(x_1(t),t)}{dt}\\ &= \frac{\partial\Phi(x_2(t),t)}{\partial t}+\frac{\partial\Phi(x_2(t),t)}{\partial x}\frac{dx_2(t)}{dt}\\ &\quad -\frac{\partial\Phi(x_1(t),t)}{\partial t}-\frac{\partial\Phi(x_1(t),t)}{\partial x}\frac{dx_1(t)}{dt}\end{aligned} \tag{1.1.26}$$

If (1.1.23) and (1.1.24) are used, (1.1.26) can be written

$$\begin{aligned}\frac{d}{dt}\int_{x_1(t)}^{x_2(t)}\Psi(x,t)dx &= \frac{\partial\Phi(x_2(t),t)}{\partial t}+\Psi(x_2(t),t)\frac{dx_2(t)}{dt}-\frac{\partial\Phi(x_1(t),t)}{\partial t}-\Psi(x_1(t),t)\frac{dx_1(t)}{dt}\\ &= \frac{\partial}{\partial t}\int_{x_1(t)}^{x_2(t)}\Psi(x,t)dx+\Psi(x_2(t),t)\frac{dx_2(t)}{dt}-\Psi(x_1(t),t)\frac{dx_1(t)}{dt}\end{aligned} \tag{1.1.27}$$

The next formal step is to interchange the order of integration on x with the partial differentiation with respect to t. The result of this final step is

$$\frac{d}{dt}\int_{x_1(t)}^{x_2(t)}\Psi(x,t)dx = \int_{x_1(t)}^{x_2(t)}\frac{\partial\Psi(x,t)}{\partial t}dx+\Psi(x_2(t),t)\frac{dx_2(t)}{dt}-\Psi(x_1(t),t)\frac{dx_1(t)}{dt} \tag{1.1.28}$$

The result (1.1.28) is a mathematical identity that holds for the function Ψ of (x,t) and functions x_1 and x_2 of t that are sufficiently smooth that the various derivatives above exist. In calculus, it sometimes goes by the name Leibnitz's rule. It is one of the standard results one needs in applied mathematics for the differentiation of integrals. We have not been precise about the smoothness assumptions sufficient to give the result (1.1.28). More rigorous derivations of this result can be found in many Calculus textbooks. References 1 and 2 contain this derivation. In any case, we are interested in this result in the special case where the functions x_1 and x_2 of t are derived from the deformation function χ. In this case, (1.1.28) immediately becomes (1.1.21)

Exercise 1.1.2

Show that (1.1.21) can be derived by a change of variables in the integral $\int_{x_1=\chi(X_1,t)}^{x_2=\chi(X_2,t)}\Psi(x,t)dx$. The first step is to write

$$\int_{x_1}^{x_2}\Psi(x,t)dx = \int_{x_1=\chi(X_1,t)}^{x_2=\chi(X_2,t)}\Psi(x,t)dx = \int_{X_1}^{X_2}\Psi(\chi(X,t),t)FdX \tag{1.1.29}$$

Differentiate (1.1.29) and rearrange the result to obtain (1.1.21).

The name "transport theorem" is suggested by viewing the last two terms in (1.1.21) as the net transported out of the spatial region $x_1 < x < x_2$ by the motion of the material.

Equation (1.1.28) and its special case (1.1.21) assume that Φ is differentiable and, thus, from (1.1.23), Ψ continuous as a function of x in $x_1 < x < x_2$. Next we wish to remove the assumption that Ψ is continuous in $x_1 < x < x_2$ and derive a generalization of (1.1.21). The derivation simply utilizes the general result (1.1.28) in two different intervals and carefully joins the results so as they apply to the interval $x_1 < x < x_2$. We assume Ψ is continuous except at a point $y(t)$ in the interval $x_1 < x < x_2$. As shown in the following figure, at the point $y(t)$ the function Ψ is allowed to undergo a discontinuity as a function of x at a fixed time t.

As indicated on the figure

$$\Psi^{-} = \lim_{x \uparrow y(t)} \Psi(x,t) \tag{1.1.30}$$

and

$$\Psi^{+} = \lim_{x \downarrow y(t)} \Psi(x,t) \tag{1.1.31}$$

We use the notation $[\Psi]$ to denote the *jump* of Ψ defined by

$$[\Psi] = \Psi^{-} - \Psi^{+} \tag{1.1.32}$$

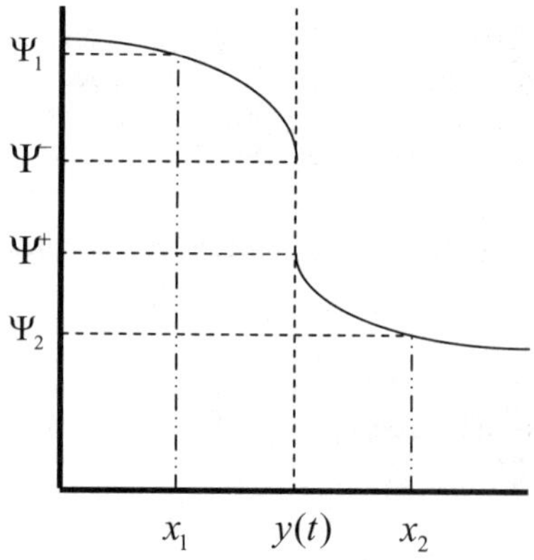

Figure 1.1.1

Exercise 1.1.3

If Ψ and Φ undergo jump discontinuities at $y(t)$ show that

$$[\Psi\Phi] = [\Psi]\Phi^- + \Psi^+[\Phi] = \Psi^-[\Phi] + [\Psi]\Phi^+$$
$$= [\Psi][\Phi] + \Psi^+[\Phi] + [\Psi]\Phi^+$$

and

$$= \frac{1}{2}\left(\Psi^+ + \Psi^-\right)[\Phi] + \frac{1}{2}[\Psi]\left(\Phi^+ + \Phi^-\right) \tag{1.1.33}$$

Exercise 1.1.4

If Ψ undergoes a jump discontinuity at $y(t)$, show that

$$\int_{x_1}^{x_2} \frac{\partial \Psi}{\partial x} dx = \Psi(x_2,t) - \Psi(x_1,t) + [\Psi] \tag{1.1.34}$$

Given a function Ψ which undergoes a jump discontinuity at $y(t)$, Reynold's theorem takes the form

$$\overline{\int_{x_1}^{x_2} \Psi(x,t)dx}^{\bullet} = \int_{x_1}^{x_2} \frac{\partial \Psi}{\partial t} dx + \Psi(x_2,t)\dot{x}(x_2,t) - \Psi(x_1,t)\dot{x}(x_1,t) + [\Psi]\dot{y} \tag{1.1.35}$$

where $\dot{y}(t)$ is the velocity of the point $y(t)$. The derivation of (1.1.35) is elementary. Because Ψ is differentiable in $x_1 < x < y(t)$ and in $y(t) < x < x_2$, (1.1.27) yields

$$\frac{d}{dt}\int_{x_1}^{y} \Psi(x,t)dx = \int_{x_1}^{y} \frac{\partial \Psi(x,t)}{\partial t} dx + \Psi^- \dot{y} - \Psi(x_1,t)\dot{x}(x_1,t) \tag{1.1.36}$$

and

$$\frac{d}{dt}\int_{y}^{x_2} \Psi(x,t)dx = \int_{y}^{x_2} \frac{\partial \Psi(x,t)}{\partial t} dx + \Psi(x_2,t)\dot{x}(x_2,t) - \Psi^+ \dot{y} \tag{1.1.37}$$

The addition of (1.1.36) and (1.1.37) yields (1.1.35). If (1.1.34) is used, (1.1.35) can also be written

$$\overline{\int_{x_1}^{x_2} \Psi(x,t)dx}^{\bullet} = \int_{x_1}^{x_2} \left(\frac{\partial \Psi}{\partial t} + \frac{\partial \Psi \dot{x}}{\partial x}\right) dx - [\Psi(\dot{x} - \dot{y})] \tag{1.1.38}$$

1.2. Balance of Mass

In this section we shall state the one dimensional form of the equation of balance of mass. This equation is the first of four fundamental principles which form the basis of continuum mechanics. The others are balance of momentum, balance of energy and the entropy inequality. These equations of balance will be discussed in subsequent sections.

We shall denote by ρ the *mass density* (mass/length) of $\mathscr{B}$ in its deformed configuration. Therefore,

$$\rho = \rho(X,t) \tag{1.2.1}$$

The corresponding quantity in the reference configuration is

$$\rho_R = \rho_R(X) \tag{1.2.2}$$

Balance of mass is the simple physical statement that the mass of the body and any of its parts are unaltered during a deformation. If an arbitrary part of the body is defined by $X_1 < X < X_2$, then it is deformed into $x_1 < x < x_2$ by the deformation, where $x_1 = \chi(X_1,t)$ and $x_2 = \chi(X_2,t)$. Balance of mass is the assertion that

$$\int_{X_1}^{X_2} \rho_R dX = \int_{x_1}^{x_2} \rho dx \tag{1.2.3}$$

for all parts of the one dimensional body $\mathscr{B}$. Because the left side of (1.2.3)is independent of t, an alternate form of balance of mass is

$$\overline{\int_{x_1}^{x_2} \rho dx}^{\bullet} = 0 \tag{1.2.4}$$

Next we shall derive the local statement of balance of mass. The statement is local in the sense that it holds at an arbitrary point X at an arbitrary time t rather than for an interval $X_1 < X < X_2$. By use of (1.1.16), (1.2.3) can be written

$$\int_{X_1}^{X_2} (\rho F - \rho_R) dX = 0 \tag{1.2.5}$$

If we assume the integrand $\rho F - \rho_R$ is a continuous function of X, the fact that (1.2.5) must hold for every interval $X_1 < X < X_2$ forces the following *local statement of balance of mass*:

$$\rho F = \rho_R \tag{1.2.6}$$

For reasons that will become clear later, we shall refer to (1.2.6) as the material form of the local statement of balance of mass. Other local statements follow by differentiation of (1.2.6). It follows from (1.2.6) that

$$\overline{\rho F}^{\bullet} = \dot{\rho} F + \rho \dot{F} = 0 \tag{1.2.7}$$

If we now use (1.1.19) and the fact that $F \neq 0$, (1.2.7) yields

$$\dot{\rho} + \rho L = 0 \tag{1.2.8}$$

Exercise 1.2.1

Show that (1.2.8) can be written in the alternate forms

$$\overline{\left(\frac{1}{\rho}\right)}^{\boldsymbol{\cdot}} = \frac{1}{\rho} L \tag{1.2.9}$$

and

$$\frac{\partial \rho}{\partial t} + \frac{\partial \rho \dot{x}}{\partial x} = 0 \tag{1.2.10}$$

Equations (1.2.8), (1.2.9) and (1.2.10) are local statements of balance of mass which hold at points where ρ and $\dot{x}$ are differentiable. If there is a point for which ρ and $\dot{x}$ undergo a jump discontinuity, we must proceed from (1.2.3) more carefully. If ρ and $\dot{x}$ suffer a jump discontinuity at $\dot{y}(t)$, it follows from (1.2.4) and (1.1.38) that

$$\int_{x_1}^{x_2} \left(\frac{\partial \rho}{\partial t} + \frac{\partial \rho \dot{x}}{\partial x} \right) dx - [\rho(\dot{x} - \dot{y})] = 0 \tag{1.2.11}$$

Equation (1.2.11) holds for all (x_1, x_2). The integrand is assumed to be continuous at all points except $\dot{y}(t)$. If (x_1, x_2) is an arbitrary interval which does not contain $\dot{y}(t)$, then (1.2.11) implies (1.2.10). Given (1.2.10), (1.2.11) then yields

$$[\rho(\dot{x} - \dot{y})] = 0 \tag{1.2.12}$$

at $x = y(t)$. The physical meaning of (1.2.12) is quite clear. It simply states that the flux of mass across the point $\dot{y}(t)$ is continuous.

Exercise 1.2.2

Use (1.2.12) and show that the jump in specific volume, $[1/\rho]$, is given by

$$\rho^{+}(\dot{x}^{+} - \dot{y})[1/\rho] = [\dot{x}] \tag{1.2.13}$$

Exercise 1.2.3

Use (1.2.10) and show that

$$\rho\dot{\Psi} = \frac{\partial\rho\Psi}{\partial t} + \frac{\partial\rho\Psi\dot{x}}{\partial x} \tag{1.2.14}$$

where Ψ is any function of (x,t).

Exercise 1.2.4

Use (1.2.14) and show that

$$\overline{\int_{x_1}^{x_2} \rho\Psi(x,t)dx}^{\,\bullet} = \int_{x_1}^{x_2} \rho\dot{\Psi}dx - [\rho\Psi(\dot{x}-\dot{y})] \tag{1.2.15}$$

1.3. Balance of Linear Momentum

In the three dimensional theory, the statement of balance of momentum consists of two parts. The first is the statement concerning the balance of linear momentum, and the second is a statement concerning the balance of angular momentum. For a one dimensional theory the concept of angular momentum does not arise.

The *linear momentum* of the part of $\mathscr{B}$ in $x_1 < x < x_2$ is

$$\int_{x_1}^{x_2} \rho\dot{x}dx$$

The rate of change of linear momentum is required to equal to the *resultant force* on the part of $\mathscr{B}$. The formal statement is written

$$\overline{\int_{x_1}^{x_2} \rho\dot{x}dx}^{\,\bullet} = f \tag{1.3.1}$$

for all parts of the body $\mathscr{B}$, where f is the resultant force acting on the part. We shall assume that f consists of two contributions and write

$$f = T(x_2,t) - T(x_1,t) + \int_{x_1}^{x_2} \rho b dx \tag{1.3.2}$$

The quantity $b(x,t)$ is called the *body force density* (body force/mass), and the integral of ρb is the *resultant body force* acting on the part of $\mathscr{B}$ in $x_1 < x < x_2$. The quantity $T(x,t)$ is a *contact force*. It results from the contact of the part of $\mathscr{B}$ in $x_1 < x < x_2$ with that not in $x_1 < x < x_2$. $T(x,t)$ is the one dimensional counterpart of stress. If $T(x,t) > 0$ (< 0) the material point is in tension (compression). If we combine (1.3.1) and (1.3.2), we obtain

$$\dot{\overline{\int_{x_1}^{x_2} \rho \dot{x} dx}} = T(x_2,t) - T(x_1,t) + \int_{x_1}^{x_2} \rho b dx \tag{1.3.3}$$

Next, we wish to deduce from (1.3.3) a local statement of balance of linear momentum. For the sake of generality, we allow T, ρ and $\dot{x}$ to suffer jump discontinuities at a point $y(t)$ in $\mathscr{B}$. It easily follows from (1.1.34) that

$$T(x_2,t) - T(x_1,t) = \int_{x_1}^{x_2} \frac{\partial T}{\partial x} dx - [T] \tag{1.3.4}$$

and, from (1.2.15), that

$$\dot{\overline{\int_{x_1}^{x_2} \rho \dot{x} dx}} = \int_{x_1}^{x_2} \rho \ddot{x} dx - [\rho \dot{x}(\dot{x} - \dot{y})] \tag{1.3.5}$$

These results allow (1.3.3) to be written

$$\int_{x_1}^{x_2} \left(\rho \ddot{x} - \frac{\partial T}{\partial x} - \rho b \right) dx - [\rho \dot{x}(\dot{x} - \dot{y})] + [T] = 0 \tag{1.3.6}$$

Since (1.3.6) must hold for all parts of $\mathscr{B}$, it follows by the same argument that produced (1.2.10) and (1.2.12) from (1.2.11) that

$$\rho \ddot{x} = \frac{\partial T}{\partial x} + \rho b \tag{1.3.7}$$

for all $x \neq y(t)$, and

$$[\rho \dot{x}(\dot{x} - \dot{y})] - [T] = 0 \tag{1.3.8}$$

at $x = y(t)$

Exercise 1.3.1

Show that an alternate form of the acceleration $\ddot{x}$ is

$$\rho \ddot{x} = \frac{\partial \rho \dot{x}}{\partial t} + \frac{\partial \rho \dot{x}^2}{\partial x} \tag{1.3.9}$$

Exercise 1.3.2

Show that (1.3.8) can be written

$$[\rho(\dot{x}-\dot{y})^2 - T] = 0 \tag{1.3.10}$$

and

$$\gamma[\dot{x}] - [T] = 0 \tag{1.3.11}$$

where $\gamma = \rho^+(\dot{x}^+ - \dot{y}) = \rho^-(\dot{x}^- - \dot{y})$.

Exercise 1.3.3

Show that

$$\gamma^2 = \frac{[T]}{[1/\rho]} \tag{1.3.12}$$

Equation (1.3.12) shows that $[T]$ and $[1/\rho]$ must have the same sign. Equation (1.3.12) is known as the *Rankine-Hugoniot* relation.

Just as (1.2.6) is the material version of balance of mass, equation (1.3.7) can be manipulated into a material version of balance of linear momentum. This material version is

$$\rho_R \ddot{x} = \frac{\partial T}{\partial X} + \rho_R b \tag{1.3.13}$$

Equation (1.3.13) follows by multiplication of (1.3.7) by F and making use of (1.2.6).

1.4. Balance of Energy

Balance of energy, or the First Law of Thermodynamics, is the statement that the rate of change of total energy equals the rate of work of the applied forces plus the rate of heat addition. The total energy includes the sum of the internal energy and the kinetic energy. If $\varepsilon(x,t)$ is the internal energy density (internal energy/mass),

$$\int_{x_1}^{x_2} \rho(\varepsilon + \frac{1}{2}\dot{x}^2)dx$$

is the *total energy* of the part of $\mathscr{B}$ in $x_1 < x < x_2$. The *rate of work* or *power* of the applied forces is

$$T(x_2,t)\dot{x}(x_2,t) - T(x_1,t)\dot{x}(x_1,t) + \int_{x_1}^{x_2} \rho\dot{x}b\,dx$$

The *rate of heat addition* arises from heat generated at points within the body $\mathscr{B}$ and from contact of the part of $\mathscr{B}$ in $x_1 < x < x_2$ with that not in $x_1 < x < x_2$. If $r(x,t)$ denotes the *heat supply density* (rate of heat addition/mass), then

$$\int_{x_1}^{x_2} \rho r dx$$

is the rate of heat addition resulting from heat generated within the body. The rate of heat addition from contact is written

$$q(x_1,t) - q(x_2,t)$$

where $q(x,t)$ is the *heat flux*. The mathematical statement which reflects balance of energy is therefore

$$\begin{aligned}\overline{\int_{x_1}^{x_2} \rho(\varepsilon + \frac{1}{2}\dot{x}^2)dx}^{\boldsymbol{\cdot}} &= T(x_2,t)\dot{x}(x_2,t) - T(x_1,t)\dot{x}(x_1,t) + \int_{x_1}^{x_2} \rho \dot{x} b dx \\ &+ q(x_1,t) - q(x_2,t) + \int_{x_1}^{x_2} \rho r dx\end{aligned} \tag{1.4.1}$$

Exercise 1.4.1

Show that (1.4.1) implies

$$\rho\overline{(\varepsilon + \frac{1}{2}\dot{x}^2)}^{\boldsymbol{\cdot}} = \frac{\partial T\dot{x}}{\partial x} - \frac{\partial q}{\partial x} + \rho \dot{x} b + \rho r \tag{1.4.2}$$

for all $x \neq y(t)$ and

$$[\rho(\varepsilon + \frac{1}{2}\dot{x}^2)(\dot{x} - \dot{y}) - T\dot{x} + q] = 0 \tag{1.4.3}$$

at $x = y(t)$.

Exercise 1.4.2

Show that when $[q] = 0$ (1.2.12) and (1.3.8) can be used to write (1.4.3) in the forms

$$[\varepsilon - \frac{T}{\rho} + \frac{1}{2}(\dot{x} - \dot{y})^2] = 0 \tag{1.4.4}$$

and

$$[\varepsilon]-\frac{1}{2}(T^{-}+T^{+})[\frac{1}{\rho}]=0 \tag{1.4.5}$$

Equation (1.4.5) is known as the *Hugoniot* relation. It is often written in terms of a different thermodynamic quantity than the internal energy density ε. The quantity that is used is the enthalpy density defined by

$$\chi=\varepsilon-\frac{T}{\rho} \tag{1.4.6}$$

Exercise 1.4.3

Given the definition (1.4.6), show that (1.4.5) takes the form

$$[\chi]+\frac{1}{2}\left(\frac{1}{\rho^{-}}+\frac{1}{\rho^{+}}\right)[T]=0 \tag{1.4.7}$$

Exercise 1.4.4

Derive a material version of (1.4.2).

Next we shall use (1.3.7) to derive from (1.4.2) a thermodynamic energy equation. If (1.3.7) is multiplied by $\dot{x}$, the result can be written

$$\rho\dot{\overline{\frac{1}{2}\dot{x}^{2}}}=\dot{x}\frac{\partial T}{\partial x}+\rho\dot{x}b \tag{1.4.8}$$

If this equation is subtracted from (1.4.2), the result is

$$\rho\dot{\varepsilon}=TL-\frac{\partial q}{\partial x}+\rho r \tag{1.4.9}$$

where the definition (1.1.17) has been used. The term $\frac{\partial T\dot{x}}{\partial x}$ in (1.4.2) arose from the rate of work of the contact forces. Since

$$\frac{\partial T\dot{x}}{\partial x}=\dot{x}\frac{\partial T}{\partial x}+TL \tag{1.4.10}$$

this rate of work decomposes into a part which changes the mechanical energy, $\dot{x}\frac{\partial T}{\partial x}$, and a part which changes the internal energy, TL. The term TL is sometimes called the *stress power.*

Exercise 1.4.5

Derive the material version of (1.4.9).

Exercise 1.4.6

On the assumption that none of the field quantities undergo jump discontinuities, show that

$$\overline{\int_{x_1}^{x_2} \rho\varepsilon dx}^{\bullet} = q(x_1,t) - q(x_2,t) + \int_{x_1}^{x_2} \rho r dx + \int_{x_1}^{x_2} TL dx \tag{1.4.11}$$

and

$$\begin{aligned}\overline{\int_{x_1}^{x_2} \rho(\frac{1}{2}\dot{x}^2) dx}^{\bullet} &= T(x_2,t)\dot{x}(x_2,t) - T(x_1,t)\dot{x}(x_1,t) + \int_{x_1}^{x_2} \rho\dot{x} b dx \\ &- \int_{x_1}^{x_2} TL dx\end{aligned} \tag{1.4.12}$$

Equations (1.4.11) and (1.4.12) show how the stress power couples the internal energy and the kinetic energy of the part of the body $\mathscr{B}$ in $x_1 < x < x_2$.

1.5. General Balance

The reader has probably noticed a formal similarity between the three balance equations discussed thus far in this chapter. Each balance equation is a special case of the following *equation of general balance:*

$$\overline{\int_{x_1}^{x_2} \rho\psi dx}^{\bullet} = \Gamma(x_2,t) - \Gamma(x_1,t) + \int_{x_1}^{x_2} \rho\varphi dx \tag{1.5.1}$$

In equation (1.5.1) the left side represents the rate of change of the amount of ψ in $x_1 < x < x_2$. The term $\Gamma(x_2,t) - \Gamma(x_1,t)$ represents the net influx of ψ, and the last term presents the supply of ψ. The following table shows the choices of ψ, Γ and φ appropriate to the equations of balance of mass, momentum and energy.

	ψ	Γ	φ
Mass	1	0	0
Momentum	$\dot{x}$	T	b
Energy	$\varepsilon+\frac{1}{2}\dot{x}^2$	$T\dot{x}-q$	$r+\dot{x}b$

Exercise 1.5.1

Derive the following local statements of the general balance:

$$\rho\dot{\psi}=\frac{\partial\Gamma}{\partial x}+\rho\varphi \tag{1.5.2}$$

for $x\neq y(t)$, and

$$[\rho\psi(\dot{x}-\dot{y})-\Gamma]=0 \tag{1.5.3}$$

for $x=y(t)$. Equation (1.5.3) is a one dimensional version of a result known as *Kotchine's theorem*.

Exercise 1.5.2

Derive a material version of (1.5.2).

1.6. The Entropy Inequality

The entropy inequality is the mathematical statement of the Second Law of Thermodynamics. In order to state this inequality, we introduce three new quantities. The *entropy density* (entropy/mass) is denoted by $\eta(x,t)$. The *entropy flux* is denoted by $h(x,t)$, and the *entropy supply density* is denoted by $k(x,t)$. These three quantities are required to obey the following *entropy inequality* or *Clausius-Duhem inequality*:

$$\overline{\int_{x_1}^{x_2}\rho\eta dx}^{\,\bullet}\geq h(x_1,t)-h(x_2,t)+\int_{x_1}^{x_2}\rho k dx \tag{1.6.1}$$

for all parts of the body $\mathscr{B}$. The *temperature* is introduced by forcing the ratio of entropy flux to heat flux to equal the ratio of entropy supply density to heat supply density. The temperature $\theta(x,t)$ is defined to be the common value of these two ratios, i.e.,

$$\theta(x,t)=\frac{q}{h}=\frac{r}{k} \tag{1.6.2}$$

We also require θ to be a *positive* number and, thus,

$$h(x,t) = \frac{q(x,t)}{\theta(x,t)} \tag{1.6.3}$$

and

$$k(x,t) = \frac{r(x,t)}{\theta(x,t)} \tag{1.6.4}$$

Given (1.6.3) and (1.6.4), (1.6.1) becomes

$$\dot{\overline{\int_{x_1}^{x_2} \rho\eta dx}} \geq \frac{q(x_1,t)}{\theta(x_1,t)} - \frac{q(x_2,t)}{\theta(x_2,t)} + \int_{x_1}^{x_2} \frac{\rho r}{\theta} dx \tag{1.6.5}$$

By a now familiar argument, (1.6.5) can be written

$$\int_{x_1}^{x_2} \left(\rho\dot{\eta} + \frac{\partial q/\theta}{\partial x} - \frac{\rho r}{\theta} \right) dx - [\rho\eta(\dot{x}-\dot{y}) + \frac{q}{\theta}] \geq 0 \tag{1.6.6}$$

for all parts of $\mathscr{B}$. The following local inequalities are valid:

$$\rho\dot{\eta} + \frac{\partial q/\theta}{\partial x} - \frac{\rho r}{\theta} \geq 0 \tag{1.6.7}$$

for $x \neq y(t)$, and

$$[\rho\eta(\dot{x}-\dot{y}) + \frac{q}{\theta}] \leq 0 \tag{1.6.8}$$

for $x = y(t)$

If (1.6.7) is written

$$\rho\dot{\eta} - \frac{q}{\theta^2}\frac{\partial \theta}{\partial x} + \frac{1}{\theta}\left(\frac{\partial q}{\partial x} - \rho r \right) \geq 0 \tag{1.6.9}$$

the term $\frac{\partial q}{\partial x} - \rho r$ can be eliminated by use of (1.4.9). The result of this elimination is

$$\rho(\theta\dot{\eta} - \dot{\varepsilon}) + TL - \frac{q}{\theta}\frac{\partial \theta}{\partial x} \geq 0 \tag{1.6.10}$$

A more convenient version of (1.6.10) results if we introduce the *Helmholtz free energy* density defined by

$$\psi = \varepsilon - \eta\theta \tag{1.6.11}$$

This definition allows (1.6.10) to be written

$$-\rho(\dot{\psi} + \eta\dot{\theta}) + TL - \frac{q}{\theta}\frac{\partial\theta}{\partial x} \geq 0 \tag{1.6.12}$$

Exercise 1.6.1

If θ_0 is a positive number, show that

$$\begin{aligned} \overline{\int_{x_1}^{x_2} \rho\left(\psi + \eta\left(\theta - \theta_0\right) + \tfrac{1}{2}\dot{\mathbf{x}}^2\right)dx}^{\,\bullet} \leq\ & T(x_2,t)\dot{x}(x_2,t) - T(x_1,t)\dot{x}(x_1,t) \\ & + q(x_1,t)\left(1 - \frac{\theta_0}{\theta(x_1,t)}\right) - q(x_2,t)\left(1 - \frac{\theta_0}{\theta(x_2,t)}\right) \\ & + \int_{x_1}^{x_2} \rho\dot{x}\cdot b\,dx + \int_{x_1}^{x_2}\left(1 - \frac{\theta_0}{\theta}\right)\rho r\,dx \end{aligned} \tag{1.6.13}$$

Equation (1.6.13) is a useful representation of the entropy inequality (1.6.5) when one wants to study the stability of certain types of bodies.

Exercise 1.6.2

Use (1.6.12) and prove that the Helmholtz free energy density cannot increase in an isothermal constant deformation process.

Exercise 1.6.3

The *enthalpy* density χ was defined by equation (1.4.6). The *Gibbs function* is defined by

$$\varsigma = \psi - \frac{T}{\rho} \tag{1.6.14}$$

Use these definitions and show that

$$\rho(\theta\dot{\eta} - \dot{\chi}) - \dot{T} - \frac{q}{\theta}\frac{\partial\theta}{\partial x} \geq 0 \tag{1.6.15}$$

and

$$-\rho(\dot{\varsigma}+\eta\dot{\theta})-\dot{T}-\frac{q}{\theta}\frac{\partial\theta}{\partial x}\geq 0 \tag{1.6.16}$$

Exercise 1.6.4

Use (1.6.16) and prove that the Gibbs function cannot increase in an isothermal constant stress process.

Exercise 1.6.5

Use the definition (1.6.11) and show that (1.4.9) can be written

$$\rho\theta\dot{\eta}=-\rho(\dot{\psi}+\eta\dot{\theta})+TL-\frac{\partial q}{\partial x}+\rho r \tag{1.6.17}$$

Exercise 1.6.6

Show that material versions of (1.6.7), (1.6.12) and (1.6.17) are

$$\rho_R\dot{\eta}+\frac{\partial q/\theta}{\partial X}-\frac{\rho_R r}{\theta}\geq 0 \tag{1.6.18}$$

$$-\rho_R(\dot{\psi}+\eta\dot{\theta})+T\dot{F}-\frac{q}{\theta}\frac{\partial\theta}{\partial X}\geq 0 \tag{1.6.19}$$

and

$$\rho_R\theta\dot{\eta}=-\rho_R(\dot{\psi}+\eta\dot{\theta})+T\dot{F}-\frac{\partial q}{\partial X}+\rho_R r \tag{1.6.20}$$

respectively.

Exercise 1.6.7

Throughout this chapter, we have developed jump expressions which govern balance of mass, momentum and energy across a jump discontinuity. We have also, with (1.6.8), developed a jump inequality which follows from the entropy inequality. It is interesting to develop material versions of these jump equations. As a first step, combine (1.2.6) and (1.2.12) and show that

$$[F^{-1}(\dot{x}-\dot{y})]=0 \tag{1.6.21}$$

The physical quantity $\dot{Y}^{\pm}=F^{-1^{\pm}}(\dot{y}-\dot{x}^{\pm})$ is the velocity in the reference configuration of the image of the spatial discontinuity. The notation $\pm$ means that the equation $\dot{Y}^{\pm}=F^{-1^{\pm}}(\dot{y}-\dot{x}^{\pm})$ is really two equations, one evaluated on the $+$ side of the discontinuity and one evaluated on the $-$ side.

In any case, it follows from (1.6.11) that conservation of mass forces $[\dot{Y}]=0$. Thus, the physical quantity $\dot{Y}$ is actually continuous across the discontinuity. Show that material versions of equations (1.3.8), (1.4.3) and (1.6.8) are

$$\rho_R\dot{Y}[\dot{x}]+[T]=0 \tag{1.6.22}$$

$$\rho_R\dot{Y}[\varepsilon+\frac{1}{2}\dot{x}^2]+[T\dot{x}]-[q]=0 \tag{1.6.23}$$

and

$$\rho_R\dot{Y}[\eta]-[\frac{q}{\theta}]\geq 0 \tag{1.6.24}$$

Exercise 1.6.8

Use the definition of $\dot{Y}$ given above and show that the jumps $[\dot{x}]$ and $[F]$ are related by

$$[\dot{x}]=-\dot{Y}[F] \tag{1.6.25}$$

Exercise 1.6.9

Derive the material version of the *Rankine-Hugoniot* relation

$$\rho_R\dot{Y}^2=\frac{[T]}{[F]} \tag{1.6.26}$$

Equation (1.6.26) is useful in the study of one dimensional shock waves in certain types of materials. It gives the velocity of the shock in the reference configuration in terms of jumps in stress and jumps in deformation. It is, in reality, the material version of (1.3.12).

Exercise 1.6.10

Show that material versions of (1.4.5) and (1.4.7) are

$$\rho_R[\varepsilon]-\frac{1}{2}(T^-+T^+)[F]=0 \tag{1.6.27}$$

and

$$\rho_R[\chi]+\frac{1}{2}\left(F^-+F^+\right)[T]=0 \tag{1.6.28}$$

As with (1.4.4) and (1.4.5), these results also assume $[q]=0$.

1.7. Example Constitutive Equations

The equations of balance are indeterminate in that they involve more variables than there are equations. This indeterminacy is to be expected since the balance equations apply to every continuous body. Experience shows that continuous bodies behave in radically different ways. There must be equations of state or *constitutive equations* which distinguish various types of materials. An important part of continuum mechanics is the study of constitutive equations. In this section we shall give examples of constitutive equations which define certain well know types of materials.

The first example is taken from gasdynamics. The material defined by the constitutive equations to be listed below is a *heat conducting compressible gas* with *constant specific heats.* The constitutive equations which define this material are

$$\varepsilon = c_v\theta + \varepsilon^+ \tag{1.7.1}$$

$$\eta = c_v \ln\theta - R\ln\rho + \eta^+ \tag{1.7.2}$$

$$T = -\pi = -\rho R\theta \tag{1.7.3}$$

and

$$q = -\kappa\frac{\partial\theta}{\partial x} \tag{1.7.4}$$

where c_v is a positive constant that represents the specific heat at constant volume, R is a positive constant that represents the gas constant, ε^+ is a constant representing the reference internal energy, η^+ is a constant representing the reference entropy, π is the one dimensional pressure and $\kappa(\theta,\rho)$ is the one dimensional thermal conductivity. The one dimensional pressure is the force of compression on the gas. It is a property of the thermal conductivity that

$$\kappa(\theta,\rho) \geq 0 \tag{1.7.5}$$

As our notation indicates, κ can depend upon θ and ρ. It follows from (1.7.1), (1.7.2) and (1.6.11) that

$$\psi = c_v\theta - \theta c_v\ln\theta + \theta R\ln\rho + \varepsilon^* - \theta\eta^* \tag{1.7.6}$$

It follows from (1.7.6), (1.7.2) and (1.7.3) that

$$\eta = -\frac{\partial\psi}{\partial\theta} \tag{1.7.7}$$

and

$$\pi = \rho^2 \frac{\partial \psi}{\partial \rho} \tag{1.7.8}$$

Therefore, the Helmholtz free energy, as a function of (θ, ρ), determines η and π. Thus, ψ is a thermodynamic potential for the material defined by (1.7.1) through (1.7.4). Given ψ, then η and π are determined by (1.7.7) and (1.7.8). The internal energy density ε is then determined from (1.6.11).

It is reasonable to question why (1.7.7) and (1.7.8) happen to hold. Other questions one could ask are why is it only the heat flux that depends upon $\frac{\partial \theta}{\partial x}$, why does q vanish when $\frac{\partial \theta}{\partial x}$ vanishes, and why must the thermal conductivity be nonnegative. In the next section we shall show that the entropy inequality places restrictions on the constitutive equations. In particular, the restrictions (1.7.5), (1.7.7) and (1.7.8) are *consequences* of the entropy inequality. How one establishes these results will be explained in the next section. It is important to note that certain of the features of (1.7.1) through (1.7.4) are *not* a consequence of the entropy inequality. For example, c_v is required to be a positive number. This requirement is a consequence of thermodynamic stability considerations.

For reference later, we shall give several other example constitutive equations. The example stress constitutive equations are the following:

1. Linear elasticity

$$T = E \frac{\partial w}{\partial X} \tag{1.7.9}$$

where E is a material constant called the one-dimensional modulus of elasticity.

2. Linear viscoelasticity (Volterra material)

$$T = E(0) \frac{\partial w(X,t)}{\partial X} + \int_0^\infty \dot{E}(s) \frac{\partial w(X,t-s)}{\partial X} ds \tag{1.7.10}$$

where $E(s)$ is the stress relaxation modulus.

3. Linear Viscous Material (Voight or Kelvin Material)

$$T = E \frac{\partial w}{\partial X} + \mu \frac{\partial \dot{w}}{\partial X} \tag{1.7.11}$$

where E and μ are material constants. The constant μ is the coefficient of viscosity.

4. Maxwellian Material

$$\tau\dot{T} + T = E\frac{\partial w}{\partial X} \tag{1.7.12}$$

where τ and E are material constants.

5. Linear Themoelasticity

$$T = E\frac{\partial w}{\partial X} - \beta(\theta - \theta_0) \tag{1.7.13}$$

where E is the isothermal modulus of elasticity and β is a constant that can be related to the coefficient of thermal expansion.

Example constitutive equations for the heat flux are the following:

1. Nonconductor

$$q = 0 \tag{1.7.14}$$

2. Fourier Heat Conductor

$$q = -\kappa\frac{\partial \theta}{\partial x} \tag{1.7.15}$$

3. Maxwell-Cattaneo Heat Conductor

$$\tau\dot{q} + q = -\kappa\frac{\partial \theta}{\partial x} \tag{1.7.16}$$

4. Gurtin-Pipkin Heat Conductor

$$q = -\int_0^\infty a(s)\frac{\partial \theta(x, t-s)}{\partial x}ds \tag{1.7.17}$$

In the formulation of any theory of material behavior there are certain principles which restrict constitutive equations. The first is a requirement of *consistency*. This requirement is that constitutive assumptions must be consistent with the axioms of balance of mass, momentum and energy and with the entropy inequality. This requirement will be the one we begin to investigate in the next section. When we consider three dimensional models, the requirements of *material frame indifference* and *material symmetry* will be used to restrict constitutive equations. As an operating procedure, we shall utilize the concept of *equipresence*. This concept states that an independent variable present in one constitutive equation should be present in all unless its presence can be shown to be in contradiction with consistency, and, for three dimensional models, material frame indifference or material symmetry.

1.8. Thermodynamic Restrictions

In this section we shall establish the type of thermodynamic restrictions described in the last section. We shall illustrate our results by examining the thermodynamic restrictions which follow for a particular set of constitutive assumptions.

By a *thermodynamic process*, we mean a set consisting of the following nine functions of (X,t): $\chi,\theta,\psi,\eta,T,q,\rho,r$ and b. The members of this set are required to obey balance of mass, balance of momentum and balance of energy. It is convenient to introduce a special symbol for the function whose value is θ and write

$$\theta=\Theta(X,t) \tag{1.8.1}$$

The constitutive equations we shall study in this section are characterized by requiring ψ,η,T and q to be determined by the *functions* Θ and χ. Formally, we shall write

$$(\psi(X,t),\eta(X,t),T(X,t),q(X,t))=f(\Theta(\cdot,\cdot),\chi(\cdot,\cdot)) \tag{1.8.2}$$

The function f is called the *response function.* An *admissible thermodynamic process* is a thermodynamic process which is consistent with (1.8.2). If we regard $\rho_R(X)$ as given, then *for every* choice of Θ and χ there exists an admissible thermodynamic process. To prove this assertion, we must construct from Θ and χ the seven remaining functions ψ,η,T,q,ρ,r and b such that balance of mass, momentum and energy are satisfied. Given (1.8.2), Θ and χ determine the four functions ψ,η,T and q. The function ρ is determined by balance of mass (1.2.6) written

$$\rho=\rho_R/F \tag{1.8.3}$$

The function b is determined by balance of linear momentum (1.3.7) written

$$b=\ddot{x}-\frac{1}{\rho}\frac{\partial T}{\partial x} \tag{1.8.4}$$

Finally, the function r is determined by balance of energy (1.6.17) written

$$r=\theta\dot{\eta}+\dot{\psi}+\eta\dot{\theta}-\frac{1}{\rho}TL+\frac{1}{\rho}\frac{\partial q}{\partial x} \tag{1.8.5}$$

The impact of the last argument is that when (1.8.2) is given, the balance equations are always satisfied no matter how we select Θ and χ.

As yet, we have not made use of the entropy inequality. The inequality (1.6.12), rewritten, is

$$-\rho(\dot{\psi}+\eta\dot{\theta})+TL-\frac{1}{\theta}qg\geq 0 \tag{1.8.6}$$

where

$$g=\frac{\partial\theta}{\partial x} \tag{1.8.7}$$

If we were to substitute (1.8.2) into (1.8.6), the resulting inequality depends, in a complicated way, on Θ and χ, and the response function f. We can view (1.8.6) as a restriction of the response function f or a restriction on the functions Θ and χ. We shall require that (1.8.6) be a restriction on f.

As an illustration, consider the case where (1.8.2) specializes to

$$\psi=\psi(\theta,g,F,\dot{F}) \tag{1.8.8}$$

$$\eta=\eta(\theta,g,F,\dot{F}) \tag{1.8.9}$$

$$T=T(\theta,g,F,\dot{F}) \tag{1.8.10}$$

and

$$q=q(\theta,g,F,\dot{F}) \tag{1.8.11}$$

These constitutive assumptions clearly contain (1.7.1) through (1.7.4) as a special case. They define a nonlinear one dimensional material that is compressible, viscous and heat conducting. Our objective is to determine how (1.8.6) restricts the functions ψ,η,T and q. First we differentiate (1.8.8) to obtain

$$\dot{\psi}=\frac{\partial\psi}{\partial\theta}\dot{\theta}+\frac{\partial\psi}{\partial g}\dot{g}+\frac{\partial\psi}{\partial F}\dot{F}+\frac{\partial\psi}{\partial\dot{F}}\ddot{F} \tag{1.8.12}$$

If this result, along with (1.8.9), (1.8.10) and (1.8.11), are substituted into (1.8.6), the result is

$$\begin{aligned}&-\rho\left(\frac{\partial\psi(\theta,g,F,\dot{F})}{\partial\theta}+\eta(\theta,g,F,\dot{F})\right)\dot{\theta}-\rho\frac{\partial\psi(\theta,g,F,\dot{F})}{\partial g}\dot{g}\\&+\left(T(\theta,g,F,\dot{F})-\rho F\frac{\partial\psi(\theta,g,F,\dot{F})}{\partial F}\right)\dot{F}F^{-1}-\rho\frac{\partial\psi(\theta,g,F,\dot{F})}{\partial\dot{F}}\ddot{F}\\&-\frac{1}{\theta}gq(\theta,g,F,\dot{F})\geq 0\end{aligned} \tag{1.8.13}$$

Equation (1.8.13) is required to hold for every choice of the functions Θ and χ. By selecting a family of functions Θ and χ each having the same θ, g, F and $\dot{F}$, the quantities $\dot{\theta}, \dot{g}$ and $\ddot{F}$ can be assigned any value. In particular they can be assigned values which violate the inequality (1.8.13) unless

$$\frac{\partial \psi(\theta, g, F, \dot{F})}{\partial \theta} + \eta(\theta, g, F, \dot{F}) = 0 \tag{1.8.14}$$

$$\frac{\partial \psi(\theta, g, F, \dot{F})}{\partial g} = 0 \tag{1.8.15}$$

and

$$\frac{\partial \psi(\theta, g, F, \dot{F})}{\partial \dot{F}} = 0 \tag{1.8.16}$$

Therefore,

$$\psi = \psi(\theta, F) \tag{1.8.17}$$

and

$$\eta = \eta(\theta, F) = -\frac{\partial \psi(\theta, F)}{\partial \theta} \tag{1.8.18}$$

Thus, η is determined by ψ and both quantities cannot depend upon g and $\dot{F}$. Given (1.8.17)and (1.8.18), (1.8.13) reduces to

$$\left(T(\theta, g, F, \dot{F}) - \rho F \frac{\partial \psi(\theta, F)}{\partial F}\right) \dot{F} F^{-1} - \frac{1}{\theta} g q(\theta, g, F, \dot{F}) \geq 0 \tag{1.8.19}$$

Because T can depend on g and q can depend on $\dot{F}$, it is not possible to conclude that the two terms in (1.8.19) are separately positive. It does follow from (1.8.19) that

$$\left(T(\theta, 0, F, \dot{F}) - \rho F \frac{\partial \psi(\theta, F)}{\partial F}\right) \dot{F} F^{-1} \geq 0 \tag{1.8.20}$$

and

$$-\frac{1}{\theta} g q(\theta, g, F, 0) \geq 0 \tag{1.8.21}$$

Equation (1.8.21) shows that when $\dot{F}=0$, the heat flux must be opposite in sign from the temperature gradient. We shall show below that $\rho F\dfrac{\partial\psi(\theta,F)}{\partial F}$ is the stress in a state of thermodynamic equilibrium. Equation (1.8.20) shows that, when $g=0$, the stress in excess of $\rho F\dfrac{\partial\psi(\theta,F)}{\partial F}$ necessarily has a nonnegative stress power.

Next we shall derive the equilibrium restrictions from (1.8.19). As a function of $(\theta,g,F,\dot{F})$, the left side of (1.8.19) is a minimum at $(\theta,0,F,0)$ for all θ and F. Because of this fact, the material defined by (1.8.8) through (1.8.11) is said to be in *thermodynamic equilibrium* when $g=\dot{F}=0$. If we define a function Φ of $(\theta,g,F,\dot{F})$ by

$$\Phi(\theta,g,F,\dot{F})=\left(T(\theta,g,F,\dot{F})-\rho F\frac{\partial\psi(\theta,F)}{\partial F}\right)\dot{F}F^{-1}-\frac{1}{\theta}gq(\theta,g,F,\dot{F})\geq 0 \tag{1.8.22}$$

then Φ is a minimum at $(\theta,0,F,0)$. Therefore,

$$\left.\frac{d\Phi(\theta,\lambda a,F,\lambda A)}{d\lambda}\right|_{\lambda=0}=0 \tag{1.8.23}$$

and

$$\left.\frac{d^2\Phi(\theta,\lambda a,F,\lambda A)}{d\lambda^2}\right|_{\lambda=0}\geq 0 \tag{1.8.24}$$

for all real numbers a and A. Since

$$\left.\frac{d\Phi(\theta,\lambda a,F,\lambda A)}{d\lambda}\right|_{\lambda=0}=\frac{\partial\Phi(\theta,0,F,0)}{\partial g}a+\frac{\partial\Phi(\theta,0,F,0)}{\partial\dot{F}}A \tag{1.8.25}$$

(1.8.23) is equivalent to

$$\frac{\partial\Phi(\theta,0,F,0)}{\partial g}=\frac{\partial\Phi(\theta,0,F,0)}{\partial\dot{F}}=0 \tag{1.8.26}$$

Equation (1.8.24) is equivalent to the requirement that the 2x2 matrix

$$\begin{bmatrix} \dfrac{\partial^2\Phi(\theta,0,F,0)}{\partial g^2} & \dfrac{\partial^2\Phi(\theta,0,F,0)}{\partial g\partial \dot{F}} \\ \dfrac{\partial^2\Phi(\theta,0,F,0)}{\partial g\partial \dot{F}} & \dfrac{\partial^2\Phi(\theta,0,F,0)}{\partial \dot{F}^2} \end{bmatrix}$$

is positive semi-definite. It easily follows from (1.8.22) and (1.8.26) that

$$q(\theta,0,F,0)=0 \tag{1.8.27}$$

and

$$T(\theta,0,F,0)=\rho F\frac{\partial\psi(\theta,F)}{\partial F} \tag{1.8.28}$$

Thus, the equilibrium heat flux must vanish and the equilibrium stress must equal $\rho F\dfrac{\partial\psi(\theta,F)}{\partial F}$.
If we define a function $T^0(\theta,F)$ by

$$T^0(\theta,F)=\rho F\frac{\partial\psi(\theta,F)}{\partial F} \tag{1.8.29}$$

and a function $T^e(\theta,g,F,\dot{F})$ by

$$T^e(\theta,g,F,\dot{F})=T(\theta,g,F,\dot{F})-T^0(\theta,F) \tag{1.8.30}$$

then T^0 is the equilibrium stress and T^e is the *dissipative* or *extra* stress. The result (1.8.28) shows that T^e vanishes in equilibrium.

Exercise 1.8.1

Calculate the elements of 2x2 matrix defined above and show that the matrix is positive semi-definite if and only if

$$\frac{\partial q(\theta,0,F,0)}{\partial g}\leq 0 \tag{1.8.31}$$

and

$$-\frac{1}{\theta}F^{-1}\frac{\partial q(\theta,0,F,0)}{\partial g}\frac{\partial T^{e}(\theta,0,F,0)}{\partial \dot{F}} \geq \frac{1}{4}\left(\frac{\partial T^{e}(\theta,0,F,0)}{\partial g}F^{-1}-\frac{1}{\theta}\frac{\partial q(\theta,0,F,0)}{\partial \dot{F}}\right)^{2} \quad (1.8.32)$$

Note, in passing, that (1.8.31) and (1.8.32) combine to yield

$$F^{-1}\frac{\partial T^{e}(\theta,0,F,0)}{\partial \dot{F}} \geq 0 \quad (1.8.33)$$

Exercise 1.8.2

Show that when the constitutive equations (1.8.8) through (1.8.11) are independent of $\dot{F}$, that (1.8.17), (1.8.18),

$$T(\theta,F)=\rho F\frac{\partial \psi(\theta,F)}{\partial F} \quad (1.8.34)$$

and

$$-\frac{1}{\theta}gq(\theta,g,F)\geq 0 \quad (1.8.35)$$

are the thermodynamic restrictions.

Exercise 1.8.3

Show that when the constitutive equations (1.8.8) through (1.8.11) are independent of g, that (1.8.17), (1.8.18),

$$q=0 \quad (1.8.36)$$

and

$$T^{e}(\theta,F,\dot{F})\dot{F}F^{-1}\geq 0 \quad (1.8.37)$$

are the thermodynamic restrictions.

Exercise 1.8.4

Express the formula (1.8.29) in terms of the variables θ and ρ and show that

$$T^0 = -\pi \tag{1.8.38}$$

where

$$\pi = \rho^2 \frac{\partial \psi(\theta, \rho)}{\partial \rho} \tag{1.8.39}$$

Exercise 1.8.5

If (1.8.18) and (1.8.29) are used, the derivative of (1.8.17) is

$$\dot{\psi} = -\eta\dot{\theta} + \frac{T^0}{\rho}\dot{F}F^{-1} \tag{1.8.40}$$

Equation (1.8.40) is called the Gibbs relation. Show that

$$\dot{\varepsilon} = \theta\dot{\eta} + \frac{T^0}{\rho}\dot{F}F^{-1} \tag{1.8.41}$$

Also, on the assumption that (1.8.18) can be solved for θ as a function of (η, F), show that

$$\theta = \frac{\partial \varepsilon(\eta, F)}{\partial \eta} \tag{1.8.42}$$

and

$$T^0(\eta, F) = \rho F \frac{\partial \varepsilon(\eta, F)}{\partial F} \tag{1.8.43}$$

Exercise 1.8.6

Use the results of Exercise 1.8.5 and show that the energy equation (1.6.17) (or (1.4.9)) reduces to

$$\rho\theta\dot{\eta} = T^e L - \frac{\partial q}{\partial x} + \rho r \tag{1.8.44}$$

Exercise 1.8.7

Show that the material version of (1.8.44) is

$$\rho_R \theta\dot{\eta} = T^e \dot{F} - \frac{\partial q}{\partial X} + \rho_R r \tag{1.8.45}$$

Exercise 1.8.8

Derive the thermodynamic restrictions for a material whose constitutive equations are

$$\psi = \psi(\theta, \dot{\theta}, g, F, \dot{F}) \tag{1.8.46}$$

$$\eta = \eta(\theta, \dot{\theta}, g, F, \dot{F}) \tag{1.8.47}$$

$$T = T(\theta, \dot{\theta}, g, F, \dot{F}) \tag{1.8.48}$$

and

$$q = q(\theta, \dot{\theta}, g, F, \dot{F}) \tag{1.8.49}$$

The material model defined by these constitutive equations could possibly yield a hyperbolic partial differential equation for θ. The model defined by (1.8.8) through (1.8.11) yields a parabolic equation which has the undesirable feature that thermal disturbances propagate with infinite speed. Will the above model yield a hyperbolic equation?

1.9. Small Departures from Thermodynamic Equilibrium

The mathematical model which results from the constitutive equations (1.8.8) through (1.8.11) is quite complicated. A less complicated model results if we assume that the departure from thermodynamic equilibrium is small. In this case we can derive approximate formulas for q and T^e. Departures from the state $(\theta, 0, F, 0)$ are measured by a positive number $\in$ defined by

$$\in^2 = g^2 + \dot{F}^2 \tag{1.9.1}$$

Given

$$T^e = T^e(\theta, g, F, \dot{F}) \tag{1.9.2}$$

and

$$q = q(\theta, g, F, \dot{F}) \tag{1.9.3}$$

we can write the following series expansions

$$T^e = \frac{\partial T^e(\theta,0,F,0)}{\partial \dot{F}} \dot{F} + \frac{\partial T^e(\theta,0,F,0)}{\partial g} g + O(\in^2) \tag{1.9.4}$$

and

$$q = \frac{\partial q(\theta,0,F,0)}{\partial g} g + \frac{\partial q(\theta,0,F,0)}{\partial \dot{F}} \dot{F} + O(\in^2) \tag{1.9.5}$$

The leading terms in both expansions vanish because of the equilibrium results (1.8.27) and (1.8.28). The coefficients in the expansions (1.9.4) and (1.9.5) correspond to material properties of the body. We shall write

$$\kappa(\theta,F) = -\frac{\partial q(\theta,0,F,0)}{\partial g} \tag{1.9.6}$$

$$\tilde{\lambda}(\theta,F) = F\frac{\partial T^e(\theta,0,F,0)}{\partial \dot{F}} \tag{1.9.7}$$

$$\alpha(\theta,F) = -F\frac{\partial q(\theta,0,F,0)}{\partial \dot{F}} \tag{1.9.8}$$

and

$$\nu(\theta,F) = \frac{\partial T^e(\theta,0,F,0)}{\partial g} \tag{1.9.9}$$

Therefore, (1.9.4)and (1.9.5)can be rewritten

$$T^e = \tilde{\lambda} L + \nu g + O(\in^2) \tag{1.9.10}$$

and

$$q = -\kappa g - \alpha L + O(\in^2) \tag{1.9.11}$$

where (1.1.19)has been used. The quantity κ is the coefficient of *thermal conductivity* and $\tilde{\lambda}$ is the *bulk* coefficient of *viscosity*. The coefficients ν and α are zero in many of the standard applications of our model and, thus, are not given names which will be familiar to the reader. Material symmetry considerations, which we have not discussed, show that ν and α must vanish for materials with a center of symmetry. This means that the constitutive equations are invariant under an inversion in the reference configuration. In any case, the material coefficients in (1.9.10) and (1.9.11)must obey the restrictions (1.8.31) and (1.8.32). These restrictions yield

$$\kappa(\theta,F) \geq 0 \tag{1.9.12}$$

and

$$\kappa(\theta,F)\tilde{\lambda}(\theta,F)\geq\frac{1}{4}\theta\left(\nu+\frac{\alpha}{\theta}\right)^2 \tag{1.9.13}$$

Because of (1.9.12), it follows from (1.9.13)that

$$\tilde{\lambda}(\theta,F)\geq 0 \tag{1.9.14}$$

Equations (1.9.12)and (1.9.14)are the classical results that the thermal conductivity and the viscosity cannot be negative.

When the remainder terms are omitted from (1.9.10) and (1.9.11), the result is a material model with *linear dissipation*. The field equations which result from utilizing these approximations are still nonlinear. In the next section we shall proceed one additional step and assume the departure from a static solution $\theta=$ const and $F-1$ is small. The resulting constitutive equations are linear and yield a set of linear governing partial differential equations.

1.10. Small Departures from Static Equilibrium

If we consider the state of constant temperature and constant deformation defined by

$$\Theta(X,t)=\theta^{+} \tag{1.10.1}$$

and

$$\chi(X,t)=X \tag{1.10.2}$$

it follows that, *in this state*, $F=1,\dot{F}=0$ and $g=0$. It immediately follows from (1.8.17), (1.8.18) and (1.8.29)that ψ, η and T^0 are constants in the state defined by (1.10.1) and (1.10.2). Also, from (1.8.27), (1.8.28), (1.8.29) and (1.8.30) it follows that q and T^e vanish. It follows from (1.3.7) that, in the state defined by (1.10.1)and (1.10.2),

$$b=0 \tag{1.10.3}$$

Likewise, the energy equation (1.8.44) yields

$$r=0 \tag{1.10.4}$$

Therefore, given (1.10.1)through (1.10.4), the field equations are identically satisfied. Such a solution is appropriately called a *static solution*. Our objective in this section is to derive the approximate constitutive and field equations which are valid near the static solution.

Departures from the static solution are measured by a positive number $\in_1$ defined by

$$\in_1^2 = (\theta - \theta^+)^2 + g^2 + (F-1)^2 + \dot{F}^2 \tag{1.10.5}$$

Note that the static solution is trivially a thermodynamic equilibrium state. In order to obtain expressions for η and T^0 which are correct up to terms of order $O(\in_1)$, we must have a representation for $\psi(\theta, F)$ correct up to terms $O(\in_1^2)$. The necessary expansion of (1.8.17) is

$$\begin{aligned}\psi(\theta,F) = \psi(\theta^+,1) &+ \frac{\partial\psi(\theta^+,1)}{\partial\theta}(\theta-\theta^+) + \frac{\partial\psi(\theta^+,1)}{\partial F}(F-1)\\ &+\frac{1}{2}\frac{\partial^2\psi(\theta^+,1)}{\partial\theta^2}(\theta-\theta^+)^2 + \frac{\partial^2\psi(\theta^+,1)}{\partial\theta\partial F}(\theta-\theta^+)(F-1)\\ &+\frac{1}{2}\frac{\partial^2\psi(\theta^+,1)}{\partial F^2}(F-1)^2 + O(\in_1^3)\end{aligned} \tag{1.10.6}$$

For the sake of a simplified notation, (1.10.6) shall be written

$$\begin{aligned}\psi(\theta,F) = \psi^+ &- \eta^+(\theta-\theta^+) + \frac{T^+}{\rho_R}(F-1)\\ &-\frac{1}{2}\frac{c_v}{\theta^+}(\theta-\theta^+)^2 - \frac{\beta}{\rho_R}(\theta-\theta^+)(F-1)\\ &+\frac{1}{2}\frac{\lambda}{\rho_R}(F-1)^2 + O(\in_1^3)\end{aligned} \tag{1.10.7}$$

where

$$\psi^+ = \psi(\theta^+,1) \tag{1.10.8}$$

$$\eta^+ = -\frac{\partial\psi(\theta^+,1)}{\partial\theta} \tag{1.10.9}$$

$$T^+ = \rho_R\frac{\partial\psi(\theta^+,1)}{\partial F} \tag{1.10.10}$$

$$c_v = -\theta^+\frac{\partial^2\psi(\theta^+,1)}{\partial\theta^2} \tag{1.10.11}$$

$$\beta = -\rho_R\frac{\partial^2\psi(\theta^+,1)}{\partial\theta\partial F} \tag{1.10.12}$$

and

$$\lambda = \rho_R \frac{\partial^2 \psi(\theta^+,1)}{\partial F^2} \tag{1.10.13}$$

From (1.8.18)and (1.8.29), η^+ and T^+ are the entropy and stress in the equilibrium state. The coefficient c_v is the *specific heat* at constant volume, λ is the *isothermal modulus of elasticity* and β is a constant that is related to the coefficient of thermal expansion. The exact relationship to the coefficient of thermal expansion is not important to us at this point. As follows from (1.2.6), ρ_R is the density in the static equilibrium state. Given (1.10.7), it follows from (1.8.18)and (1.8.29)that

$$\eta = \eta^+ + \frac{c_v}{\theta^+}(\theta - \theta^+) + \frac{\beta}{\rho_R}(F-1) + O(\epsilon_1^2) \tag{1.10.14}$$

and

$$T^0 = T^+ - \beta(\theta - \theta^+) + \lambda(F-1) + O(\epsilon_1^2) \tag{1.10.15}$$

Next, we need expressions for T^e and q valid near the static solution. Recalling that both T^e and q must vanish whenever g and $\dot{F}$ are zero, we obtain from (1.9.2) and (1.9.3)

$$T^e = \frac{\partial T^e(\theta^+,0,1,0)}{\partial g} g + \frac{\partial T^e(\theta^+,0,1,0)}{\partial \dot{F}} \dot{F} + O(\epsilon_1^2) \tag{1.10.16}$$

and

$$q = \frac{\partial q(\theta^+,0,1,0)}{\partial g} g + \frac{\partial q(\theta^+,0,1,0)}{\partial \dot{F}} \dot{F} + O(\epsilon_1^2) \tag{1.10.17}$$

By use of (1.8.7) and (1.1.11),

$$\begin{aligned} g &= \frac{\partial \theta}{\partial x} = \frac{\partial \theta}{\partial X} F^{-1} = \frac{\partial \theta}{\partial X} \frac{1}{1+(F-1)} \\ &= \frac{\partial \theta}{\partial X} + O(\epsilon_1^2) \end{aligned} \tag{1.10.18}$$

This result, along with (1.9.6) through (1.9.9) allows equations (1.10.16) and (1.10.17) to be written

$$T^e = \nu^+ \frac{\partial \theta}{\partial X} + \tilde{\lambda}^+ \dot{F} + O(\epsilon_1^2) \tag{1.10.19}$$

and

$$q=-\kappa^{+}\frac{\partial\theta}{\partial X}-\alpha^{+}\dot{F}+O(\epsilon_1^2) \tag{1.10.20}$$

where

$$\nu^{+}=\nu(\theta^{+},1) \tag{1.10.21}$$

$$\tilde{\lambda}^{+}=\tilde{\lambda}(\theta^{+},1) \tag{1.10.22}$$

$$\kappa^{+}=\kappa(\theta^{+},1) \tag{1.10.23}$$

and

$$\alpha^{+}=\alpha(\theta^{+},1) \tag{1.10.24}$$

Equations (1.10.14), (1.10.15), (1.10.19) and (1.10.20) are the basis for our linear constitutive equations. If we simply drop the remainder terms in these equations, the following *linear constitutive equations* are obtained:

$$\eta=\eta^{+}+\frac{c_v}{\theta^{+}}(\theta-\theta^{+})+\frac{\beta}{\rho_R}(F-1) \tag{1.10.25}$$

$$T^{0}=T^{+}-\beta(\theta-\theta^{+})+\lambda(F-1) \tag{1.10.26}$$

$$T^{e}=\nu^{+}\frac{\partial\theta}{\partial X}+\tilde{\lambda}^{+}\dot{F} \tag{1.10.27}$$

and

$$q=-\kappa^{+}\frac{\partial\theta}{\partial X}-\alpha^{+}\dot{F} \tag{1.10.28}$$

If follows from (1.9.12), (1.9.13) and (1.10.14) that

$$\kappa^{+}\geq 0 \tag{1.10.29}$$

$$\frac{\kappa^{+}\tilde{\lambda}^{+}}{\theta^{+}}\geq\frac{1}{4}(\nu^{+}+\alpha^{+}/\theta^{+})^{2} \tag{1.10.30}$$

and

$$\tilde{\lambda}^{+} \geq 0 \tag{1.10.31}$$

The field equations which result from these linear constitutive equations follow by substitution into the material statements of balance of momentum and energy. Because of (1.10.3), balance of momentum (1.3.13) yields

$$\rho_R \ddot{x} = \lambda \frac{\partial F}{\partial X} + \tilde{\lambda}^{+} \frac{\partial \dot{F}}{\partial X} - \beta \frac{\partial \theta}{\partial X} + \nu^{+} \frac{\partial^2 \theta}{\partial X^2} \tag{1.10.32}$$

The material statement of balance of energy for our material is (1.8.45). This equation repeated here, is

$$\rho_R \theta \dot{\eta} = T^{e} \dot{F} - \frac{\partial q}{\partial X} + \rho_R r \tag{1.10.33}$$

If (1.10.4), (1.10.25), (1.10.27) and (1.10.28) are used, the energy equation (1.10.33) reduces to

$$\rho_R \frac{\theta}{\theta^{+}} c_v \dot{\theta} + \theta \beta \dot{F} = \kappa^{+} \frac{\partial^2 \theta}{\partial X^2} + \alpha^{+} \frac{\partial \dot{F}}{\partial X} + \left(\nu^{+} \frac{\partial \theta}{\partial X} + \tilde{\lambda}^{+} \dot{F} \right) \dot{F} \tag{1.10.34}$$

Note that (1.10.34) is not a linear partial differential equation because of the products $\theta\dot{\theta}, \theta\dot{F}, \left(\frac{\partial \theta}{\partial X}\right)\dot{F}$ and $\dot{F}\dot{F}$. If (1.10.34) is linearized, the result is

$$\rho_R c_v \dot{\theta} + \theta^{+} \beta \dot{F} = \kappa^{+} \frac{\partial^2 \theta}{\partial X^2} + \alpha^{+} \frac{\partial \dot{F}}{\partial X} \tag{1.10.35}$$

Since we have formally linearized our constitutive equations, it is reasonable to utilize the linear partial differential equation (1.10.35) rather than (1.10.34) as the equation governing balance of energy. The coupled partial differential equations (1.10.32) and (1.10.35), along with suitable initial and boundary data, determine χ and θ for our model.

Exercise 1.10.1

As in elementary thermodynamics, it is sometimes convenient to use independent variables other than θ and F. Solve (1.10.25) for $\theta - \theta^{+}$ in terms of $\eta - \eta^{+}$ and $F - 1$ and use this result to eliminate θ in favor of η in (1.10.26). The result should be

$$T^{0} = T^{+} + \lambda^{*}(F - 1) - \frac{\theta^{+} \beta}{c_v}(\eta - \eta^{+}) \tag{1.10.36}$$

where

$$\lambda^* = \lambda + \frac{\theta^+}{\rho_R c_v}\beta^2 \tag{1.10.37}$$

Recalling that λ is the isothermal modulus of elasticity, the coefficient λ^* is the *isentropic modulus of elasticity.*

Exercise 1.10.2

Express ε in terms of η and F and show that

$$\begin{aligned}\varepsilon = \varepsilon^+ + \theta^+(\eta-\eta^+) + \frac{T^+}{\rho_R}(F-1) + \frac{1}{2}\frac{\theta^+}{c_v}(\eta-\eta^+)^2 \\ -\frac{\theta^+}{\rho_R c_v}\beta(\eta-\eta^+)(F-1) + \frac{1}{2}\frac{\lambda^*}{\rho_R}(F-1)^2\end{aligned} \tag{1.10.38}$$

and, from (1.8.42),

$$\theta = \theta^+ + \frac{\theta^+}{c_v}(\eta-\eta^+) - \frac{\theta^+}{\rho_R c_v}\beta(F-1) \tag{1.10.39}$$

Also show that (1.8.43) yields (1.10.36).

1.11. Some Features of the Linear Model

It is interesting to investigate certain features of the model formulated in Section 1.10. For simplicity, we shall assume that the material is such that none of the constitutive equations depend upon $\dot{F}$. This special case is achieved by taking

$$\tilde{\lambda}^+ = \alpha^+ = 0 \tag{1.11.1}$$

This assumption implies, from (1.10.30), that

$$\nu^+ = 0 \tag{1.11.2}$$

Therefore, T^e vanishes and the constitutive equations reduce to

$$\eta = \eta^+ + \frac{c_v}{\theta^+}(\theta-\theta^+) + \frac{\beta}{\rho_R}\frac{\partial w}{\partial X} \tag{1.11.3}$$

$$T = T^{+} - \beta(\theta - \theta^{+}) + \lambda \frac{\partial w}{\partial X} \tag{1.11.4}$$

and

$$q = -\kappa \frac{\partial \theta}{\partial X} \tag{1.11.5}$$

where (1.1.14) and (1.1.15) have been used to express the results in terms of the displacement gradients. In order to simplify the notation, we have written κ rather than κ^{+} for the thermal conductivity. Of course from (1.9.12), κ is restricted by

$$\kappa \geq 0 \tag{1.11.6}$$

The field equations in our special case follow from (1.11.1), (1.11.2), (1.10.32) and (1.10.35). They are

$$\rho_R \frac{\partial^2 w}{\partial t^2} = \lambda \frac{\partial^2 w}{\partial X^2} - \beta \frac{\partial \theta}{\partial X} \tag{1.11.7}$$

and

$$\rho_R c_v \frac{\partial \theta}{\partial t} + \theta^{+} \beta \frac{\partial^2 w}{\partial t \partial X} = \kappa \frac{\partial^2 \theta}{\partial X^2} \tag{1.11.8}$$

where the displacement has been introduced from (1.1.15).

Exercise 1.11.1

Show that for the special case of a nonconductor $(\kappa = 0)$ that (1.11.8) yields $\dot{\eta} = 0$.

Exercise 1.11.2

Show that for a nonconductor with initially uniform entropy that the displacement is a solution of the following wave equation:

$$\frac{\partial^2 w}{\partial t^2} = a^{*2} \frac{\partial^2 w}{\partial X^2} \tag{1.11.9}$$

where

$$a^{*2} = \frac{\lambda^{*}}{\rho_R} \tag{1.11.10}$$

Because (1.11.9) is a wave equation, a^* is called the *isentropic wave speed.*

Exercise 1.11.3

Show that in the limit of very large conductivity $(\kappa \to \infty)$ that suitable boundary conditions can always be prescribed such that $\theta = const$. Note that in this case the displacement satisfies the following wave equation:

$$\frac{\partial^2 w}{\partial t^2} = a^2 \frac{\partial^2 w}{\partial X^2} \tag{1.11.11}$$

where

$$a^2 = \frac{\lambda}{\rho_R} \tag{1.11.12}$$

is the isothermal wave speed squared.

Exercise 1.11.4

The definitions (1.11.10) and (1.11.12) presume that both λ and λ^* are non-negative quantities. Show that if $c_v > 0$, the isentropic wave speed is larger than the isothermal wave speed. Calculate the ratio a^*/a for the case where the material is air modeled as a heat conducting compressible perfect gas with constant specific heats.

Certain properties of the solutions to (1.11.7) and (1.11.8) are revealed by deriving the single partial differential equation obeyed by $w(X,t)$. This equation, which is also obeyed by $\Theta(X,t)$ is the result of expanding the following 2×2 operator determinant.

$$\begin{vmatrix} \rho_R \dfrac{\partial^2}{\partial t^2} - \lambda \dfrac{\partial^2}{\partial X^2} & \beta \dfrac{\partial}{\partial X} \\ \theta^+ \beta \dfrac{\partial^2}{\partial t \partial X} & \rho_R c_v \dfrac{\partial}{\partial t} - \kappa \dfrac{\partial^2}{\partial X^2} \end{vmatrix} = 0 \tag{1.11.13}$$

If this determinant is expanded and rearranged, the result is the following fourth order partial differential equation:

$$\frac{\partial}{\partial t}\left(\frac{\partial^2 w}{\partial t^2} - a^{*2} \frac{\partial^2 w}{\partial X^2}\right) - \frac{\kappa}{\rho_R c_v} \frac{\partial^2}{\partial X^2}\left(\frac{\partial^2 w}{\partial t^2} - a^2 \frac{\partial^2 w}{\partial X^2}\right) = 0 \tag{1.11.14}$$

where a and a^* are the wave speeds defined by (1.11.12) and (1.11.10), respectively. It is convenient to define a characteristic time τ_κ by

$$\tau_\kappa = \frac{\kappa}{\lambda c_v} \tag{1.11.15}$$

and rewrite (1.11.14)as

$$\frac{\partial}{\partial t}\left(\frac{\partial^2 w}{\partial t^2} - a^{*2}\frac{\partial^2 w}{\partial X^2}\right) - a^2\tau_\kappa \frac{\partial^2}{\partial X^2}\left(\frac{\partial^2 w}{\partial t^2} - a^2\frac{\partial^2 w}{\partial X^2}\right) = 0 \tag{1.11.16}$$

The characteristic time τ_κ can be used to assign meaning to the expressions "short time" and "long time". For fixed t, the dimensionless time t/τ_κ is large if τ_κ is small and is small if τ_κ is large. A *short time* approximation to (1.11.16) is a solution valid for small t/τ_κ. A long time approximation to (1.11.16) is a solution valid for large t/τ_κ. For large τ_κ the second term in (1.11.16) dominates and, as a result, disturbances propagate with the isothermal wave speed. Likewise, for small τ_κ the first term in (1.11.16) dominates and disturbances propagate with the isentropic wave speed. It is reasonable to expect that short time approximations to the solutions of (1.11.16) approach solutions of (1.11.11) and long time approximations approach solutions of (1.11.9). The intuitive argument which is sometimes used to support this assertion is that for short times the heat conduction has not yet influenced the material, and the material acts as if it is in an isothermal process. For long times, the dissipative effects of heat conduction have taken place and, after this, the material acts as if it is in an isentropic process.

It is helpful when one tries to get some general feelings about the behavior of the one dimensional material defined by (1.11.16) to look at some representative numerical values. The following table is adapted from one in Ref. 3.

Properties of Four Metals at $20^o C$

Quantity	Aluminum	Copper	Iron	Lead
a (m/sec)	6320	4360	5800	2140
a^* (m/sec)	6432	4396	5801	2217
τ_κ (sec)	$2.15(10)^{-12}$	$5.78(10)^{-12}$	$.571(10)^{-12}$	$5.24(10)^{-12}$

Table 1.11.1

To the extent that these metals are typical, the small characteristic time τ_κ would cause solutions of (1.11.16) to behave like long term approximations.

Exercise 1.11.5

Calculate the three parameters a, a^* and τ_κ for the two materials defined by the following table:

Quantity	Air (as a Perfect Gas)	AL 2024-T3
$\theta^+\ (^oC)$	20	20
$\rho_R\,(kg/m^3)$	1.205	2770
$c_v\ (J/kg^oK)$	717.4	963
$\kappa\,(W/m^oK)$	.0257	190.5
λ (Pa)	101400	108308270700
$\beta\ (Pa/^o\,K)$	346	4876

Additional insight into the behavior of the material described by (1.11.7) and (1.11.8) can be obtained by investigating the propagation of harmonic waves defined by solutions of the form

$$w(X,t) = de^{-\varphi X}e^{i(\omega/q)(X-qt)} \tag{1.11.17}$$

and

$$\theta(X,t)-\theta^+ = fe^{-\varphi X}e^{i(\omega/q)(X-qt)} \tag{1.11.18}$$

In (1.11.17) and (1.11.18), it is understood that the real part of the assumed solution is used. The quantity ω is a real number and represents the *frequency* of the harmonic wave. The quantity q is a real number which represents the *phase velocity* of the wave. The quantity φ is a real number called the *attenuation coefficient*. The coefficients d and f are complex numbers which correspond to the amplitudes of the wave. The *wave number* is defined by

$$k = \frac{\omega}{q} + i\varphi \tag{1.11.19}$$

This definition allows (1.11.17) and (1.11.18) to be written

$$w(X,t) = de^{i(kX-\omega t)} \tag{1.11.20}$$

and

$$\theta(X,t)-\theta^+ = fe^{i(kX-\omega t)} \tag{1.11.21}$$

In order that (1.11.20) and (1.11.21) represent a solution of (1.11.7) and (1.11.8), k and ω cannot be independent. Since ω is a prescribed real number, k will be determined as a function of ω. Such a relation is known as the *dispersion relation*.

Exercise 1.11.6

Substitute (1.11.20) into (1.11.16) and show that the dispersion relation $k(\omega)$ is a solution of

$$a^2\tau_\kappa k^2(a^2k^2-\omega^2)-i\omega(a^{*2}k^2-\omega^2)=0 \tag{1.11.22}$$

Equation (1.11.22) is a quadratic in k^2. Therefore, there are four modes of propagation of the form (1.11.20). Since high frequency corresponds to short time, the phase velocity in the high frequency approximation for two of the modes is easily shown to be $+a$ and $-a$. Likewise, the low frequency approximation yields phase velocities $+a^*$ and $-a^*$.

In circumstances where the two limiting cases of high and low frequency does not apply, the phase velocities and the attenuation coefficients depend upon the frequency. A dispersion relation which yields a frequency dependent phase velocity is called a *dispersive mode*. The approximations just cited correspond to the nondispersive limits of the solutions $k(\omega)$. The other modes are dispersive and, roughly speaking, correspond to the propagation of the thermal disturbance.

The explicit formulas for k implied by (1.11.22) are complicated but can be derived. The details can be found in the work of P. Chadwick in Ref. 3

Exercise 1.11.7

Show that the roots of (1.11.22) can be written

$$\frac{2ak}{\sqrt{\omega}}=\pm\left\{\sqrt{\omega+i\omega_\kappa\left(\frac{a^*}{a}\right)^2+(1+i)\sqrt{2\omega_\kappa\omega}}\pm\sqrt{\omega+i\omega_\kappa\left(\frac{a^*}{a}\right)^2-(1+i)\sqrt{2\omega_\kappa\omega}}\right\} \tag{1.11.23}$$

where $\omega_\kappa = 1/\tau_\kappa$.

In equation, (1.11.23) the first pair of $\pm$ symbols corresponds to waves propagating in the $\pm x$ direction. We shall only consider this $+$ case below. The second pair of $\pm$ symbols correspond to two possible modes of propagation.

If one adopts the properties of air as given in Exercise 1.11.5 above, equation (1.11.23) can be used to calculate the phase velocity and the attenuation coefficient as a function of frequency for each mode. The results of this calculation yield the following plots.

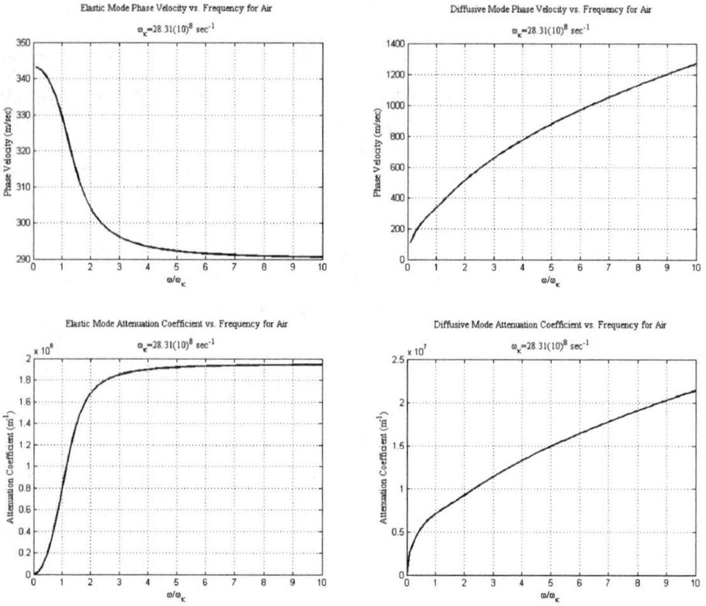

For the elastic mode, the first figure shows how the phase velocity begins at a^* and, as the frequency is increased, approaches a. For this same mode, the attenuation coefficient begins at zero and grows, as the frequence is increased, to a constant value. It is possible to use (1.11.22) to derive the following high frequency approximation for the attenuation coefficient in this case. The result is

$$\lim_{\frac{\omega}{\omega_K}\to\infty} \varphi(\omega) = \frac{\omega_K}{2a}\frac{a^{*2}-a^2}{a^2} \tag{1.11.24}$$

Chadwick gives approximations for $k(\omega)$. One family of approximations to (1.11.23) involves the assumption that a and a^* are close. The formal way this approximation is generated is to expand (1.11.23) for small values of the ratio $\dfrac{a^{*2}-a^2}{a^2}$. It is readily established from (1.10.37), (1.11.10) and (1.11.12) that $\dfrac{a^{*2}-a^2}{a^2} = \dfrac{\theta^+}{\rho_R^2 c_v}\beta^2$. Therefore, small values of $\dfrac{a^{*2}-a^2}{a^2}$ correspond to small values of the coefficient β. It is β which couples the equation of motion (1.11.7) and the energy equation (1.11.8). Table 1.11.1 shows that for the metals given, the isothermal wave speed and the isentropic wave speeds are very close. This type of approximation would not be good for air, where, the ratio a^{*2}/a^2 is approximately the ratio of the specific heat at constant pressure to the specific heat at constant volume. If air is modeled as a perfect gas, this ratio is 1.4. Exercise 1.11.4

asks that the ratio a^*/a be calculated in the case where air is modeled as a perfect gas with constant specific heats.

The other family of approximations is one where the ratio ω/ω_κ is small. Table 1.11.1 shows that τ_κ is small, and, thus, $\omega_\kappa = 1/\tau_\kappa$ is large. The characteristic time for air, as calculated in Exercise 1.11.5, is approximately $3.5(10)^{-10}$ sec. As a result, an approximation based upon the assumption that the corresponding ratio ω/ω_κ is small appears to be broadly useful. Chadwick gives the following results for the two phase velocities and the associated attenuation coefficients.

$$q_{(1)} = a^*\left(1 - \frac{(a^{*2}/a^2 - 1)(7 - 3a^{*2}/a^2)}{8(a^{*2}/a^2)^4}(\omega/\omega_\kappa)^2 + O\left((\omega/\omega_\kappa)^4\right)\right) \tag{1.11.25}$$

$$q_{(2)} = \frac{a^2}{a^*}\sqrt{2(\omega/\omega_\kappa)}\left(1 - \frac{(a^{*2}/a^2 - 1)}{2(a^{*2}/a^2)^2}(\omega/\omega_\kappa) + O\left((\omega/\omega_\kappa)^2\right)\right) \tag{1.11.26}$$

$$\varphi_{(1)} = \frac{\omega_\kappa}{a^*}\left(\frac{(a^{*2}/a^2 - 1)}{2(a^{*2}/a^2)^2}(\omega/\omega_\kappa)^2 + O\left((\omega/\omega_\kappa)^4\right)\right) \tag{1.11.27}$$

and

$$\varphi_{(2)} = \frac{a^*\omega_\kappa}{a^2}\sqrt{\omega/2\omega_\kappa}\left(1 - \frac{(a^{*2}/a^2 - 1)}{2(a^{*2}/a^2)^2}(\omega/\omega_\kappa) + O\left((\omega/\omega_\kappa)^2\right)\right) \tag{1.11.28}$$

These approximate expressions display the two modes of propagation. The first mode depends upon the frequency through second order terms, while the second mode has a stronger dependence. It is the first mode that propagates with a phase velocity near the isentropic speed a^*. Additional discussion of plane harmonic waves in thermoelastic materials can be found in references 4, 5 and 6.

In the last four sections, we have examined the constitutive equations of a special one dimensional material. Our thermodynamic results, along with the approximations introduced in Sections 9 and 10, allowed a brief consideration of the behavior of waves propagating in the one dimensional material. If one were interested in the behavior of a broader class of initial boundary value problems, one would need some sort of existence and uniqueness theorem. Uniqueness theorems in elasticity usually arise from some sort of energy argument. An illustration, in the context of the one dimensional material being discussed here, is provided by the following exercise.

Exercise 1.11.8

Use (1.11.4), (1.11.5), (1.11.7) and (1.11.8) and show that

$$\begin{aligned}\frac{\partial}{\partial t}\int_a^b\left[\frac{1}{2}\frac{\rho_R c_v}{\theta^+}(\theta-\theta^+)^2+\frac{1}{2}\lambda\left(\frac{\partial w}{\partial X}\right)^2+\rho_R\frac{1}{2}\left(\frac{\partial w}{\partial t}\right)^2\right]dX\\ =(T(b,t)-T^+)\frac{\partial w(b,t)}{\partial t}-(T(a,t)-T^+)\frac{\partial w(a,t)}{\partial t}\\ -\left(\frac{\theta(b,t)-\theta^+}{\theta^+}\right)q(b,t)+\left(\frac{\theta(a,t)-\theta^+}{\theta^+}\right)q(a,t)\\ -\int_a^b\frac{\kappa}{\theta^+}\left(\frac{\partial\theta}{\partial X}\right)^2dX\end{aligned} \tag{1.11.29}$$

where a and b are fixed positions.

Exercise 1.11.9

Consider a boundary-initial value problem for (1.11.7) and (1.11.8) where the velocity and heat flux vanish at each end of the material. Show that

$$\frac{\partial}{\partial t}\int_a^b\left[\frac{1}{2}\frac{\rho_R c_v}{\theta^+}(\theta-\theta^+)^2+\frac{1}{2}\lambda\left(\frac{\partial w}{\partial X}\right)^2+\frac{1}{2}\rho_R\left(\frac{\partial w}{\partial t}\right)^2\right]dX\le 0 \tag{1.11.30}$$

Exercise 1.11.10

Assume that $c_v>0$ and $\lambda>0$. For initial conditions, assume that

$$\theta(X,0)=\theta^+ \tag{1.11.31}$$

$$\frac{\partial w(X,0)}{\partial X}=0 \tag{1.11.32}$$

and

$$\frac{\partial w(X,0)}{\partial t}=0 \tag{1.11.33}$$

Show that (1.11.30) implies that

$$\theta(X,t)=\theta^+ \tag{1.11.34}$$

$$\frac{\partial w(X,t)}{\partial X}=0 \tag{1.11.35}$$

and

$$\frac{\partial w(X,t)}{\partial t} = 0 \tag{1.11.36}$$

for all $t \geq 0$. The argument used in this exercise provides a uniqueness theorem for solutions of (1.11.7) and (1.11.8). Given the assumptions made in the problem, it is necessary that (1.11.34), (1.11.35) and (1.11.36) hold. Notice that $c_v > 0$ and $\lambda > 0$ were among the assumptions made. These stability results are not consequences of the second law of thermodynamics.

References

1. SOKOLNIKOFF, I. S., *Advanced Calculus,* McGraw-Hill, New York (1939).
2. SOKOLNIKOFF, I. S., and E. S. SOKOLNIKOFF, *High Mathematics for Engineers and Physicists*, McGraw-Hill, New York (1941).
3. CHADWICK, P., *Thermoelasticity, The Dynamic Theory*, Progress in Solid Mechanics, Vol. 1, (ed. I. N. Sneddon), North Holland, Amsterdam (1960).
4. NOWINSKI, J. L., *Theory of Thermoelasticity with Applications*, Sijthoff & Noordhoff, Alphen aan den Rijn (1978).
5. Boley, D. A., and J. H. Weiner, Theory of Thermal Stresses, John Wiley & Sons, New York (1960).
6. PARKUS, H., *Thermoelasticity*, Blaisdell, London (1968).

Bibliography

1. TRUESDELL, C., and R. A. TOUPIN, *The Classical Field Theories*, Handbuch der Physik, Vol.III/l (ed. S. Flugge). Springer-Verlag, Berlin (1960).
2. TRUESDELL, C., and W. NOLL, *The Non-Linear Field Theories of Mechanics*, Handbuch der Physik,Vol. III/3 (ed. S. Flugge), Springer-Verlag, Berlin (1965).
3. WANG, C.-C., and C. TRUESDELL, *Introduction to Rational Elasticity*, Noordhoff, Leyden (1973).
4. TRUESDELL, C., *Rational Thermodynamics*, Second Edition, Springer-Verlag, Berlin (1984).

2

Kinematics of Motion

This chapter is concerned with the three dimensional kinematics of motion and strain. The material presented represents the three dimensional generalization of the material contained in Section 1.1. After the introduction of the concepts of deformation, velocity, acceleration and deformation gradient in Sections 2.1 and 2.2, Section 2.3 contains a discussion of the transformation of linear, surface and volume elements which are induced by a deformation. This discussion leads naturally to a discussion of nonlinear strain kinematics in Section 2.4 and linear, or infinitesimal, strain kinematics in Section 2.5.

2.1. Bodies and Deformations

Given the limited mathematical background established in the Appendix A, it is convenient here to regard a body $\mathscr{B}$ to be a primitive concept. Precise mathematical definitions of a body can be found, for example, in Reference 1, 2 and 3. Roughly speaking a body is a set $\mathscr{B}$ endowed with a topological and differentiable structure. In addition, a body is endowed with a family of functions, called *configurations*, which map $\mathscr{B}$ into a subset of $\mathscr{E}$. The structure of $\mathscr{B}$ is such that it is meaningful to require configurations to be diffeomorphisms. In addition, bodies are required to be connected. We shall denote the elements of $\mathscr{B}$, called particles, by X. If $\boldsymbol{\chi}$ is a configuration, then

$$\mathbf{x}=\boldsymbol{\chi}(X) \tag{2.1.1}$$

is the position in $\mathscr{E}$ occupied by X in $\mathscr{B}$.

Definition: A *motion* of the body $\mathscr{B}$ is a one parameter family of configurations.

The parameter is the time t and a motion is written

$$\mathbf{x}=\boldsymbol{\chi}(X,t) \tag{2.1.2}$$

for all (X, t) in $\mathscr{B}\times(-\infty,\infty)$. Since a configuration is a diffeomorphism, for each t the motion $\boldsymbol{\chi}$ has a smooth inverse such that

$$X=\boldsymbol{\chi}^{-1}(\mathbf{x},t) \tag{2.1.3}$$

for all $(\mathbf{x},t)$ in $\boldsymbol{\chi}(\mathscr{B})\times(-\infty,\infty)$. The point $\mathbf{x}$ is the *place occupied* by the particle X at the time t. Physical observations can never be made on a body except in some region of physical space $\mathscr{E}$. In many cases it is convenient to reflect this fact by use of a fixed configuration, called a *reference configuration*. We shall denote this fixed configuration by $\boldsymbol{\kappa}$. The configuration $\boldsymbol{\kappa}$ may be, but need not be, a configuration actually occupied by the body in the course of its motion. The position of X in $\boldsymbol{\kappa}$ will be denoted by $\mathbf{X}$. Thus,

$$X = \boldsymbol{\kappa}(X) \tag{2.1.4}$$

Figure 2.1.1 shows the geometric arrangement reflected in equations (2.1.2) and (2.1.4).

The function $\boldsymbol{\chi}_{\kappa}$ defined by

$$\mathbf{x} = \boldsymbol{\chi}\left(\boldsymbol{\kappa}^{-1}(X),t\right) \tag{2.1.5a}$$

$$\equiv \boldsymbol{\chi}_{\kappa}(X,t) \tag{2.1.5b}$$

for all (X,t) in $\boldsymbol{\kappa}(\mathscr{B}) \times (-\infty,\infty)$ is called the *deformation function* relative to the reference configuration $\boldsymbol{\kappa}$. As the above notation suggests and the definition (2.1.5) shows, the deformation function depends upon the choice of the reference configuration.

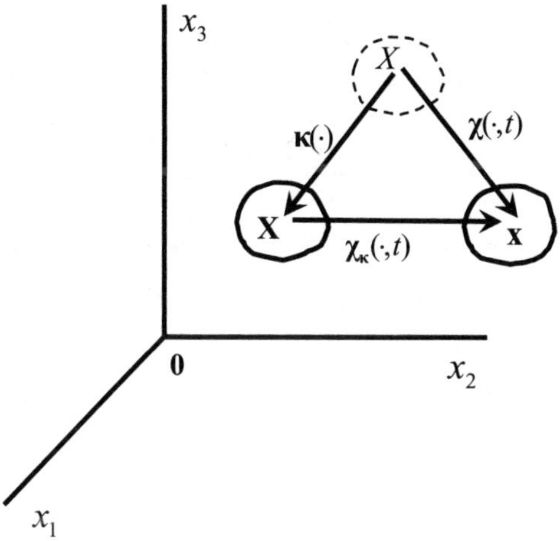

Figure 2.1.1

The coordinates of the point $\mathbf{X}$, denoted by (X_1, X_2, X_3), are called the *material coordinates* of the particle at $\mathbf{X}$, while the coordinates of the point $\mathbf{x}$, denoted by (x_1, x_2, x_3) are called the *spatial coordinates* of the particle at $\mathbf{x}$. As a matter of convention, the subscripts on the material coordinates will always be in Latin majuscules and those on the spatial coordinates will be Latin minuscules. Therefore, in components, (2.1.2) and (2.1.5*b*) can be written

$$x_i = \chi_i(X,t) \tag{2.1.6}$$

and

$$x_i = \chi_{\kappa_i}(X_1, X_2, X_3, t) \equiv \chi_{\kappa_i}(X_I, t) \tag{2.1.7}$$

where

$$x_i = (\mathbf{x} - \mathbf{0}) \cdot \mathbf{i}_i \tag{2.1.8}$$

and

$$X_I = (\mathbf{X} - \mathbf{0}) \cdot \mathbf{i}_I \tag{2.1.9}$$

In the following sections it will be assumed, without comment, that the diffeomorphisms $\boldsymbol{\chi}$ and $\boldsymbol{\chi}_\kappa$ have sufficient smoothness in order to allow for the existence of any derivatives that appear.

2.2. Velocity, Acceleration and Deformation Gradients

Given a motion $\boldsymbol{\chi}$, there are several kinematic quantities which can be calculated. In this section several of these quantities are defined.

Definition: The *velocity* of the particle *X*, written $\dot{\mathbf{x}}$ is defined by

$$\dot{\mathbf{x}} = \frac{\partial \chi(X,t)}{\partial t} \tag{2.2.1}$$

It follows from (2.1.5) and (2.2.1) that

$$\dot{\mathbf{x}} = \frac{\partial \chi_\kappa(\mathbf{X},t)}{\partial t} \tag{2.2.2}$$

Of course, the velocity computed by (2.2.2) does not depend upon the special configuration $\boldsymbol{\kappa}$.

Definition: The *acceleration* of the particle X, written $\ddot{\mathbf{x}}$, is defined by

$$\ddot{\mathbf{x}} = \frac{\partial^2 \boldsymbol{\chi}(X,t)}{\partial t} \tag{2.2.3}$$

Also, it is true that

$$\ddot{\mathbf{x}} = \frac{\partial^2 \chi_\kappa(\mathbf{X},t)}{\partial t^2} \tag{2.2.4}$$

Definition: The *displacement* of X relative to the configuration $\boldsymbol{\kappa}$ is defined by

$$\mathbf{w} = \chi_\kappa(\mathbf{X},t) - \mathbf{X} \tag{2.2.5}$$

Any time dependent scalar, vector, or tensor field Ψ can be regarded as a function of (X, t), $(\mathbf{X}, t)$ or $(\mathbf{x}, t)$ whenever the motion $\mathbf{x}=\boldsymbol{\chi}(X,t)=\boldsymbol{\chi}_{\kappa}(\mathbf{X},t)$ is given. If we regard Ψ to be a function of $(\mathbf{X}, t)$, we are using *material* variables. If we regard Ψ to be function of $(\mathbf{x}, t)$, we are using *spatial* variables. For the sake of notational simplicity, it is convenient to use the same symbol for the three possible functions whose value is Ψ. For example, the following equations should not be confusing:

$$\Psi=\Psi(X,t) \tag{2.2.6a}$$

$$\Psi=\Psi(\mathbf{X},t) \tag{2.2.6b}$$

and

$$\Psi=\Psi(\mathbf{x},t) \tag{2.2.6c}$$

It should be remembered also that the function Ψ in (2.2.6*b*) will depend upon the configuration $\boldsymbol{\kappa}$.

Definition: The *material derivative* of Ψ, written $\dot{\Psi}$ is defined by

$$\dot{\Psi}=\frac{\partial\Psi(X,t)}{\partial t}=\frac{\partial\Psi(\mathbf{X},t)}{\partial t} \tag{2.2.7}$$

It follows from (2.2.5) and the above definitions that

$$\dot{\mathbf{w}}=\dot{\mathbf{x}} \tag{2.2.8}$$

and

$$\ddot{\mathbf{w}}=\ddot{\mathbf{x}} \tag{2.2.9}$$

The material derivative can be expressed in terms of $\Psi(\mathbf{x},t)$. By the chain rule, it follows that

$$\dot{\Psi}=\frac{\partial\Psi(\mathbf{x},t)}{\partial t}+\left(grad\,\Psi(\mathbf{x},t)\right)\dot{\mathbf{x}} \tag{2.2.10}$$

In components, (2.2.10) is

$$\dot{\Psi}_{j_1 j_2\cdots j_p}=\frac{\partial\Psi_{j_1 j_2\cdots j_p}(\mathbf{x},t)}{\partial t}+\frac{\partial\Psi_{j_1 j_2\cdots j_p}(\mathbf{x},t)}{\partial x_k}\dot{x}_k \tag{2.2.11}$$

As an example of (2.2.10), consider the velocity, $\dot{\mathbf{x}}$. Then, the acceleration can be written

$$\ddot{\mathbf{x}}=\frac{\partial \dot{\mathbf{x}}(\mathbf{x},t)}{\partial \mathbf{t}}+\left(grad\,\dot{\mathbf{x}}(\mathbf{x},t)\right)\dot{\mathbf{x}} \tag{2.2.12}$$

or

$$\ddot{x}_j=\frac{\partial \dot{x}_j(\mathbf{x},t)}{\partial t}+\frac{\partial \dot{x}_j(\mathbf{x},t)}{\partial x_k}\dot{x}_k \tag{2.2.13}$$

In (2.2.10) and (2.2.12) the gradient with respect to spatial coordinates (x_1, x_2, x_3) has been denoted by "grad". The gradient with respect to material coordinates will be denoted by "GRAD". The divergence with respect to spatial coordinates will be denoted by "div", and the divergence with respect to material coordinates will be denoted by "Div".

Definition: The *deformation gradient* at (**X**, t) is a linear transformation in $\mathscr{L}(\mathscr{V};\mathscr{V})$ defined by

$$\mathbf{F}=GRAD\,\chi_\kappa(\mathbf{X},t) \tag{2.2.14}$$

In those cases where no confusion can arise, **F**(**X** *t*) will be written **F**(*t*) or, simply, **F**.

Since the functions χ and $\boldsymbol{\kappa}$ are diffeomorphisms, the composition of χ and $\boldsymbol{\kappa}^{-1}$, which is $\boldsymbol{\chi}_\kappa$, is a diffeomophism. The differentiability of $\boldsymbol{\chi}_\kappa$ has been used in the definition (2.2.14).

The fact that $\boldsymbol{\chi}_\kappa$ and $\boldsymbol{\chi}_\kappa^{-1}$ are one-to-one has an important and clear physical significance. It omits the possibility of a material point at $\mathbf{X}$ being mapped into more than one point $\mathbf{x}$, and, conversely. It is a well known theorem of general topology that a homeomorphism maps connected sets into connected sets. [Ref. 4]. If this theorem is applied to the function $\boldsymbol{\chi}_\kappa$, it follows that a point is mapped into a point, a line into a line, a surface into a surface and a region into a region.

Exercise 2.1.1

Show that

$$\mathbf{F}(\mathbf{X},t)\,grad\,\boldsymbol{\chi}_\kappa^{-1}(x,t)=grad\,\boldsymbol{\chi}_\kappa^{-1}(\mathbf{x},t)\,\mathbf{F}(\mathbf{X},t)=\mathbf{I} \tag{2.2.15}$$

Equation (2.2.15) shows that grad $\boldsymbol{\chi}_\kappa^{-1}(\mathbf{x},t)$ is the inverse of the linear transformation **F**. In the following, this linear transformation will be denoted by $\mathbf{F}^{-1}$. It follows from (2.2.15) and (A.5.22) that

$$\det \mathbf{F}\neq 0 \tag{2.2.16}$$

for all (**X**, *t*) in $\boldsymbol{\chi}(\mathscr{B})\times(-\infty,\infty)$.

The component representation of the deformation gradient $\mathbf{F}$ follows from (A.8.14), (2.1.5*b*) and (2.2.14). The result shall be written

$$\mathbf{F}=\frac{\partial x_i}{\partial X_J}\mathbf{i}_i\otimes\mathbf{i}_J \tag{2.2.17}$$

In addition, if the component version of $\mathbf{X}=\boldsymbol{\chi}_{\boldsymbol{\kappa}}^{-1}(\mathbf{x},t)$ is written

$$X_J=\chi_{\kappa J}^{-1}(x_1,x_2,x_3,t) \tag{2.2.18}$$

we can write

$$\mathbf{F}^{-1}=\frac{\partial X_J}{\partial x_i}\mathbf{i}_J\otimes\mathbf{i}_i \tag{2.2.19}$$

Definition: The *displacement gradient* at (**X**, *t*) is a linear transformation in $\mathscr{L}(\mathscr{V};\mathscr{V})$ defined by

$$\mathbf{H}(\mathbf{X},t)=\text{GRAD}\,\mathbf{w}(\mathbf{X},t) \tag{2.2.20}$$

It easily follows from (2.2.5) and (2.2.14) that

$$\mathbf{H}=\mathbf{F}-\mathbf{I} \tag{2.2.21}$$

Definition: The *velocity gradient* at (**x**, *t*) is a linear transformation in $\mathscr{L}(\mathscr{V};\mathscr{V})$ defined by

$$\mathbf{L}(\mathbf{x},t)=\text{grad}\,\dot{\mathbf{x}}(\mathbf{x},t) \tag{2.2.22}$$

Often, we shall write $\mathbf{L}$ or $\mathbf{L}(t)$ for $\mathbf{L}(\mathbf{x},t)$.

The component representation for $\mathbf{L}$ can be shown to be

$$\mathbf{L}=\frac{\partial \dot{x}_j}{\partial x_k}\mathbf{i}_j\otimes\mathbf{i}_k \tag{2.2.23}$$

Given (2.2.14) and (2.2.22), we shall prove that

$$\dot{\mathbf{F}}=\mathbf{L}\mathbf{F} \tag{2.2.24}$$

The argument necessary to establish (2.2.24) is as follows:

$$\dot{\mathbf{F}} = \overline{\text{GRAD}\,\boldsymbol{\chi}_\kappa(\mathbf{X},t)}^{\cdot} = \text{GRAD}\frac{\partial \boldsymbol{\chi}_\kappa(\mathbf{X},t)}{\partial t}$$

$$= grad\,\dot{\mathbf{x}}\,\text{GRAD}\,\boldsymbol{\chi}_\kappa(\mathbf{X},t)$$

$$= \mathbf{LF}$$

Exercise 2.1.2

Show that

$$tr\,\mathbf{L} = div\,\dot{\mathbf{x}} \tag{2.2.25}$$

and

$$\dot{\overline{\mathbf{F}^{-1}}} = -\mathbf{F}^{-1}\mathbf{L} \tag{2.2.26}$$

Equation (2.2.24) implies that

$$\dot{\overline{(\det \mathbf{F})}} = (\det \mathbf{F})\,\text{tr}\,\mathbf{L} \tag{2.2.27}$$

Equation (2.2.27) follows by the following argument.

By the definition (A.5.19), we can write

$$(\det \mathbf{F})\mathbf{u}\cdot(\mathbf{v}\times\mathbf{w}) = \mathbf{Fu}\cdot(\mathbf{Fv}\times\mathbf{Fw}) \tag{2.2.28}$$

where **u**, **v** and **w** are arbitrary vectors in $\mathscr{V}$. If (2.2.28) is differentiated, the result is

$$\begin{aligned}\dot{\overline{(\det \mathbf{F})}}\,\mathbf{u}\cdot(\mathbf{v}\times\mathbf{w}) &= \dot{\mathbf{F}}\mathbf{u}\cdot(\mathbf{Fv}\times\mathbf{Fw}) + \mathbf{Fu}\cdot(\dot{\mathbf{F}}\mathbf{v}\times\mathbf{Fw}) + \mathbf{Fu}\cdot(\mathbf{Fv}\times\dot{\mathbf{F}}\mathbf{w})\\ &= \mathbf{LFu}\cdot(\mathbf{Fv}\times\mathbf{Fw}) + \mathbf{Fu}\cdot(\mathbf{LFv}\times\mathbf{Fw}) + \mathbf{Fu}\cdot(\mathbf{Fv}\times\mathbf{LFw})\\ &= (\text{tr}\,\mathbf{L})\mathbf{Fu}\cdot(\mathbf{Fv}\times\mathbf{Fw})\end{aligned}$$

where (A.5.36) and (2.2.24) have been used. Because $\mathbf{u}$, $\mathbf{v}$ and $\mathbf{w}$ are arbitrary the result (2.2.27) is obtained.

Exercise 2.1.3

Use (2.2.27) and show that

$$\frac{\partial(\det \mathbf{F})}{\partial \mathbf{F}} = (\det \mathbf{F})\mathbf{F}^{-1^T} \tag{2.2.29}$$

It is also true that

$$\operatorname{Div}\left(\left(\det \mathbf{F}\right)\mathbf{F}^{-1^T}\right)=\mathbf{0} \tag{2.2.30}$$

and

$$\operatorname{div}\left(\left(\det \mathbf{F}^{-1}\right)\mathbf{F}^{T}\right)=\mathbf{0} \tag{2.2.31}$$

The proof of these results is rather complicated and will be summarized in the following exercises.

Exercise 2.1.4

Show that (2.2.30) is equivalent to the component formula

$$\frac{\partial}{\partial \mathrm{X}_J}\left[\left(\det \mathbf{F}\right)\frac{\partial \mathrm{X}_J}{\partial x_j}\right]=\mathbf{0} \tag{2.2.32}$$

Exercise 2.1.5

Use equation (A.5.27) and show that

$$\varepsilon_{QJK}\left(\det \mathbf{F}\right)\frac{\partial \mathrm{X}_Q}{\partial x_q}=\varepsilon_{qjk}\frac{\partial x_j}{\partial \mathrm{X}_J}\frac{\partial x_k}{\partial \mathrm{X}_K} \tag{2.2.33}$$

Exercise 2.1.6

Multiply equation (2.2.33) by ε_{RJK} and show that

$$\left(\det \mathbf{F}\right)\frac{\partial \mathrm{X}_R}{\partial x_q}=\frac{1}{2}\varepsilon_{RJK}\,\varepsilon_{qjk}\frac{\partial x_j}{\partial \mathrm{X}_J}\frac{\partial x_k}{\partial \mathrm{X}_K} \tag{2.2.34}$$

Exercise 2.1.7

Use (2.2.34) and prove the validity of (2.2.30).The proof of (2.2.31) follows an identical argument with **x** and **X** interchanged.

Exercise 2.2.8

Use (2.2.31) and show that

$$\operatorname{div}((\det \mathbf{F}^{-1})\mathbf{F}\mathbf{u}) = (\det \mathbf{F}^{-1})\operatorname{Div}\mathbf{u} \tag{2.2.35}$$

for an arbitrary vector field $\mathbf{u}$.

2.3. Transformation of Linear, Surface and Volume Elements

It follows from (2.2.14) that

$$d\mathbf{x} = \mathbf{F}\, d\mathbf{X} \tag{2.3.1}$$

The vector $d\mathbf{X}$ at $\mathbf{X}$ represents an infinitesimal segment of material in the reference configuration, and the vector $d\mathbf{x}$ at $\mathbf{x}$ represents an infinitesimal segment of material in the deformed configuration. Equation (2.3.1) represents the transformation law for *linear elements* of material under the deformation $\mathbf{x} = \boldsymbol{\chi}_\kappa(\mathbf{X}, t)$.

If dV is a material element of volume at $\mathbf{X}$ and dv denotes its image under the mapping $\mathbf{x} = \chi_\kappa(\mathbf{X}, t)$ then we wish to show that

$$dv = |\det \mathbf{F}|\, dV \tag{2.3.2}$$

By definition, the volume elements dV and dv are given by

$$dV = |d\mathbf{X}_1 \cdot (d\mathbf{X}_2 \times d\mathbf{X}_3)| \tag{2.3.3}$$

and

$$dv = |d\mathbf{x}_1 \cdot (d\mathbf{x}_2 \times d\mathbf{x}_3)| \tag{2.3.4}$$

where $d\mathbf{x}_1 = \mathbf{F}\, d\mathbf{X}_1, d\mathbf{x}_2 = \mathbf{F}\, d\mathbf{X}_2$ and $d\mathbf{x}_3 = \mathbf{F}\, d\mathbf{X}_3$. Therefore, by (A.5.19),

$$\begin{aligned} dv &= |\mathbf{F}\, d\mathbf{X}_1 \cdot (\mathbf{F}\, d\mathbf{X}_2 \times \mathbf{F}\, d\mathbf{X}_3)| \\ &= |(\det \mathbf{F})\, d\mathbf{X}_1 \cdot (d\mathbf{X}_2 \times d\mathbf{X}_3)| = |\det \mathbf{F}|\, dV \end{aligned}$$

Equation (2.3.2) shows us again the physical importance of (2.2.16). By (2.3.2), we see that (2.2.16) is in reality a statement about the permanence of the material.

Next, we wish to obtain a formula that relates a material element of *area* at $\mathbf{X}$ to its image at $\mathbf{x}$. It will not be possible to discuss here, in a careful fashion, the idea of an oriented surface element. However, we can obtain the desired result by the following argument. If $d\mathbf{S}$ denotes a material element of area at $\mathbf{X}$, we require that

$$dV = d\mathbf{X}_1 \cdot d\mathbf{S} \tag{2.3.5}$$

If we compare (2.3.3) and (2.3.5), it is seen that $d\mathbf{S}$ must be given by

$$d\mathbf{S} = \pm\, d\mathbf{X}_2 \times d\mathbf{X}_3 \tag{2.3.6}$$

where the plus or minus sign is used, depending upon the orientation of the vectors $d\mathbf{X}_2$ *and* $d\mathbf{X}_3$ relative to $d\mathbf{X}_1$. A formula similar to (2.3.5) can be written to define $d\mathbf{s}$, the image of $d\mathbf{S}$, at $\mathbf{x}$. It is

$$dv = d\mathbf{x}_1 \cdot d\mathbf{s} \tag{2.3.7}$$

where $d\mathbf{x}_1 = \mathbf{F}d\mathbf{X}_1$. Therefore, by (2.3.2),

$$d\mathbf{x}_1 \cdot d\mathbf{s} = \left|\det \mathbf{F}\right| d\mathbf{X}_1 \cdot d\mathbf{S} \tag{2.3.8}$$

or

$$\left(\mathbf{F}\, d\mathbf{X}_1\right)\cdot d\mathbf{s} = d\mathbf{X}_1 \cdot \left(\mathbf{F}^{\mathrm{T}} d\mathbf{s}\right) = \left|det\, \mathbf{F}\right| d\mathbf{X}_1 \cdot d\mathbf{S} \tag{2.3.9}$$

Therefore, equation (2.3.9) can be written,

$$d\mathbf{X}_1 \cdot \left(\mathbf{F}^{\mathrm{T}} d\mathbf{s} - \left|\det \mathbf{F}\right| d\mathbf{S}\right) = 0 \tag{2.3.10}$$

If we now regard $d\mathbf{X}_1$ as arbitrary, equation (2.3.10) yields the *transformation law for material surface elements*,

$$d\mathbf{s} = \left|\det \mathbf{F}\right| \mathbf{F}^{-1^{\mathrm{T}}} d\mathbf{S} \tag{2.3.11}$$

The next formulas we shall discuss are expressions for differentiating volume, surface and line integrals. If $\mathscr{P}$ is a part of the body $\mathscr{B}$, then $\chi(\mathscr{P},t)$ is the region occupied by $\mathscr{P}$ at the time t. It is a material region in the sense that it moves with the part $\mathscr{P}$ such that every point within $\chi(\mathscr{P},t)$ has the material velocity $\dot{\mathbf{x}}$. We shall denote the surface of $\chi(\mathscr{P},t)$ by $\partial\chi(\mathscr{P},t)$. The vector $d\mathbf{s}$ denotes the outward drawn vector element of area. The geometric arrangement is shown in the figure below.

For a volume integral the following result, known as *Reynold's Theorem*, shall be established:

$$\overline{\int_{\chi(\mathscr{P},t)} \Psi dv} = \int_{\chi(\mathscr{P},t)} \frac{\partial \Psi}{\partial t} dv + \oint_{\partial\chi(\mathscr{P},t)} \Psi \dot{\mathbf{x}} \cdot d\mathbf{s} \tag{2.3.12}$$

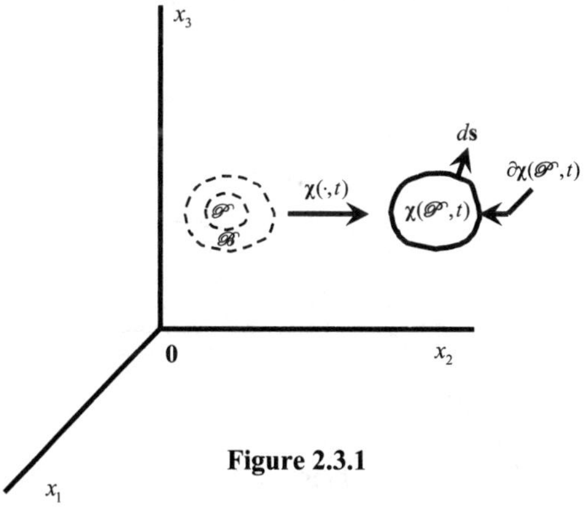

Figure 2.3.1

In order to prove (2.3.12), we first utilize (2.3.2) and write

$$\dot{\overline{\int_{\chi(\mathscr{P},t)} \Psi dv}} = \dot{\overline{\int_{\kappa(\mathscr{P})} \Psi \left|\det \mathbf{F}\right| dV}} = \int_{\kappa(\mathscr{P})} \dot{\overline{\left(\Psi \left|\det \mathbf{F}\right|\right)}} dV$$

$$= \int_{\kappa(\mathscr{P})} \left(\dot{\Psi} \left|\det \mathbf{F}\right| + \Psi \dot{\overline{\left|\det \mathbf{F}\right|}}\right) dV$$

By use of (2.2.27) and (2.3.2), this result can be rewritten

$$\dot{\overline{\int_{\chi(\mathscr{P},t)} \Psi dv}} = \int_{\chi(\mathscr{P},t)} \left(\dot{\Psi} + \Psi \operatorname{tr} \mathbf{L}\right) dv \tag{2.3.13}$$

If we now use (2.2.10) and (2.2.25), it is seen that

$$\dot{\Psi} + \Psi \operatorname{tr} \mathbf{L} = \frac{\partial \Psi}{\partial t} + \operatorname{div}\left(\Psi \dot{\mathbf{x}}\right) \tag{2.3.14}$$

If (2.3.14) is substituted into (2.3.13), the divergence term can be converted, by use of (A.8.27), to the surface integral which appears in (2.3.12).

Equation (2.3.12) is the three dimensional generalization of (1.1.21). The reader should note that the derivation of (2.3.12) is the three dimensional version of the solution to Exercise 1.1.2.

Exercise 2.1.8

If $\mathscr{S}(t)$ denotes a *material surface*, i.e., a surface that always consists of the same material points, show that

$$\dot{\overline{\int_{\mathscr{S}(t)} \Psi\, d\mathbf{s}}} = \int_{\mathscr{S}(t)} \left(\dot{\Psi}\, d\mathbf{s} + \left(\operatorname{tr} \mathbf{L}\right) \Psi\, d\mathbf{s} - \Psi \mathbf{L}^T d\mathbf{s}\right) \tag{2.3.15}$$

The geometric arrangement appropriate to this exercise is illustrated in the Fig. 2.3.2.

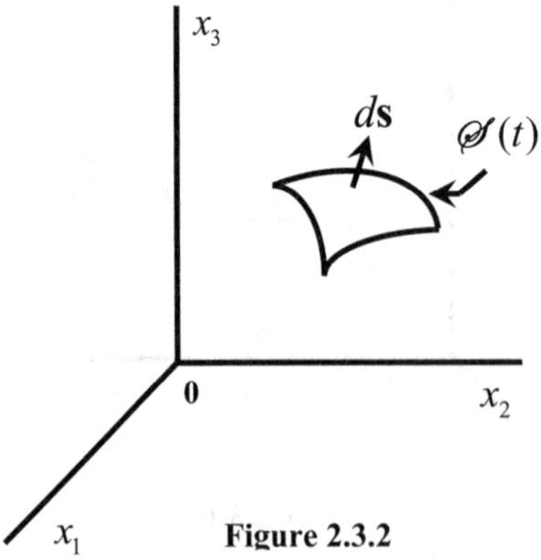

Figure 2.3.2

Exercise 2.1.9

If $\mathscr{L}(t)$ denotes a material curve, i.e., a curve that always consists of the same material points, show that

$$\overline{\int_{\mathscr{L}(t)} \Psi\, d\mathbf{x}}^{\,\boldsymbol{\cdot}} = \int_{\mathscr{L}(t)} \left(\dot{\Psi}\, d\mathbf{x} + \Psi\, \mathbf{L}\, d\mathbf{x} \right) \tag{2.3.16}$$

Exercise 2.1.10

The *circulation* about a closed curve $\mathscr{L}(t)$ is defined by

$$\Gamma\left(\mathscr{L}(t)\right) = \oint_{\mathscr{L}(t)} \dot{\mathbf{x}} \cdot d\mathbf{x} \tag{2.3.17}$$

Show that

$$\overline{\Gamma\left(\mathscr{L}(t)\right)}^{\,\boldsymbol{\cdot}} = \oint_{\mathscr{L}(t)} \ddot{\mathbf{x}} \cdot d\mathbf{x} \tag{2.3.18}$$

Equation (2.3.18) is useful in discussing the kinematics of vorticity and in proving classical theorems regarding circulation.[Ref. 5]

It is important to recognize that (2.3.12) can be generalized to the point where it is divorced from the idea that $\chi(\mathscr{P},t)$ is a material region. In fact, if $\mathscr{R}(t)$ denotes a region in $\mathscr{E}$ and if points on the boundary of $\mathscr{R}(t)$, $\partial\mathscr{R}(t)$ move with velocity $\mathbf{v}(\mathbf{x}, t)$, then it is possible to show that

$$\frac{d}{dt}\int_{\mathscr{R}(t)} \Psi\, dv = \int_{\mathscr{R}(t)} \frac{\partial \Psi}{\partial t} dv + \oint_{\partial\mathscr{R}(t)} \Psi \mathbf{v} \cdot d\mathbf{s} \tag{2.3.19}$$

[Ref.6, Sect.112]

The derivation of (2.3.12) assumed, among other things, that Ψ was differentiable throughout $\chi(\mathscr{P},t)$. It is useful to have a generalization of (2.3.12) valid when Ψ suffers a *jump* discontinuity on a surface $\Sigma(t)$ in $\chi(\mathscr{P},t)$. The generalization we seek is the three dimensional generalization of (1.1.31). Consider Fig. 2.3.3.

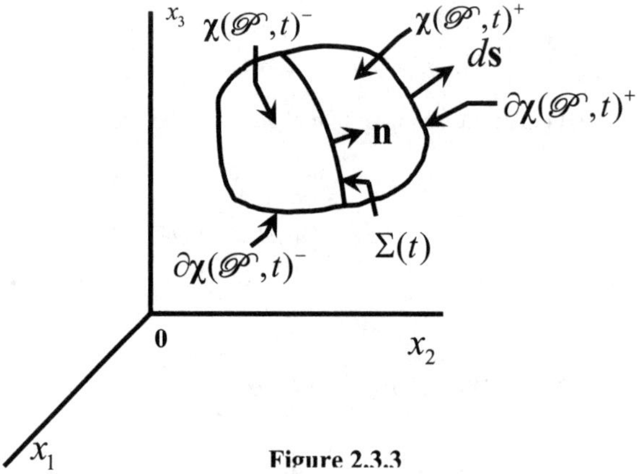

Figure 2.3.3

The surface $\Sigma(t)$ divides $\chi(\mathscr{P},t)$. into the two regions shown, $\chi(\mathscr{P},t)^+$ and $\chi(\mathscr{P},t)^-$. It is important to note that $\chi(\mathscr{P},t)^+$ and $\chi(\mathscr{P},t)^-$ are not material volumes. This assertion becomes clear when one recognizes that the surface $\Sigma(t)$ dividing $\chi(\mathscr{P},t)$. is not fixed in $\chi(\mathscr{P},t)$. The unit normal $\mathbf{n}$ to $\Sigma(t)$ is directed into $\chi(\mathscr{P},t)^+$. The surface $\Sigma(t)$ is allowed to be moving and its *normal velocity* is written $\mathbf{n}u_n$. If we assume Ψ is differentiable in both $\chi(\mathscr{P},t)^-$ and $\chi(\mathscr{P},t)^+$, then Ψ approaches limiting values Ψ^- and Ψ^+ as $\Sigma(t)$ is approached from $\chi(\mathscr{P},t)^-$ and $\chi(\mathscr{P},t)^+$, respectively. Equation (2.3.19) can be used to obtain

$$\frac{d}{dt}\int_{\chi(\mathscr{P},t)^+}\Psi\,dv=\int_{\chi(\mathscr{P},t)^+}\frac{\partial\Psi}{\partial t}dv+\int_{\partial\chi(\mathscr{P},t)^+}\Psi\dot{\mathbf{x}}\cdot d\mathbf{s}-\int_{\Sigma(t)}\Psi^+u_n\,d\sigma \tag{2.3.20}$$

for $\chi(\mathscr{P},t)^+$, and

$$\frac{d}{dt}\int_{\chi(\mathscr{P},t)^-}\Psi\,dv=\int_{\chi(\mathscr{P},t)^-}\frac{\partial\Psi}{\partial t}dv+\int_{\partial\chi(\mathscr{P},t)^-}\Psi\dot{\mathbf{x}}\cdot d\mathbf{s}+\int_{\Sigma(t)}\Psi^-u_n\,d\sigma \tag{2.3.21}$$

for $\chi(\mathscr{P},t)^-$. In (2.3.20) and (2.3.21), $d\sigma$ is the element of area of the surface $\Sigma(t)$. The sum of (2.3.20) and (2.3.21) yields

$$\overline{\int_{\chi(\mathscr{P},t)}\Psi\,dv}^{\,\cdot}=\int_{\chi(\mathscr{P},t)}\frac{\partial\Psi}{\partial t}dv+\oint_{\partial\chi(\mathscr{P},t)}\Psi\dot{\mathbf{x}}\cdot d\mathbf{s}+\int_{\Sigma(t)}[\Psi]u_n d\sigma \tag{2.3.22}$$

where the jump $[\Psi]$ is defined by (A.8.31). Equation (2.3.22) generalizes (1.1.35). By use of (A.8.30), (2.3.22) can be written

$$\overline{\int_{\chi(\mathscr{P},t)} \Psi\, dv}^{\,\cdot} = \int_{\chi(\mathscr{P},t)} \left(\frac{\partial \Psi}{\partial t} + \operatorname{div}(\Psi \dot{\mathbf{x}}) \right) dv - \int_{\Sigma(t)} [\Psi(\dot{\mathbf{x}} \cdot \mathbf{n} - u_n)] d\sigma \tag{2.3.23}$$

If we now use (2.3.14), the result (2.3.23) can be rewritten in the form

$$\overline{\int_{\chi(\mathscr{P},t)} \Psi\, dv}^{\,\cdot} = \int_{\chi(\mathscr{P},t)} \left(\dot{\Psi} + \Psi \operatorname{tr} \mathbf{L} \right) dv - \int_{\Sigma(t)} [\Psi(\dot{\mathbf{x}} \cdot \mathbf{n} - u_n)] d\sigma \tag{2.3.24}$$

The reader should recognize (2.3.23) as the three dimensional generalization of (1.1.38).

Exercises 2.3.4-2.3.8:

Kinematics of Rotating Coordinates. Consider the geometric arrangement shown in the Fig. 2.3.4. The point $\mathbf{c} = \mathbf{c}(t)$ is the origin of a translating and rotating coordinate system with orthonormal basis $\left\{\mathbf{i}_1^*, \mathbf{i}_2^*, \mathbf{i}_3^*\right\}$. Since $\left\{\mathbf{i}_1^*, \mathbf{i}_2^*, \mathbf{i}_3^*\right\}$ is a basis for $\mathscr{V}$, we can write

$$\mathbf{i}_j^* = Q_{kj} \mathbf{i}_k \tag{2.3.25}$$

where $\left[Q_{kj} \right]$ is an *orthogonal* time-dependent matrix.

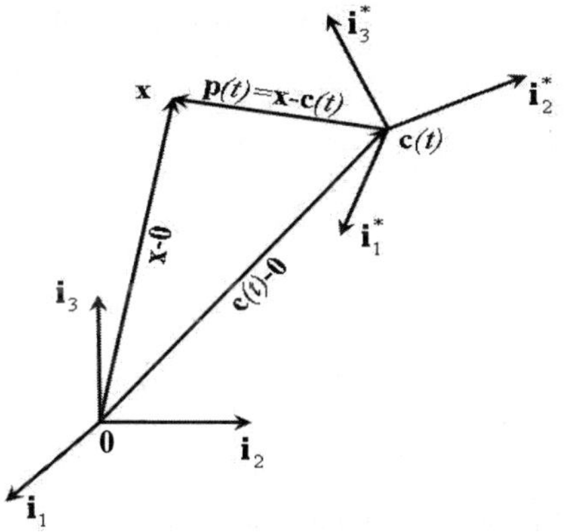

Figure 2.3.4

Exercise 2.1.11

Show that

$$\frac{d\mathbf{i}_j^*}{dt} = Z_{kj}(t) \mathbf{i}_k^* \tag{2.3.26}$$

where $\left[Z_{kj}(t)\right]$ is a *skew-symmetric* matrix defined by

$$Z_{kj}(t)=Q_{sk}\frac{dQ_{sj}(t)}{dt} \tag{2.3.27}$$

Exercise 2.1.12

Define $\mathbf{Z}(t)$ in $\mathscr{L}(\mathscr{V};\mathscr{V})$ by

$$\mathbf{Z}(t)=Z_{kj}(t)\mathbf{i}_k^*\otimes\mathbf{i}_j^* \tag{2.3.28}$$

and show that

$$\frac{d\mathbf{i}_j^*}{dt}=\mathbf{Z}\mathbf{i}_j^* \tag{2.3.29}$$

The linear transformation $\mathbf{Z}$ measures the angular velocity of $\left\{\mathbf{i}_1^*,\mathbf{i}_2^*,\mathbf{i}_3^*\right\}$ with respect to $\left\{\mathbf{i}_1,\mathbf{i}_2,\mathbf{i}_3\right\}$.

Exercise 2.1.13

If $\boldsymbol{\omega}$ is the angular velocity vector of the basis $\left\{\mathbf{i}_1^*,\mathbf{i}_2^*,\mathbf{i}_3^*\right\}$ defined by

$$\mathbf{Z}\mathbf{v}=\boldsymbol{\omega}\times\mathbf{v} \tag{2.3.30}$$

for all $\mathbf{v}$ in $\mathscr{V}$, show that

$$\omega_k=-\tfrac{1}{2}\varepsilon_{ksj}Z_{sj} \tag{2.3.31}$$

and

$$\frac{d\mathbf{i}_j^*}{dt}=\boldsymbol{\omega}\times\mathbf{i}_j^* \tag{2.3.32}$$

Exercise 2.1.14

If $\mathbf{u}=\mathbf{u}(\mathbf{X},t)$ is a time dependent vector field, show that

$$\dot{\mathbf{u}}=\frac{\delta\mathbf{u}}{\delta t}+\boldsymbol{\omega}\times\mathbf{u} \tag{2.3.33}$$

where $\delta\mathbf{u}/\delta t$ is the material derivation of $\mathbf{u}$ seen by an observer in the translating and rotating coordinate system, i.e., if $\mathbf{u}=u_j^*\mathbf{i}_j^*$ then

$$\frac{\delta \mathbf{u}}{\delta t} = \dot{u}_j^* \mathbf{i}_j^*$$

Exercise 2.1.15

From the figure above, we have

$$\mathbf{x} - \mathbf{0} = \mathbf{p}(t) + \mathbf{c}(t) - \mathbf{0} \tag{2.3.34}$$

The velocity $\dot{\mathbf{x}}$ associated with the motion $\mathbf{x} = \boldsymbol{\chi}_\kappa(\mathbf{X}, t)$ is thus given by

$$\dot{\mathbf{x}} = \frac{\delta \mathbf{p}}{\delta t} + \boldsymbol{\omega} \times \mathbf{p} + \dot{\mathbf{c}}(t) \tag{2.3.35}$$

where we have used (2.3.33). Show that

$$\mathbf{L} = \operatorname{grad}\left(\frac{\delta \mathbf{p}}{\delta t}\right) + \mathbf{Z} \tag{2.3.36}$$

2.4. Strain Kinematics

In this section, the kinematics of *local strain* are discussed. The discussion is "local" in the sense that the kinematical ideas apply to the deformation of infinitesimal linear elements of material. The ideas introduced here will aid in understanding certain quantities that will appear later when special types of materials are considered.

If the polar decomposition theorem, Theorem A.5.12, is applied to $\mathbf{F}$ we have

$$\mathbf{F} = \mathbf{R}\mathbf{U} \tag{2.4.1a}$$

$$= \mathbf{V}\mathbf{R} \tag{2.4.1b}$$

where $\mathbf{R}$ is orthogonal and $\mathbf{U}$ and $\mathbf{V}$ are symmetric and positive definite. The tensor $\mathbf{R}$ is called the *rotation tensor*, while $\mathbf{U}$ and $\mathbf{V}$ are the *right* and *left stretch tensors*, respectively. The positive definite and symmetric tensors $\mathbf{B}$ and $\mathbf{C}$, defined by

$$\mathbf{B} = \mathbf{F}\mathbf{F}^T \tag{2.4.2}$$

and

$$\mathbf{C} = \mathbf{F}^T\mathbf{F} \tag{2.4.3}$$

are the left and right *Cauchy- Green tensors*, respectively.

It follows from (2.4.1) and (2.3.1) that

$$d\mathbf{x} = \mathbf{RU}\, d\mathbf{X} \tag{2.4.4a}$$

$$= \mathbf{VR}\, d\mathbf{X} \tag{2.4.4b}$$

The meaning of (2.4.4a) should be clear. Given the material element $d\mathbf{X}$, $\mathbf{U}d\mathbf{X}$ is the material element stretched by $\mathbf{U}$. The orthogonal linear transformation $\mathbf{R}$ then rotates the stretched material element into the deformed element $d\mathbf{x}$. Equation (2.4.4b) represents the same deformation, but with the rotation and stretch taking place in a reverse order. Therefore, $\mathbf{U}$ and $\mathbf{V}$ measure the stretch (or strain) of a linear element, while $\mathbf{R}$ measures its rotation. It easily follows from (2.4.1), (2.4.2) and (2.4.3) that

$$\mathbf{V} = \mathbf{RUR}^T \tag{2.4.5}$$

$$\mathbf{B} = \mathbf{RCR}^T \tag{2.4.6}$$

$$\mathbf{B} = \mathbf{V}^2 \tag{2.4.7}$$

and

$$\mathbf{C} = \mathbf{U}^2 \tag{2.4.8}$$

As observed in the proof of Theorem A.5.11, $\mathbf{U}$ and $\mathbf{V}$ have the same eigenvalues. Thus, they correspond to the same stretch. If $\mathbf{n}$ is an eigenvector of $\mathbf{U}$, then we know that $\mathbf{Rn}$ is the corresponding eigenvector of $\mathbf{V}$. Therefore, $\mathbf{V}$ and $\mathbf{U}$ correspond to the same stretch but in different directions related by the orthogonal linear transformation $\mathbf{R}$.

Exercise 2.4.1

Show that

$$\mathbf{V}^n = \mathbf{RU}^n\mathbf{R}^T \tag{2.4.9}$$

where n is an integer.

It follows from (2.4.7) and (2.4.8) that $\mathbf{B}$ and $\mathbf{C}$ are also measures of the strain. An equivalent method of reaching this conclusion is to consider the following argument. The squared lengths of the linear element at $\mathbf{X}$ and $\mathbf{x}$ are

$$dS^2 = d\mathbf{X} \cdot d\mathbf{X} \tag{2.4.10}$$

and

$$ds^2 = d\mathbf{x} \cdot d\mathbf{x}$$

The difference in these squared lengths is clearly a measure of the strain. It is a simple calculation, based upon (2.3.1), to show that

$$ds^2 - dS^2 = d\mathbf{X}\cdot\left(\left(\mathbf{C}-\mathbf{I}\right)d\mathbf{X}\right) = d\mathbf{x}\cdot\left(\left(\mathbf{I}-\mathbf{B}^{-1}\right)d\mathbf{x}\right) \tag{2.4.11}$$

Equations (2.3.11) indicate that $\mathbf{C}$ and $\mathbf{B}$ measure the strain. The tensor

$$\mathbf{E} = \tfrac{1}{2}\left(\mathbf{C}-\mathbf{I}\right) \tag{2.4.12}$$

is called the *material strain tensor* or the *Lagrangian strain tensor*.

Exercise 2.4.2

Show that

$$\mathbf{C} = \mathbf{I}+\mathbf{H}+\mathbf{H}^T+\mathbf{H}^T\mathbf{H} \tag{2.4.13}$$

$$\mathbf{B} = \mathbf{I}+\mathbf{H}+\mathbf{H}^T+\mathbf{H}\mathbf{H}^T \tag{2.4.14}$$

and

$$\mathbf{E} = \tfrac{1}{2}\left(\mathbf{H}+\mathbf{H}^T+\mathbf{H}^T\mathbf{H}\right) \tag{2.4.15}$$

Exercise 2.4.3

Show that

$$I_C = 3 + 2I_E \tag{2.4.16}$$

$$II_C = 3 + 4I_E + 4II_E \tag{2.4.17}$$

and

$$III_C = 1 + 2I_E + 4II_E + 8III_E \tag{2.4.18}$$

where I_C, II_C, III_C are the fundamental invariants defined by (A.5.44) through (A.5.46).

Given the various strain measures introduced in this section, their material derivatives are *rate of strain* measures. Given the deformation gradient $\mathbf{F}$, its derivative $\dot{\mathbf{F}}$ measures the *rate of deformation*. If (2.2.24) is written

$$\mathbf{L} = \dot{\mathbf{F}}\mathbf{F}^{-1} \tag{2.4.19}$$

it follows that **L** also measures the rate of deformation. More precisely, it follows from (2.4.19) that

$$\mathbf{L}=\dot{\mathbf{F}}\Big|_{\mathbf{F}=\mathbf{I}} \tag{2.4.20}$$

and, thus, **L** is the rate of deformation measured with respect to a configuration which instantaneously coincides with the reference configuration.

Given the decompositions (2.4.1), it is natural to use (2.4.20) and attempt to decompose **L** into a part which measures rate of rotation and a part which measures rate of deformation. From (2.4.1a),

$$\dot{\mathbf{F}}=\mathbf{R}\dot{\mathbf{U}}+\dot{\mathbf{R}}\mathbf{U} \tag{2.4.21}$$

When **F** = **I**, it follows from the polar decomposition theorem that **U** = **I** and **R** = **I**. Therefore, (2.4.21) yields

$$\dot{\mathbf{F}}\Big|_{\mathbf{F}=\mathbf{I}}=\dot{\mathbf{U}}\Big|_{\mathbf{F}=\mathbf{I}}+\dot{\mathbf{R}}\Big|_{\mathbf{F}=\mathbf{I}} \tag{2.4.22}$$

Therefore, in the state which instantaneously coincides with the reference configuration, $\dot{\mathbf{U}}\Big|_{\mathbf{F}=\mathbf{I}}$ measures the rate of deformation and $\dot{\mathbf{R}}\Big|_{\mathbf{F}=\mathbf{I}}$ measures the rate of rotation. Because **R** is orthogonal,

$$\mathbf{R}\mathbf{R}^T=\mathbf{I} \tag{2.4.23}$$

The derivative of (2.4.23) yields

$$\dot{\mathbf{R}}\mathbf{R}^T=-(\dot{\mathbf{R}}\mathbf{R}^T)^T \tag{2.4.24}$$

Therefore, $\dot{\mathbf{R}}\mathbf{R}^T$ is a skew symmetric linear transformation, and, thus,

$$\dot{\mathbf{R}}\Big|_{\mathbf{F}=\mathbf{I}}=-\dot{\mathbf{R}}^T\Big|_{\mathbf{F}=\mathbf{I}} \tag{2.4.25}$$

Because **U** is symmetric, $\dot{\mathbf{U}}$ is symmetric and, thus,

$$\dot{\mathbf{U}}\Big|_{\mathbf{F}=\mathbf{I}}=\dot{\mathbf{U}}^T\Big|_{\mathbf{F}=\mathbf{I}} \tag{2.4.26}$$

Therefore, (2.4.22) is a decomposition of $\dot{\mathbf{F}}\Big|_{\mathbf{F}=\mathbf{I}}$ into its symmetric and skew symmetric parts. Because of (2.4.20), (2.4.22) is the decomposition of **L** into its symmetric and skew symmetric parts. Therefore

$$\tfrac{1}{2}\left(\mathbf{L}+\mathbf{L}^T\right)=\dot{\mathbf{U}}\Big|_{\mathbf{F}=\mathbf{I}} \tag{2.4.27}$$

and

$$\tfrac{1}{2}\left(\mathbf{L}-\mathbf{L}^T\right)=\dot{\mathbf{R}}\Big|_{\mathbf{F}=\mathbf{I}} \tag{2.4.28}$$

We shall use the symbols **D** and **W** for the symmetric and skew symmetric parts of **L**, respectively. These quantities are, of course, defined by

$$\mathbf{D}=\tfrac{1}{2}\left(\mathbf{L}+\mathbf{L}^T\right) \tag{2.4.29}$$

and

$$\mathbf{W}=\tfrac{1}{2}\left(\mathbf{L}-\mathbf{L}^T\right) \tag{2.4.30}$$

The linear transformation **D** measures the rate of strain with respect to a state which instantaneously coincides with the reference configuration, and **W** measures the rate of rotation with respect to the same state. **D** is called the *stretching* tensor and **W** is called the *spin* tensor. **D** is sometimes called the *rate of strain* tensor, and the vector formed from the skew symmetric **W** is called the *vorticity* vector.

Exercise 2.4.4

Show that

$$\dot{\mathbf{C}}=2\mathbf{F}^T\mathbf{D}\mathbf{F} \tag{2.4.31}$$

Exercise 2.4.5

Show that

$$\mathbf{D}=\tfrac{1}{2}\mathbf{R}\left(\dot{\mathbf{U}}\mathbf{U}^{-1}+\mathbf{U}^{-1}\dot{\mathbf{U}}\right)\mathbf{R}^T \tag{2.4.32}$$

and

$$\mathbf{W}=\dot{\mathbf{R}}\mathbf{R}+\tfrac{1}{2}\mathbf{R}\left(\dot{\mathbf{U}}\mathbf{U}^{-1}-\mathbf{U}^{-1}\dot{\mathbf{U}}\right)\mathbf{R}^T \tag{2.4.33}$$

Equations (2.4.32) and (2.4.33) show that **D** is not a pure rate of stretching and **W** is not a pure rate of rotation when these rates are measured with respect to a fixed configuration. If the body is rigid, $\mathbf{U}=\mathbf{I}$ for all **X**. Equation (2.4.32) yields $\mathbf{D}=\mathbf{0}$ for a rigid body, and (2.4.33) yields $\mathbf{W}=\dot{\mathbf{R}}\mathbf{R}^T$. The term $\dot{\mathbf{R}}\mathbf{R}^T$ represents the *angular velocity* of the rigid body.

Exercise 2.4.6

Show that, for a rigid body with velocity $\dot{\mathbf{x}}_o$ at the point $\mathbf{x}_o$, the velocity of any point **x** in the body is given by

$$\dot{\mathbf{x}}=\dot{\mathbf{x}}_o+\dot{\mathbf{R}}\mathbf{R}^T\left(\mathbf{x}-\mathbf{x}_o\right) \tag{2.4.34}$$

Equation (2.4.34) shows that the velocity of a rigid body can always be represented by a translation of a point $\mathbf{x}_o$ in the body plus a rotation about that point.

Exercise 2.4.7

Show that

$$\mathbf{D}=\tfrac{1}{2}\left[\operatorname{grad}\left(\frac{\delta\mathbf{p}}{\delta t}\right)+\operatorname{grad}\left(\frac{\delta\mathbf{p}}{\delta t}\right)^T\right] \tag{2.4.35}$$

and

$$\mathbf{W}=\tfrac{1}{2}\left[\operatorname{grad}\left(\frac{\delta\mathbf{p}}{\delta t}\right)-\operatorname{grad}\left(\frac{\delta\mathbf{p}}{\delta t}\right)^T\right]+\mathbf{Z}, \tag{2.4.36}$$

where $\mathbf{p}$ is the position vector introduced in (2.3.34). These results show that the spin tensor is sensitive to rigid rotations while stretching tensor is not.

Exercise 2.4.8

For certain types of viscoelastic fluids an important kinematic quantity is the *nth Rivlin-Ericksen* tensor, $\mathbf{A}_n$, for $n=1,2,\ldots$, defined by

$$\frac{d^n\mathbf{C}}{dt^n}=\mathbf{F}^T\mathbf{A}_n\mathbf{F} \tag{2.4.37}$$

Show that

$$\mathbf{A}_1=2\mathbf{D} \tag{2.4.38}$$

and

$$\mathbf{A}_2=\operatorname{grad}\ddot{\mathbf{x}}+\left(\operatorname{grad}\ddot{\mathbf{x}}\right)^T+2\mathbf{L}^T\mathbf{L} \tag{2.4.39}$$

Exercise 2.4.9

If $\mathbf{F}(\mathbf{X}, t)$ is orthogonal for every $\mathbf{X}$ and for all t and if $\mathbf{F}$ is differentiable, show that $\mathbf{F}$ is independent of $\mathbf{X}$ and that the body is rigid.

Exercise 2.4.10

Consider the homogeneous deformation of the form

$$\mathbf{x}=\mathbf{F}\mathbf{X} \tag{2.4.40}$$

where **F** is the deformation gradient defined in Exercise A.5.22 of Appendix A. Given a material element in the shape of a unit cube aligned with the coordinate axes, construct the deformed element which results from the deformation (2.4.40).

Exercise 2.4.11

Show that

$$\operatorname{div}((\det \mathbf{F}^{-1})\mathbf{V}) = (\det \mathbf{F}^{-1})\operatorname{Div}\mathbf{R} \tag{2.4.41}$$

One approach to working this exercise is to choose $\mathbf{u} = \mathbf{R}^T\mathbf{v}$ in (2.2.35) where $\mathbf{v}$ is an arbitrary vector.

2.5. Infinitesimal Strain Kinematics

If the displacement gradients are, in some sense, small, the results of Section 2.4 are simplified considerably. This section is concerned with a discussion of the kinematics of strain in this special circumstance. The results are important in the classical theory of linear elasticity.

The magnitude of a deformation is measured by a positive number $\in$ defined by

$$\in = \sqrt{\operatorname{tr}\left(\mathbf{H}\mathbf{H}^T\right)} \equiv \|\mathbf{H}\| \tag{2.5.1}$$

In components, $\in$ is defined by

$$\in = \sqrt{\frac{\partial w_j}{\partial X_J}\frac{\partial w_j}{\partial X_J}} \tag{2.5.2}$$

Equation (2.5.1) defines the *norm* of a linear transformation **H** in terms of the trace operation. It is easy to show that this definition of norm is consistent with the definition (A.3.1), and it has the properties summarized in Theorem A.3.1.

It follows from (2.5.1) that when $\in$ is small every component of **H** is necessarily small. We shall denote by $O(\in)$ any function of **H** with the property that

$$\|O(\in)\| < M \in \tag{2.5.3}$$

as $\in \rightarrow 0$, where M is any positive number. If $\in \ll 1$, the deformation is said to be *small* or *infinitesimal*.

It is a consequence of (2.5.1) that

$$\mathbf{H} = O(\in) \tag{2.5.4}$$

and

$$\mathbf{H}^T = O(\in) \tag{2.5.5}$$

Since

$$O(\in^n)O(\in^m) = O(\in^{n+m}) \tag{2.5.6}$$

for positive integers *n* and *m*, it follows from (2.5.4) and (2.5.5) that

$$\mathbf{H}^T\mathbf{H} = O(\in^2) \tag{2.5.7}$$

and

$$\mathbf{H}\mathbf{H}^T = O(\in^2) \tag{2.5.8}$$

If $\tilde{\mathbf{E}}$ denotes the symmetric linear transformation defined by

$$\tilde{\mathbf{E}} = \frac{1}{2}(\mathbf{H} + \mathbf{H}^T) \tag{2.5.9}$$

it follows from (2.4.13), (2.4.14), (2.4.15), (2.5.7) and (2.5.8) that

$$\mathbf{C} = \mathbf{I} + 2\tilde{\mathbf{E}} + O(\in^2) \tag{2.5.10}$$

$$\mathbf{B} = \mathbf{I} + 2\tilde{\mathbf{E}} + O(\in^2) \tag{2.5.11}$$

and

$$\mathbf{E} = \tilde{\mathbf{E}} + O(\varepsilon^2) \tag{2.5.12}$$

Equations (2.5.10) through (2.5.12) show that $\tilde{\mathbf{E}}$ is a measure of the strain whenever terms like $O(\in^2)$ are neglected. It is for this reason that $\tilde{\mathbf{E}}$ is called the *infinitesimal strain measure*.

It is also possible to derive approximate expressions for **U**, **V**, and **R** that are valid for infinitesimal deformations. These expressions are

$$\mathbf{U} = \mathbf{I} + \tilde{\mathbf{E}} + O(\in^2) \tag{2.5.13}$$

$$\mathbf{V} = \mathbf{I} + \tilde{\mathbf{E}} + O(\in^2) \tag{2.5.14}$$

and

$$\mathbf{R}=\mathbf{I}+\tilde{\mathbf{R}}+O\left(\in^2\right) \tag{2.5.15}$$

where

$$\tilde{\mathbf{R}}=\frac{1}{2}\left(\mathbf{H}-\mathbf{H}^T\right) \tag{2.5.16}$$

Equations (2.5.13) and (2.5.14) follow from (2.4.7) and (2.4.8) by extracting the square root of (2.5.10) and (2.5.11). Equations (2.5.15) follows from (2.4.1a), written in the form

$$\begin{aligned}\mathbf{R} &= \mathbf{F}\mathbf{U}^{-1}=(\mathbf{I}+\mathbf{H})\left[\mathbf{I}+\tilde{\mathbf{E}}+O\left(\in^2\right)\right]^{-1}=(\mathbf{I}+\mathbf{H})\left[\mathbf{I}-\tilde{\mathbf{E}}+O\left(\in^2\right)\right]\\ &=\mathbf{I}+\mathbf{H}-\tilde{\mathbf{E}}+O\left(\in^2\right)=\mathbf{I}+\mathbf{H}-\frac{1}{2}\left(\mathbf{H}+\mathbf{H}^T\right)+O\left(\in^2\right)\\ &=\mathbf{I}+\frac{1}{2}\left(\mathbf{H}-\mathbf{H}^T\right)+O\left(\in^2\right)=\mathbf{I}+\tilde{\mathbf{R}}+O\left(\in^2\right)\end{aligned}$$

The kinematical meaning of $\tilde{\mathbf{R}}$ should be clear from (2.5.15). It is the linear correction to the rotation $\mathbf{R}=\mathbf{I}$. This is the reason $\tilde{\mathbf{R}}$ is called the *infinitesimal rotation tensor*. It follows from (2.5.9) and (2.5.16) that

$$\mathbf{H}=\tilde{\mathbf{E}}+\tilde{\mathbf{R}} \tag{2.5.17}$$

Equation (2.5.17) is a decomposition of the infinitesimal deformation into the sum of an infinitesimal strain and an infinitesimal rotation.

Exercise 2.1.16

Show that

$$\det\mathbf{F}=1+\operatorname{tr}\mathbf{H}+O\left(\in^2\right) \tag{2.5.18}$$

$$=1+\operatorname{tr}\tilde{\mathbf{E}}+O\left(\in^2\right) \tag{2.5.19}$$

$$I_C=3+2\operatorname{tr}\tilde{\mathbf{E}}+O\left(\in^2\right) \tag{2.5.20}$$

$$II_C=3+4\operatorname{tr}\tilde{\mathbf{E}}+O\left(\in^2\right) \tag{2.5.21}$$

and

$$III_C=1+2\operatorname{tr}\tilde{\mathbf{E}}+O\left(\in^2\right) \tag{2.5.22}$$

Equation (2.5.18) shows that $\operatorname{tr}\mathbf{H}=\operatorname{tr}\tilde{\mathbf{E}}$ measures volume deformations in the case of small deformations. In linear elasticity, $\operatorname{tr}\tilde{\mathbf{E}}$ is called the *dilatation*.

References

1. TRUESDELL, C. and W. NOLL, *The Non-Linear Field Theories of Mechanics*, Handbuch der Physik, Vol. III/3 (ed. S. Flugge). Springer-Verlag (1965).
2. WANG, C.-C., On the Geometric Structure of Simple Bodies, A Mathematical Foundation for the Theory of Continuous Distributions of Dislocations, *Arch. Rational Mech. Anal.*, **27**, 33-94 (1967).
3. NOLL, W., Materially Uniform Simple Bodies with Inhomogeneities, *Arch. Rational Mech. Anal.*, **27**, 1-32 (1967).
4. SIMMONS, G. F., *Introduction to Topology and Modern Analysis*, McGraw-Hill, New York (1963).
5. TRUESDELL, C. and R. A. TOUPIN, *The Classical Field Theories*, Handbuch der Physik, Vol. III/l (ed. S. Flugge), Springer-Verlag, Berlin (1960).
6. MILNE, E. A., *Vectorial Mechanics*, Interscience, New York (1948).

Bibliography

7. WANG, C.-C., and C. TRUESDELL, *Introduction to Rational Elasticity*, Noordhoff, Leyden (1973).
8. TRUESDELL, C., *A First Course in Rational Continuum Mechanics*, Vol. 1, Academic Press, New York (1977).
9. GURTIN, M. E., *An Introduction to Continuum Mechanics*, Academic Press, New York (1981)
10. CHADWICK, P., *Continuum Mechanics*, George Allen & Unwin Ltd., London (1976).
11. SPENCER, A. J. M., *Continuum Mechanics*, Longman, London (1980).
12. MALVERN, L. E., *Introduction to the Mechanics of a Continuous Medium*, Prentice-Hall, Englewood Cliffs (1969).
13. LEIGH, D. C., *Nonlinear Continuum Mechanics*, McGraw-Hill, New York (1968).
14. JAUNZEMIS, W., *Continuum Mechanics*, Macmillan, New York, (1967).
15. FUNG, Y. C., A *First Course in Continuum Mechanics* (2nd Ed.), Prentice-Hall, Englewood Cliffs (1977).
16. ZHONG-HENG, GAO, Rates of Stretch Tensors, *Journal of Elasticity*, **14,** 263-267 (1984).
17. HOGER, A. and D. E. CARLSON, Determination of the Stretch and Rotations in the Polar Decomposition of the Deformation Gradient, *Quarterly of Appl. Math.*, **42**, 113-117 (1984).
18. HOGER, A. and D. E. CARLSON, On the Derivative of the Square Root of a Tensor and Guo's Rate Theorems, *Journal of Elasticity*, **14**, 329-336 (1984).

3

Equations of Balance

This chapter is concerned with the three dimensional statements of the equations of balance. As in Chapter I, the discussion will include statements of balance of mass, linear momentum, energy and the entropy inequality. In addition, we shall discuss balance of moment of momentum. This balance equation is trivial in one dimension.

3.1. Balance of Mass

The body $\mathscr{B}$ is assumed to be endowed with a nonnegative scalar property known as the *mass*. For our purposes, the idea of mass will be introduced by assuming the existence of a positive valued function on $\mathscr{B}$ called the mass density. The *mass density*, or simply the density, of X in $\mathscr{B}$ at the time t is given by

$$\rho = \rho(X,t) \tag{3.1.1}$$

By use of (2.1.2), we can regard ρ to be a function of $(\mathbf{x},t)$. In this case, we shall write

$$\rho = \rho(\mathbf{x},t) \tag{3.1.2}$$

where, for notational simplicity, we have used the same symbol for the two different functions which appear in (3.1.1) and (3.1.2). Also, when the independent variables are $(\mathbf{X},t)$ we shall again write

$$\rho = \rho(\mathbf{X},t) \tag{3.1.3}$$

The above notational convention will not cause any confusion since the context will always indicate which function is intended.

If $\boldsymbol{\kappa}$ is the reference configuration, the density of $\mathscr{B}$ in $\boldsymbol{\kappa}$ will be written

$$\rho_R = \rho_R(\mathbf{X}) \tag{3.1.4}$$

The *mass* of a part $\mathscr{P}$ of $\mathscr{B}$ in the configuration χ at the time t is

$$m(\mathscr{P},t) = \int_{\chi(\mathscr{P},t)} \rho \, dv \tag{3.1.5}$$

Axiom (Balance of Mass). For all parts $\mathscr{P}$ in $\mathscr{B}$,

$$m(\mathscr{P},t) = \text{const} \tag{3.1.6}$$

or, equivalently,

$$\overline{m(\mathscr{P},t)}^{\cdot} = 0 \tag{3.1.7}$$

Note that $m(\cdot,t)$ is a set *function*; i.e., the domain of $m(\cdot,t)$ is a collection of sets. Each element of the domain represents a part of $\mathscr{B}$. Clearly, the axiom of balance of mass is equivalent to the following statements:

1.
$$m(\mathscr{P},t) = m(\mathscr{P},\tau) \tag{3.1.8}$$

for $t \neq \tau$,

2.
$$\overline{\int_{\chi(\mathscr{P},t)} \rho \, dv}^{\cdot} = 0 \tag{3.1.9}$$

and

3.
$$\int_{\chi(\mathscr{P},t)} \rho \, dv = \int_{\kappa(\mathscr{P})} \rho_R \, dV \tag{3.1.10}$$

for all $\mathscr{P}$ in $\mathscr{B}$.

If we use (2.3.2), equation (3.1.10) can be written

$$\int_{\kappa(\mathscr{P})} \left(\rho \left|\det \mathbf{F}\right| - \rho_R\right) dV = 0 \tag{3.1.11}$$

Since $\mathscr{P}$ is an arbitrary part of $\mathscr{B}$, (3.1.11) is equivalent to

$$\rho \left|\det \mathbf{F}\right| = \rho_R \tag{3.1.12}$$

for all $(\mathbf{X},t)$ in $\boldsymbol{\kappa}(\mathscr{B}) \times (-\infty,\infty)$. Equation (3.1.12) is a *local* version of the axiom of balance of mass. When ρ and $\mathbf{F}$ are differentiable functions, it follows from (3.1.12) and (3.1.4) that

$$\overline{\rho \left|\det \mathbf{F}\right|}^{\cdot} = 0 \tag{3.1.13}$$

Exercise 3.1.1

Show that (3.1.13) is equivalent to the following equations:

$$\rho\overline{\left(\frac{1}{\rho}\right)}^{\cdot} = \operatorname{div}\dot{\mathbf{x}} \tag{3.1.14}$$

$$\dot{\rho} + \rho \operatorname{div}\dot{\mathbf{x}} = 0 \tag{3.1.15}$$

and

$$\frac{\partial \rho}{\partial t} + \operatorname{div}\rho\dot{\mathbf{x}} = 0 \tag{3.1.16}$$

Exercise 3.1.2

Derive (3.1.16) from (3.1.9) by use of (2.3.12).

Equations (3.1.12), (3.1.14), (3.1.15) and (3.1.16) summarize the usual local statements of the axiom of balance of mass. Equation (3.1.12) is the form usually used in solid mechanics while the others are common in fluid mechanics. All of these equations hold at points where ρ, $\dot{\mathbf{x}}$, and $\mathbf{F}$ are differentiable.

Exercise 3.1.3

Start with (3.1.16) and derive an integral statement of the axiom of balance of mass for an arbitrary region $\mathscr{R}(t)$. The answer is

$$\frac{d}{dt}\int_{\mathscr{R}(t)} \rho\, dv = -\oint_{\partial\mathscr{R}(t)} \rho(\dot{\mathbf{x}} - \mathbf{v})\cdot d\mathbf{s} \tag{3.1.17}$$

where $\mathbf{v}$ denotes the velocity of a point on the surface $\partial\mathscr{R}(t)$. What do the terms in (3.1.17) represent physically?

In deriving (3.1.14), (3.1.15) and (3.1.16), we assumed that ρ, $\dot{\mathbf{x}}$, and $\mathbf{F}$ are differentiable within $\chi(\mathscr{P},t)$. From our one dimensional discussion in Chapter I, we know that in certain physical problems this is not the case. Across a three dimensional shock wave ρ, $\dot{\mathbf{x}}$, and $\mathbf{F}$ are not continuous. They undergo jump discontinuities. A typical problem in continuum mechanics is to compute the properties of these discontinuities. In order to relax the assumptions leading to (3.1.14), (3.1.15) and (3.1.16), we shall now consider the case when ρ, $\dot{\mathbf{x}}$, and $\mathbf{F}$ are differentiable at all points except on a surface $\Sigma(t)$, where ρ, $\dot{\mathbf{x}}$, and $\mathbf{F}$ suffer jump discontinuities. It follows from (3.1.9) and (2.3.23) that

$$\int_{\chi(\mathscr{P},t)}\left(\frac{\partial\rho}{\partial t}+\operatorname{div}\rho\dot{\mathbf{x}}\right)dv=\int_{\Sigma(t)}[\rho(\dot{x}_n-u_n)]d\sigma \tag{3.1.18}$$

where

$$\dot{x}_n=\dot{\mathbf{x}}\cdot\mathbf{n} \tag{3.1.19}$$

is the normal component of $\dot{\mathbf{x}}$ at $\Sigma(t)$. The arbitrary nature of $\chi(\mathscr{P},t)$ allows us to conclude from (3.1.18) that (3.1.16) holds at points where ρ and $\dot{\mathbf{x}}$ are differentiable. In addition, (3.1.18) shows that

$$\int_{\Sigma(t)}[\rho(\dot{x}_n-u_n)]d\sigma=0 \tag{3.1.20}$$

Equation (3.1.20) yields

$$[\rho(\dot{x}_n-u_n)]=0 \tag{3.1.21}$$

at points $\mathbf{x}$ on $\Sigma(t)$. Equation (3.1.21) follows from (3.1.20) and the arbitrary nature of $\Sigma(t)$. The physical meaning of (3.1.21) should be clear. It simply states that the mass flux per unit of area is continuous across $\Sigma(t)$. The reader has no doubt noticed the formal similarity of the derivation of (3.1.21) and its one dimensional version (1.2.12).

Exercise 3.1.4

Show that the three dimensional generalization of (1.2.14) is

$$\rho\dot{\psi}=\frac{\partial\rho\psi}{\partial t}+\operatorname{div}(\rho\psi\dot{\mathbf{x}}) \tag{3.1.22}$$

Exercise 3.1.5

Show that the three dimensional generalization of (1.2.15) is

$$\overline{\int_{\chi(\mathscr{P},t)}\rho\psi\,dv}^{\,\cdot}=\int_{\chi(\mathscr{P},t)}\rho\dot{\psi}\,dv-\int_{\Sigma(t)}[\rho\psi(\dot{x}_n-u_n)]d\sigma \tag{3.1.23}$$

Exercise 3.1.6

Use (2.3.36) and show that

$$\operatorname{div}\dot{\mathbf{x}}=\operatorname{div}\left(\frac{\delta\mathbf{p}}{\delta t}\right) \tag{3.1.24}$$

Since $\overline{(1/\rho)}^{\cdot} = \delta(1/\rho)/\delta t$, (3.1.24) and (3.1.14) combine to yield

$$\rho \frac{\delta(1/\rho)}{\delta t} = \operatorname{div}\left(\frac{\delta \mathbf{p}}{\delta t}\right) \tag{3.1.25}$$

which is the local statement of balance of mass appropriate to an observer in a rotating and translating coordinate system.

Exercise 3.1.7

Show that (3.1.25) can be written

$$\left.\frac{\partial \rho}{\partial t}\right|_{\mathbf{p}} + \operatorname{div}\left(\rho \frac{\delta \mathbf{p}}{\delta t}\right) = 0 \tag{3.1.26}$$

where $\left.\frac{\partial \rho}{\partial t}\right|_{\mathbf{p}}$ denotes the time derivative of $\rho = \rho(\mathbf{p},t)$ at constant $\mathbf{p}$.

Exercise 3.1.8

The body $\mathscr{B}$ is *incompressible* if $\dot{\rho} = 0$ for all X in $\mathscr{B}$. Show that in this case

$$|\det \mathbf{F}| = 1 \tag{3.1.27}$$

and

$$\operatorname{div} \dot{\mathbf{x}} = 0 \tag{3.1.28}$$

3.2. Balance of Linear Momentum

In this section, the axiom of balance of linear momentum is stated and discussed.

The *linear momentum* of the part $\mathscr{P}$ of the body $\mathscr{B}$ is

$$\mathbf{k}(\mathscr{P},t) = \int_{\chi(\mathscr{P},t)} \rho \dot{\mathbf{x}}\, dv \tag{3.2.1}$$

The *center of mass* of the part $\mathscr{P}$ of the body $\mathscr{B}$ is

$$\mathbf{x}_c(\mathscr{P},t) = \frac{1}{m(\mathscr{P},t)} \int_{\chi(\mathscr{P},t)} \rho \mathbf{x}\, dv \tag{3.2.2}$$

where $m(\mathscr{P},t)$ is given by (3.1.5).

Exercise 3.2.1

Show that the linear momentum of the part $\mathscr{P}$, no matter what deformation it undergoes, equals the linear momentum of a mass point of mass $m(\mathscr{P},t)$ located at the center of mass moving with the velocity of the center of mass, i.e., show that

$$\mathbf{k}(\mathscr{P},t) = m(\mathscr{P},t)\dot{\mathbf{x}}_c \tag{3.2.3}$$

It should be stressed that (3.2.3) depends strongly on the fact that the mass within $\chi(\mathscr{P},t)$ is constant. The concept of center of mass is of little value in those cases where the mass is not constant.

The statement of the axiom of balance of linear momentum takes the concept of *force* as fundamental. If $\mathbf{f}(\mathscr{P},t)$ denotes the *resultant force* acting on the part $\mathscr{P}$ in $\mathscr{B}$ at the time t, the axiom of balance of linear momentum is the following:

Axiom (Balance of Linear Momentum) For all parts $\mathscr{P}$ in $\mathscr{B}$,

$$\dot{\mathbf{k}}(\mathscr{P},t) = \mathbf{f}(\mathscr{P},t) \tag{3.2.4}$$

Equation (3.2.4) will be referred to as the *equation of motion* of $\mathscr{P}$. Also, note that this axiom is stated for the material occupying the volume $\chi(\mathscr{P},t)$. The important property of $\chi(\mathscr{P},t)$ is that is contains a constant mass. It will be shown in an exercise how (3.2.4) is modified when a region with variable mass is used.

Equation (3.2.4) is usually referred to as "Newton's second law," although it was stated explicitly for the first time by Euler [Ref. 1]. Equivalent forms of (3.2.4) are

$$\overline{\int_{\chi(\mathscr{P},t)} \rho\dot{\mathbf{x}}\, dv}^{\,\cdot} = \mathbf{f}(\mathscr{P},t) \tag{3.2.5}$$

and

$$m(\mathscr{P},t)\ddot{\mathbf{x}}_c = \mathbf{f}(\mathscr{P},t) \tag{3.2.6}$$

Exercise 3.2.2

Show that if (3.2.6) is written with respect to the translating and rotating system, it takes the form

$$m(\mathcal{P},t)\left(\ddot{\mathbf{c}}+\frac{\delta^2\mathbf{p}_c}{\delta t^2}+\dot{\boldsymbol{\omega}}\times\mathbf{p}_c+2\boldsymbol{\omega}\times\frac{\delta\mathbf{p}_c}{\delta t}+\boldsymbol{\omega}\times(\boldsymbol{\omega}\times\mathbf{p}_c)\right)=\mathbf{f}(\mathcal{P},t) \tag{3.2.7}$$

where $\mathbf{p}_c$ denotes the position vector to $\mathbf{x}_c$ from $\mathbf{c}(t)$. Identify the terms in (3.2.7).

Assume that the body $\mathscr{B}$ and its motion are given.

Definition. A *system of forces* for the body in motion is characterized by the following conditions [Ref. 2, Sect. 16]:

1. For every t in $(-\infty,\infty)$ there exists a vector field $\mathbf{b}(\cdot,t)$ defined on $\chi(\mathscr{P},t)$.
2. For every t in $(-\infty,\infty)$ and for every $\mathscr{P}$ in $\mathscr{B}$ there exists a vector field $\mathbf{t}(\cdot,\mathscr{P},t)$ defined on $\partial\chi(\mathscr{P},t)$.

The vector field $\mathbf{b}(\cdot,t)$ is called the external *body force density* on $\mathscr{B}$. The vector $\mathbf{f}_b(\mathscr{P},t)$ defined by

$$\mathbf{f}_b(\mathscr{P},t)=\int_{\chi(\mathscr{P},t)}\rho\mathbf{b}\,dv \tag{3.2.8}$$

is the *resultant external body force* exerted on $\mathscr{P}$ at the time t. The vector field $\mathbf{t}(\cdot,\mathscr{P},t)$ is called the *stress vector* and represents the contact force acting on $\mathscr{P}$ at the time t. The *resultant contact force* $\mathbf{f}_c(\mathscr{P},t)$ exerted on $\mathscr{P}$ at the time t is defined by

$$\mathbf{f}_c(\mathscr{P},t)=\oint_{\partial\chi(\mathscr{P},t)}\mathbf{t}(\mathbf{x},\mathscr{P},t)\,ds \tag{3.2.9}$$

where $ds=\|d\mathbf{s}\|$. Physically, $\mathbf{f}_c(\mathscr{P},t)$ is the resultant force exerted by the material outside of $\mathscr{P}$ which is in contact with the surface $\partial\chi(\mathscr{P},t)$. The resultant force $\mathbf{f}(\mathscr{P},t)$ exerted on $\mathscr{P}$ in $\mathscr{B}$ at the time t is defined by

$$\mathbf{f}(\mathscr{P},t)=\mathbf{f}_b(\mathscr{P},t)+\mathbf{f}_c(\mathscr{P},t) \tag{3.2.10}$$

Under certain rather formal mathematical assumptions, one can prove that there exists a vector field $\mathbf{t}(\cdot,\mathbf{n},t)$, where $\mathbf{n}$ is the unit exterior normal to $\partial\chi(\mathscr{P},t)$, such that

$$\mathbf{t}(\mathbf{x},\mathscr{P},t)=\mathbf{t}(\mathbf{x},\mathbf{n},t) \tag{3.2.11}$$

for all $\mathbf{x}$ on $\partial\chi(\mathscr{P},t)$ and t in $(-\infty,\infty)$. The mathematical argument leading to (3.2.11) can be found in Ref. 3. Additional discussion relevant to this argument can be found in Refs. 4, 5, and 6.

By use of (3.2.8), (3.2.9), (3.2.10) and (3.2.11), the axiom (3.2.5) can be written

$$\overline{\int_{\chi(\mathscr{P},t)} \rho\dot{\mathbf{x}}\,dv} = \oint_{\partial\chi(\mathscr{P},t)} \mathbf{t}(\mathbf{x},\mathbf{n},t)\,ds + \int_{\chi(\mathscr{P},t)} \rho\mathbf{b}\,dv \tag{3.2.12}$$

The *normal stress* at $\mathbf{x}$ on $\partial\chi(\mathscr{P},t)$ is

$$\mathbf{t}(\mathbf{x},\mathbf{n},t)\cdot\mathbf{n}$$

Since $\mathbf{t}(\mathbf{x},\mathbf{n},t)$ is the local effect of the material outside of $\chi(\mathscr{P},t)$ on the material inside of $\chi(\mathscr{P},t)$, if $\mathbf{t}(\mathbf{x},\mathbf{n},t)\cdot\mathbf{n}$ is negative, the normal stress is *compressive* and, if $\mathbf{t}(\mathbf{x},\mathbf{n},t)\cdot\mathbf{n}$ is positive, the normal stress is a *tension*. The projection of $\mathbf{t}(\mathbf{x},\mathbf{n},t)$ in the direction of a plane normal to $\mathbf{n}$ is the *shear stress*. If $\mathbf{t}(\mathbf{x},\mathbf{n},t)$ is parallel to $\mathbf{n}$, and if the factor of proportionality is independent of $\mathbf{n}$, the stress is *hydrostatic*.

Next, we shall show that if $\mathbf{t}(\mathbf{x},\mathbf{n},t)$ is a continuous function of $\mathbf{x}$, there exists a linear transformation $\mathbf{T}$, depending upon $\mathbf{x}$ and t, such that all stress vectors $\mathbf{t}(\mathbf{x},\mathbf{n},t)$ at $\mathbf{x}$ and t are determined by

$$\mathbf{t}(\mathbf{x},\mathbf{n},t) = \mathbf{T}(\mathbf{x},t)\mathbf{n} \tag{3.2.13}$$

This result is known as *Cauchy's theorem*, and it asserts that $\mathbf{t}(\mathbf{x},\mathbf{n},t)$ depends upon the surface orientation only in a *linear* fashion. The tensor $\mathbf{T}$ is called the *stress tensor*. The proof of (3.2.13) is based on a preliminary result which is stated as follows:

If d denotes a characteristic dimension of $\chi(\mathscr{P},t)$ and if $\rho\ddot{\mathbf{x}}$ and $\rho\mathbf{b}$ are bounded for all $\chi(\mathscr{P},t)$, then

$$\lim_{d\to 0}\frac{1}{d^2}\oint_{\partial\chi(\mathscr{P},t)} \mathbf{t}(\mathbf{x},\mathbf{n},t)\,ds = \mathbf{0} \tag{3.2.14}$$

Because d is the characteristic dimension of $\chi(\mathscr{P},t)$, then $\chi(\mathscr{P},t)$ is proportional to d^3 and $\partial\chi(\mathscr{P},t)$ is proportional to d^2. The proportionality factors depend only on the shape of $\chi(\mathscr{P},t)$. If we select $\chi(\mathscr{P},t)$ such that it does not contain a surface of discontinuity, we can use (3.1.23) with $\psi = \dot{\mathbf{x}}$ to write (3.2.12) in the form

$$\oint_{\partial\chi(\mathscr{P},t)} \mathbf{t}(\mathbf{x},\mathbf{n},t)\,ds = \int_{\chi(\mathscr{P},t)} \rho(\ddot{\mathbf{x}}-\mathbf{b})\,dv \tag{3.2.15}$$

If we divide (3.2.15) by d^2 and let $d\to 0$, we immediately obtain (3.2.14) since, by assumption, $\rho\ddot{\mathbf{x}}$ and $\rho\mathbf{b}$ are bounded. Returning to the proof of Cauchy's Theorem, consider the tetrahedron shown in the following figure. This is tetrahedron has three of its faces parallel to the coordinate planes through $\mathbf{x}$ and the fourth with normal $\mathbf{n}$. If $d\mathbf{s}$ is the element of area for the slant face, the element of area for the face perpendicular to the coordinate axis $\mathbf{i}_j$ is $ds_j = (\mathbf{n}\cdot\mathbf{i}_j)ds$. If (3.2.14) is applied to the tetrahedron, it follows that

$$\mathbf{t}(\mathbf{x},\mathbf{n},t)+\sum_{j=1}^{3}\mathbf{t}(\mathbf{x},-\mathbf{i}_j,t)(\mathbf{n}\cdot\mathbf{i}_j)=\mathbf{0}$$

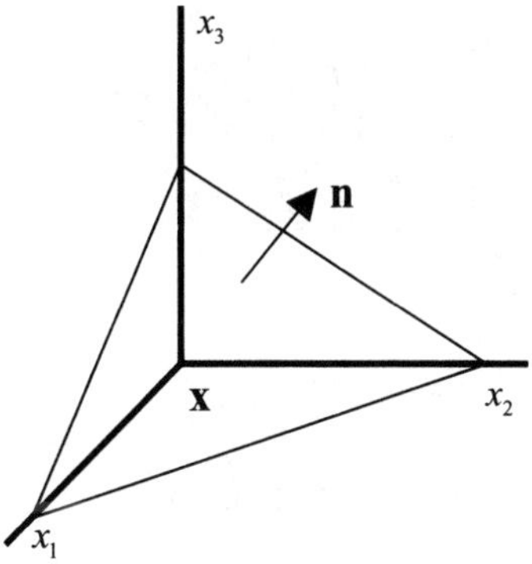

Figure 3.2.1

Therefore,

$$\mathbf{t}(\mathbf{x},\mathbf{n},t)=-\left(\sum_{j=1}^{3}\mathbf{t}(\mathbf{x},-\mathbf{i}_j,t)\otimes\mathbf{i}_j\right)\mathbf{n}\equiv\mathbf{T}(\mathbf{x},t)\mathbf{n}$$

By use of (A.4.3), the component representation for the stress tensor can be written

$$\mathbf{T}=T_{jk}\mathbf{i}_j\otimes\mathbf{i}_k \tag{3.2.16}$$

Exercise 3.2.3

Use (3.2.13) and (3.2.16) and show that the components of $\mathbf{T}$, T_{jk}, represents the force per unit of area in the j th direction of the k th face of a unit cube. The following figure illustrates the results of this exercise.

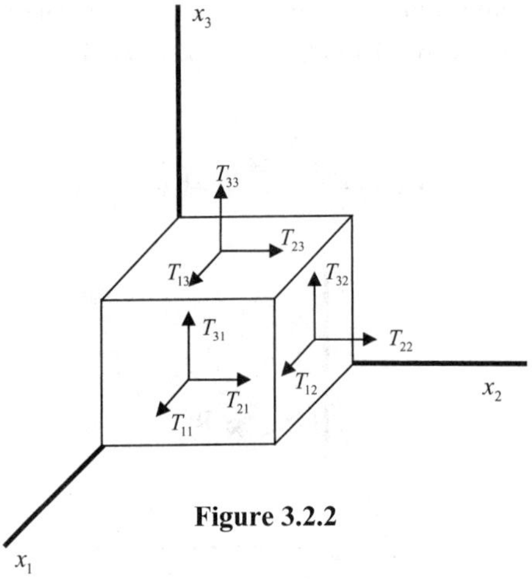

Figure 3.2.2

If equation (3.2.13) is substituted into (3.2.12), the result can be written

$$\dot{\overline{\int_{\chi(\mathscr{P},t)} \rho \dot{\mathbf{x}}\, dv}} = \oint_{\partial\chi(\mathscr{P},t)} \mathbf{T}(\mathbf{x},t)\, d\mathbf{s} + \int_{\chi(\mathscr{P},t)} \rho \mathbf{b}\, dv \tag{3.2.17}$$

If we assume $\chi(\mathscr{P},t)$ does not contain a surface of discontinuity, we can follow the same procedure used to derive (3.2.15) from (3.2.12) and deduce from (3.2.17) that

$$\int_{\chi(\mathscr{P},t)} \rho \ddot{\mathbf{x}}\, dv = \oint_{\partial\chi(\mathscr{P},t)} \mathbf{T}(\mathbf{x},t)\, d\mathbf{s} + \int_{\chi(\mathscr{P},t)} \rho \mathbf{b}\, dv \tag{3.2.18}$$

If the surface integral in (3.2.18) is converted into a volume integral by use of (A.8.29), it follows that

$$\int_{\chi(\mathscr{P},t)} \left(\rho \ddot{\mathbf{x}} - \operatorname{div}\mathbf{T} - \rho \mathbf{b}\right) dv = \mathbf{0} \tag{3.2.19}$$

Because $\chi(\mathscr{P},t)$ is arbitrary, (3.2.19) implies the following local statement of balance of linear momentum:

$$\rho \ddot{\mathbf{x}} = \operatorname{div}\mathbf{T} + \rho \mathbf{b} \tag{3.2.20}$$

This result holds at points $\mathbf{x}$ not on a surface of discontinuity. The component version of (3.2.20) is

$$\rho \ddot{x}_j = \frac{\partial T_{jk}}{\partial x_k} + \rho b_j \tag{3.2.21}$$

Exercise 3.2.4

Show that the following equations are equivalent statements of the local statement of balance of linear momentum.

1. $$\rho \ddot{\mathbf{x}} = \operatorname{div}\mathbf{T} + \rho \mathbf{b} \tag{3.2.22}$$

2. $$\rho\left(\frac{\partial \dot{\mathbf{x}}}{\partial t} + (\operatorname{grad}\dot{\mathbf{x}})\dot{\mathbf{x}}\right) = \operatorname{div}\mathbf{T} + \rho \mathbf{b} \tag{3.2.23}$$

3. $$\frac{\partial \rho \dot{\mathbf{x}}}{\partial t} + \operatorname{div}(\rho \dot{\mathbf{x}} \otimes \dot{\mathbf{x}}) = \operatorname{div}\mathbf{T} + \rho \mathbf{b} \tag{3.2.24}$$

4. $$\rho\left[\frac{\partial \dot{\mathbf{x}}}{\partial t} + \operatorname{grad}\left(\frac{\dot{\mathbf{x}}^2}{2}\right) + 2\mathbf{W}\dot{\mathbf{x}}\right] = \operatorname{div}\mathbf{T} + \rho \mathbf{b} \tag{3.2.25}$$

Exercise 3.2.5

Show that, for the material in $\mathscr{R}(t)$, the balance equation for linear momentum is

$$\frac{d}{dt}\int_{\mathscr{R}(t)} \rho \dot{\mathbf{x}}\, dv = -\oint_{\partial\mathscr{R}(t)} \rho \dot{\mathbf{x}}(\dot{\mathbf{x}} - \mathbf{v}) \cdot d\mathbf{s} + \oint_{\partial\mathscr{R}(t)} \mathbf{T}\, d\mathbf{s} + \int_{\mathscr{R}(t)} \rho \mathbf{b}\, dv \tag{3.2.26}$$

What do the terms in (3.2.26) represent physically?

Equation (3.2.20) represents the form of the equation of balance of linear momentum that occurs frequently in textbooks on fluid mechanics. While (3.2.20) applies to any continuum material, in theories of nonlinear elasticity it is useful to derive a *material* form of the axiom of balance of linear momentum. This result, derived below, is expressed in terms of the first *Piola-Kirchhoff* stress tensor. This tensor is given the symbol $\mathbf{T}_R$, and is defined by

$$\mathbf{T}_R = |\det \mathbf{F}|\mathbf{T}\mathbf{F}^{-1^T} \tag{3.2.27}$$

Exercise 3.2.6

Show that

$$\mathbf{T}\, d\mathbf{s} = \mathbf{T}_R\, d\mathbf{S} \tag{3.2.28}$$

This results shows that $\mathbf{T}_R$ is the stress tensor measured per unit of undeformed area.

The local statement of the material form of the axiom of balance of linear momentum will now be shown to be

$$\rho_R \ddot{\mathbf{x}} = \mathrm{Div}\mathbf{T}_R + \rho_R \mathbf{b} \tag{3.2.29}$$

The derivation of (3.2.29) follows by first multiplication of (3.2.20) by $|\det \mathbf{F}|$ and making use of (3.1.12). The result is

$$\rho_R \ddot{\mathbf{x}} = |\det \mathbf{F}| \mathrm{div}\mathbf{T} + \rho_R \mathbf{b} \tag{3.2.30}$$

By use of (2.2.30), it follows that

$$|\det \mathbf{F}| \mathrm{div}\mathbf{T} = \mathrm{Div}\left(|\det \mathbf{F}| \mathbf{T}\mathbf{F}^{-1^T}\right) = \mathrm{Div}\mathbf{T}_R \tag{3.2.31}$$

which yields the desired result. The reader should compare the derivation of (3.2.29) with its one dimensional version (1.3.13). In some applications the *second Piola-Kirchhoff* stress tensor is introduced. It is defined by

$$\tilde{\mathbf{T}} = \mathbf{F}^{-1}\mathbf{T}_R \tag{3.2.32}$$

Exercise 3.2.7

Show that the following jump equation governs linear momentum across a singular surface $\Sigma(t)$.

$$[\rho(\dot{x}_n - u_n)\dot{\mathbf{x}}] - [\mathbf{T}]\mathbf{n} = \mathbf{0} \tag{3.2.33}$$

Equation (3.2.33) is the three dimensional generalization of (1.3.8).

Exercise 3.2.8

Show that (3.2.33) can also be written

$$[\rho(\dot{\mathbf{x}} - u_n\mathbf{n}) \otimes (\dot{\mathbf{x}} - u_n\mathbf{n}) - \mathbf{T}]\mathbf{n} = \mathbf{0} \tag{3.2.34}$$

and

$$\gamma[\dot{\mathbf{x}}] - [\mathbf{T}]\mathbf{n} = \mathbf{0} \tag{3.2.35}$$

where $\gamma = \rho^+\left(\dot{x}_n^+ - u_n\right) = \rho^-\left(\dot{x}_n^- - u_n\right)$. Equations (3.2.34) and (3.2.35) generalize (1.3.10) and (1.3.11), respectively.

Exercise 3.2.9

Show that when $\mathbf{T}$ is given by $\mathbf{T} = -\pi\mathbf{I}$, where π is the pressure, that (3.2.34) is equivalent to

$$[\pi + \rho(\dot{x}_n - u_n)^2] = 0 \tag{3.2.36}$$

and

$$\mathbf{n} \times [\dot{\mathbf{x}}] = \mathbf{0} \tag{3.2.37}$$

if $\gamma \neq 0$. Equation (3.2.37) shows that the tangential component of $\dot{\mathbf{x}}$ is continuous across the discontinuity whenever $\mathbf{T} = -\pi\mathbf{I}$ and $\gamma \neq 0$.

Exercise 3.2.10

Show that

$$\rho\left(\ddot{\mathbf{c}} + \frac{\delta^2\mathbf{p}}{\delta t^2} + \dot{\boldsymbol{\omega}} \times \mathbf{p} + 2\boldsymbol{\omega} \times \frac{\delta\mathbf{p}}{\delta \mathbf{t}} + \boldsymbol{\omega} \times (\boldsymbol{\omega} \times \mathbf{p})\right) = \operatorname{div}\mathbf{T} + \rho\mathbf{b} \tag{3.2.38}$$

Equation (3.2.38) is the equation of motion which is utilized in geophysical continuum mechanics.

3.3. Balance of Angular Momentum

The *moment of momentum* or *angular momentum* of a part of $\mathscr{P}$ of $\mathscr{B}$ about the origin $\mathbf{0}$ is defined by

$$\mathbf{h}_o(\mathscr{P}, t) = \int_{\chi(\mathscr{P}, t)} \rho(\mathbf{x} - \mathbf{0}) \times \dot{\mathbf{x}}\, dv \tag{3.3.1}$$

It is clear from (3.3.1) that the moment of momentum has been assumed to be due entirely to the velocity $\dot{\mathbf{x}}$. This is certainly the most common case. However, there are theories of *polar materials* from which $\mathbf{h}_o(\mathscr{P}, t)$ contains a part due to an intrinsic spin moment of momentum. We shall not consider such theories in this text [Refs. 2, 7].

Axiom (Balance of Moment of Momentum) The rate of change of moment of momentum of an arbitrary part $\mathscr{P}$ of $\mathscr{B}$ about the origin $\mathbf{0}$ is equal to the resultant of the torques about the origin. If $\mathbf{l}_o(\mathscr{P}, t)$ denotes this resultant torque, then

$$\dot{\mathbf{h}}_o(\mathscr{P}, t) = \mathbf{l}_o(\mathscr{P}, t) \tag{3.3.2}$$

It is assumed here that $\mathbf{l}_0(\mathscr{P},t)$ is due to the body and contact forces. Therefore, *surface couples* and *body couples* are not allowed. With this assumption, $\mathbf{l}_o(\mathscr{P},t)$ is given by

$$\mathbf{l}_o(\mathscr{P},t)=\oint_{\partial\chi(\mathscr{P},t)}(\mathbf{x}-\mathbf{0})\times\mathbf{T}(\mathbf{x},t)\,ds+\int_{\chi(\mathscr{P},t)}\rho(\mathbf{x}-\mathbf{0})\times\mathbf{b}\,dv \tag{3.3.3}$$

Therefore, the axiom of balance of angular momentum can be written

$$\overline{\int_{\chi(\mathscr{P},t)}\rho(\mathbf{x}-\mathbf{0})\times\dot{\mathbf{x}}\,dv}^{\,\boldsymbol{\cdot}}=\oint_{\partial\chi(\mathscr{P},t)}(\mathbf{x}-\mathbf{0})\times\mathbf{T}(\mathbf{x},t)\,ds+\int_{\chi(\mathscr{P},t)}\rho(\mathbf{x}-\mathbf{0})\times\mathbf{b}\,dv \tag{3.3.4}$$

It should be clear to the reader that one could use (A.5.17) and (A.5.18) to rewrite (3.3.4) as an equivalent tensor equation. In this equivalent statement, moment of momentum and torque are treated as skew-symmetric linear transformations.

Exercise 3.3.1

Show that, when linear momentum is balanced, the balance of angular momentum with respect to one fixed point is necessary and sufficient for the balance of angular momentum with respect to any other fixed point.

Exercise 3.3.2

If

$$\mathbf{h}_a(\mathscr{P},t)=\int_{\chi(\mathscr{P},t)}\rho(\mathbf{x}-\mathbf{x}_a)\times\dot{\mathbf{x}}\,dv$$

is the angular momentum about the variable point $\mathbf{x}_a$, show that the balance equation for $\mathbf{h}_a(\mathscr{P},t)$ is

$$\dot{\mathbf{h}}_a(\mathscr{P},t)=\mathbf{l}_a(\mathscr{P},t)-m(\mathscr{P},t)\dot{\mathbf{x}}_a\times\dot{\mathbf{x}}_c \tag{3.3.5}$$

where $\mathbf{l}_a(\mathscr{P},t)$ is the resultant of moments taken about $\mathbf{x}_a$ and $\dot{\mathbf{x}}_a$ is the velocity of the point $\mathbf{x}_a$. In what cases does the above result reduce to $\dot{\mathbf{h}}_a(\mathscr{P},t)=\mathbf{l}_a(\mathscr{P},t)$?

Exercise 3.3.3

If

$$\mathbf{h}_a^r(\mathscr{P},t)=\int_{\chi(\mathscr{P},t)}\rho(\mathbf{x}-\mathbf{x}_a)\times(\dot{\mathbf{x}}-\dot{\mathbf{x}}_a)\,dv$$

is the *relative angular momentum* about the variable point $\mathbf{x}_a$, show that the balance equation for $\mathbf{h}_a^r(\mathscr{P},t)$ is

$$\dot{\mathbf{h}}_a^r(\mathscr{P},t)=\mathbf{l}_a(\mathscr{P},t)+m(\mathscr{P},t)\ddot{\mathbf{x}}_a\times(\mathbf{x}_c-\mathbf{x}_a) \tag{3.3.6}$$

where $\ddot{\mathbf{x}}_a$ is the acceleration of the point $\mathbf{x}_a$. In what cases does the above result reduce to $\dot{\mathbf{h}}_a^r(\mathscr{P},t)=\mathbf{l}_a(\mathscr{P},t)$?

Exercise 3.3.4

Use the results of Exercise 3.3.3 and derive Euler's equations for a rigid body $\mathscr{B}$.

Next, we wish to derive a local version of the axiom (3.3.4). This derivation requires that we define the operation of the cross product of a vector with a linear transformation. If $\mathbf{v}$ is in $\mathscr{V}$ and $\mathbf{A}$ is in $\mathscr{L}(\mathscr{V};\mathscr{V})$ then $\mathbf{v}\times\mathbf{A}$ is a linear transformation defined by

$$(\mathbf{v}\times\mathbf{A})\mathbf{u}=\mathbf{v}\times(\mathbf{A}\mathbf{u}) \tag{3.3.7}$$

for all $\mathbf{u}$ in $\mathscr{V}$. Equation (3.3.7) can be used to derive the following component formula:

$$\mathbf{v}\times\mathbf{A}=\varepsilon_{jqm}\upsilon_q A_{mk}\mathbf{i}_j\otimes\mathbf{i}_k \tag{3.3.8}$$

Given (3.3.4), we can follow a now standard argument and show that

$$\rho\overline{\left((\mathbf{x}-\mathbf{0})\times\dot{\mathbf{x}}\right)}^{\boldsymbol{\cdot}}=\operatorname{div}\left((\mathbf{x}-\mathbf{0})\times\mathbf{T}\right)+\rho(\mathbf{x}-\mathbf{0})\times\mathbf{b} \tag{3.3.9}$$

at points $\mathbf{x}$ not on a singular surface.

It is a fact that equation (3.3.9) and balance of linear momentum, (3.2.20), are equivalent if and only if the stress tensor is symmetric, i.e.

$$\mathbf{T}=\mathbf{T}^T \tag{3.3.10}$$

The proof of this assertion depends upon the identities

$$\overline{\left((\mathbf{x}-\mathbf{0})\times\dot{\mathbf{x}}\right)}^{\boldsymbol{\cdot}}=(\mathbf{x}-\mathbf{0})\times\ddot{\mathbf{x}} \tag{3.3.11}$$

and

$$\operatorname{div}\left((\mathbf{x}-\mathbf{0})\times\mathbf{T}\right)=(\mathbf{x}-\mathbf{0})\times\operatorname{div}\mathbf{T}+\varepsilon_{jmk}T_{km}\mathbf{i}_j \tag{3.3.12}$$

Equation (3.3.11) follows directly, since

$$\overline{(\mathbf{x}-\mathbf{0})}^{\cdot}\times\dot{\mathbf{x}}=\dot{\mathbf{x}}\times\dot{\mathbf{x}}=\mathbf{0}$$

Equation (3.3.12) is most easily obtained by expressing the left side in components, differentiating the result, and then writing the answer in direct notation. In components, the left side of (3.3.12) is

$$\operatorname{div}\left((\mathbf{x}-\mathbf{0})\times\mathbf{T}\right)=\frac{\partial\left(\varepsilon_{jsk}x_sT_{km}\right)}{\partial x_m}\mathbf{i}_j \tag{3.3.13}$$

Therefore,

$$\operatorname{div}\left((\mathbf{x}-\mathbf{0})\times\mathbf{T}\right)=\varepsilon_{jsk}x_s\frac{\partial T_{km}}{\partial x_m}i_j+\varepsilon_{jsk}\frac{\partial x_s}{\partial x_m}T_{km}i_j \tag{3.3.14}$$

If the identity

$$\frac{\partial x_s}{\partial x_m}=\delta_{sm} \tag{3.3.15}$$

is used, (3.3.14) reduces to (3.3.12). By use of (3.3.11) and (3.3.12), it follows that (3.3.9) can be written

$$(\mathbf{x}-\mathbf{0})\times\left(\rho\ddot{\mathbf{x}}-\operatorname{div}\mathbf{T}-\rho\mathbf{b}\right)=\varepsilon_{jmk}T_{km}\mathbf{i}_j \tag{3.3.16}$$

It is easy to conclude that the right side of (3.3.16) is zero if and only if $\mathbf{T}$ is symmetric. Thus, if $\mathbf{T}$ is symmetric, (3.3.16) implies (3.2.20). Conversely, if (3.2.20) is used in (3.3.16), then $\mathbf{T}$ must be symmetric. The result just established is a theorem proven by *Cauchy* in 1827. This theorem is no longer true if the material is a *polar material*, i.e., a material with intrinsic spin, surface couples, and body couples. From our standpoint, it is an extremely useful result, since we can avoid the use of the balance equation for moment of momentum in the complicated form (3.3.9) by replacing it with (3.3.10). In terms of the first Piola-Kirchhoff stress tensor $\mathbf{T}_R$ defined by (3.2.27), the result (3.3.10) can be written

$$\mathbf{T}_R\mathbf{F}^T=\mathbf{F}\mathbf{T}_R^T \tag{3.3.17}$$

thus $\mathbf{T}_R$ is not symmetric. An elementary calculation shows that $\tilde{\mathbf{T}}$ defined by (3.2.32) is symmetric.

No additional information is obtained from (3.3.4) by allowing a surface of discontinuity to occur in the material. The jump equation for angular momentum turns out to be equivalent to (3.2.33).

3.4. Balance of Energy (First Axiom of Thermodynamics)

In this section the axiom of balance of energy is considered. In addition to the quantities previously introduced in this chapter, the balance of energy axiom introduces the ideas of *internal energy* and *heat*. As in Chapter I, these two concepts are regarded as fundamental undefined properties of the material.

The *kinetic energy* of a part $\mathscr{P}$ of $\mathscr{B}$ at the time t is defined by

$$T(\mathscr{P},t)=\int_{\chi(\mathscr{P},t)}\frac{1}{2}\rho\dot{\mathbf{x}}^2 dv \tag{3.4.1}$$

If ε denotes the internal energy density (i.e., the internal energy/mass), the *internal energy* of $\mathscr{P}$ in $\mathscr{B}$ at the time t is

$$E(\mathscr{P},t)=\int_{\chi(\mathscr{P},t)}\rho\varepsilon\, dv \tag{3.4.2}$$

The *total energy* of $\mathscr{P}$ in $\mathscr{B}$ at the time t is the sum of $T(\mathscr{P},t)$ and $E(\mathscr{P},t)$. The *power* (rate of work) of the applied forces acting on $\mathscr{P}$ in $\mathscr{B}$ at the time t is

$$P(\mathscr{P},t)=\oint_{\partial\chi(\mathscr{P},t)}\dot{\mathbf{x}}\cdot(\mathbf{T}\,d\mathbf{s})+\int_{\chi(\mathscr{P},t)}\rho\dot{\mathbf{x}}\cdot\mathbf{b}\,dv \tag{3.4.3}$$

The *rate of heat addition* to $\mathscr{P}$ is regarded as arising from the generation of heat within the body and from contact of $\mathscr{P}$ with the material outside of $\mathscr{P}$ (conduction). Thus, if $Q(\mathscr{P},t)$ denotes the rate of heat addition to $\mathscr{P}$ at the time t, we can write

$$Q(\mathscr{P},t)=-\oint_{\partial\chi(\mathscr{P},t)}q(\mathbf{x},\mathscr{P},t)\,ds+\int_{\chi(\mathscr{P},t)}\rho r(\mathbf{x},t)\,dv \tag{3.4.4}$$

The quantity q in (3.4.4) represents the rate of heat flow per unit of area across $\partial\chi(\mathscr{P},t)$. It is positive when heat is being removed from $\mathscr{P}$. The quantity r in (3.4.4) represents the rate of generation of heat at $\mathbf{x}$ in $\chi(\mathscr{P},t)$ at the time t. It is called the *heat supply* density. Equation (3.4.4) has the same interpretation for heat as (3.2.10) has for resultant force.

Axiom (The First Axiom of Thermodynamics). For all parts $\mathscr{P}$ in $\mathscr{B}$,

$$\overline{E(\mathscr{P},t)+T(\mathscr{P},t)}^{\,\boldsymbol{\cdot}}=P(\mathscr{P},t)+Q(\mathscr{P},t) \tag{3.4.5}$$

In terms of the definitions (3.4.1) through (3.4.4), the axiom (3.4.5) takes the form

$$\overline{\int_{\chi(\mathscr{P},t)} \rho\left(\varepsilon+\frac{1}{2}\dot{\mathbf{x}}^2\right) dv}^{\,\boldsymbol{\cdot}} = \oint_{\partial\chi(\mathscr{P},t)} \dot{\mathbf{x}}\cdot(\mathbf{T}\,d\mathbf{s}) - \oint_{\partial\chi(\mathscr{P},t)} q(\mathbf{x},\mathscr{P},t)\,ds + \int_{\chi(\mathscr{P},t)} \rho(\dot{\mathbf{x}}\cdot\mathbf{b}+r)\,dv \tag{3.4.6}$$

By use of (3.3.10), (3.1.23) and (A.8.27), (3.4.6) can be written

$$\oint_{\partial\chi(\mathscr{P},t)} q(\mathbf{x},\mathscr{P},t)\,ds = \int_{\chi(\mathscr{P},t)} \left(\operatorname{div}(\mathbf{T}\dot{\mathbf{x}}) + \rho\dot{\mathbf{x}}\cdot\mathbf{b} + \rho r - \rho\overline{\left(\varepsilon+\frac{1}{2}\dot{\mathbf{x}}^2\right)}^{\,\boldsymbol{\cdot}}\right) dv \tag{3.4.7}$$

when one assumes $\chi(\mathscr{P},t)$ does not contain a surface of discontinuity. The same type of argument which produced (3.2.11) and (3.2.13) can be used to show that

$$q(\mathbf{x},\mathscr{P},t) = q(\mathbf{x},\mathbf{n},t) = \mathbf{q}(\mathbf{x},t)\cdot\mathbf{n} \tag{3.4.8}$$

The vector $\mathbf{q}(\mathbf{x},t)$ is known as the *heat flux vector*. Equation (3.4.8) is a result first established by *Stokes*. With (3.4.8) and (3.3.10), (3.4.6) and (3.4.7) become

$$\overline{\int_{\chi(\mathscr{P},t)} \rho\left(\varepsilon+\frac{1}{2}\dot{\mathbf{x}}^2\right) dv}^{\,\boldsymbol{\cdot}} = \oint_{\partial\chi(\mathscr{P},t)} (\mathbf{T}\dot{\mathbf{x}}-\mathbf{q})\cdot d\mathbf{s} + \int_{\chi(\mathscr{P},t)} \rho(\dot{\mathbf{x}}\cdot\mathbf{b}+r)\,dv \tag{3.4.9}$$

and

$$\int_{\chi(\mathscr{P},t)} \left(\rho\overline{\left(\varepsilon+\frac{1}{2}\dot{\mathbf{x}}^2\right)}^{\,\boldsymbol{\cdot}} - \operatorname{div}(\mathbf{T}\dot{\mathbf{x}}) - \rho\dot{\mathbf{x}}\cdot\mathbf{b} - \rho r + div\,\mathbf{q}\right) dv = 0 \tag{3.4.10}$$

Equation (3.4.10) immediately yields the following local statement of the axiom of balance of energy:

$$\rho\overline{\left(\varepsilon+\frac{1}{2}\dot{\mathbf{x}}^2\right)}^{\,\boldsymbol{\cdot}} = \operatorname{div}(\mathbf{T}\dot{\mathbf{x}}) - \operatorname{div}\mathbf{q} + \rho\dot{\mathbf{x}}\cdot\mathbf{b} + \rho r \tag{3.4.11}$$

Of course, (3.4.11) generalizes (1.4.2). This result holds at points $\mathbf{x}$ not at a surface of discontinuity.

Exercise 3.4.1

Show that the following forms of (3.4.11) are equivalent:

1. $$\rho\overline{\left(\varepsilon+\frac{1}{2}\dot{\mathbf{x}}^2\right)}^{\cdot}=\operatorname{div}\left(\mathbf{T}\dot{\mathbf{x}}\right)-\operatorname{div}\mathbf{q}+\rho\dot{\mathbf{x}}\cdot\mathbf{b}+\rho r \tag{3.4.12}$$

2. $$\rho\frac{\partial}{\partial t}\left(\varepsilon+\frac{1}{2}\dot{\mathbf{x}}^2\right)+\rho\left(\operatorname{grad}\left(\varepsilon+\frac{1}{2}\dot{\mathbf{x}}^2\right)\right)\cdot\dot{\mathbf{x}}=\operatorname{div}\left(\mathbf{T}\dot{\mathbf{x}}\right)-\operatorname{div}\mathbf{q}+\rho\dot{\mathbf{x}}\cdot\mathbf{b}+\rho r \tag{3.4.13}$$

and

3. $$\frac{\partial}{\partial t}\rho\left(\varepsilon+\frac{1}{2}\dot{\mathbf{x}}^2\right)+\operatorname{div}\left(\rho\left(\varepsilon+\frac{1}{2}\dot{\mathbf{x}}^2\right)\dot{\mathbf{x}}\right)=\operatorname{div}\left(\mathbf{T}\dot{\mathbf{x}}\right)-\operatorname{div}\mathbf{q}+\rho\dot{\mathbf{x}}\cdot\mathbf{b}+\rho r \tag{3.4.14}$$

Exercise 3.4.2

Assume that the body force $\mathbf{b}$ is given by $\mathbf{b}=-\operatorname{grad}v$, where $v(\mathbf{x},t)$ is the potential energy. Show that (3.4.11) can be written

$$\rho\overline{\left(\varepsilon+v+\frac{1}{2}\dot{\mathbf{x}}^2\right)}^{\cdot}=\operatorname{div}\left(\mathbf{T}\dot{\mathbf{x}}\right)-\operatorname{div}\mathbf{q}+\rho r+\rho\frac{\partial v}{\partial t} \tag{3.4.15}$$

Exercise 3.4.3

Show that, for the material in $\mathscr{R}(t)$, the balance of energy equation the energy is

$$\begin{aligned}\frac{d}{dt}\int_{\mathscr{R}(t)}\rho\left(\varepsilon+\frac{1}{2}\dot{\mathbf{x}}^2\right)dv=&-\oint_{\partial\mathscr{R}(t)}\rho\left(\varepsilon+\frac{1}{2}\dot{\mathbf{x}}^2\right)\left(\dot{\mathbf{x}}-\mathbf{v}\right)\cdot d\mathbf{s}\\&+\oint_{\partial\mathscr{R}(t)}\left(\mathbf{T}\dot{\mathbf{x}}-\mathbf{q}\right)\cdot d\mathbf{s}\\&+\int_{\mathscr{R}(t)}\rho\left(\dot{\mathbf{x}}\cdot\mathbf{b}+r\right)dv\end{aligned} \tag{3.4.16}$$

What do the terms in (3.4.16) represent physically?

Exercise 3.4.4

Show that the material form of the axiom of balance of energy is

$$\rho_R\overline{\left(\varepsilon+\frac{1}{2}\dot{\mathbf{x}}^2\right)}^{\cdot}=\operatorname{Div}\left(\mathbf{T}_R^T\dot{\mathbf{x}}\right)-\operatorname{Div}\mathbf{q}_R+\rho_R\dot{\mathbf{x}}\cdot\mathbf{b}+\rho_R r \tag{3.4.17}$$

where

$$\mathbf{q}_R = \left|\det \mathbf{F}\right| \mathbf{F}^{-1}\mathbf{q} \tag{3.4.18}$$

As in Section 1.4, it is possible to derive a balance equation for the internal energy density alone by "removing the kinetic energy" from (3.4.11) with the equation of motion (3.2.20). It is convenient to call the resulting equation a *thermodynamic energy equation*. However, this does not imply that the internal energy density is not influenced by things that might be attributed physically to mechanical origins. It must be stressed that the thermodynamic energy equation does not represent a new axiom. It will be derived from (3.4.11) and (3.2.20). It follows from (3.2.20) that

$$\rho\dot{\mathbf{x}}\cdot\ddot{\mathbf{x}} = \frac{1}{2}\rho\overline{\left(\dot{\mathbf{x}}^2\right)}^{\cdot} = \dot{\mathbf{x}}\cdot\operatorname{div}\mathbf{T} + \rho\dot{\mathbf{x}}\cdot\mathbf{b} \tag{3.4.19}$$

It should be clear that (3.4.19) can be used to subtract the kinetic energy and the rate of work of the body force from (3.4.11). The result is

$$\rho\dot{\varepsilon} = \operatorname{div}\left(\mathbf{T}\dot{\mathbf{x}}\right) - \dot{\mathbf{x}}\cdot\operatorname{div}\mathbf{T} - \operatorname{div}\mathbf{q} + \rho r \tag{3.4.20}$$

Equation (3.4.20) shows that the internal energy density ε is changed because of the heat addition and what is left over from $\operatorname{div}\left(\mathbf{T}\dot{\mathbf{x}}\right) - \dot{\mathbf{x}}\cdot\operatorname{div}\mathbf{T}$. Important physical information can be obtained by examination of the quantities $\operatorname{div}\left(\mathbf{T}\dot{\mathbf{x}}\right)$ and $\dot{\mathbf{x}}\cdot\operatorname{div}\mathbf{T}$. It should be clear that

1. $\operatorname{div}\left(\mathbf{T}\dot{\mathbf{x}}\right) =$ rate of work/vol of the surface forces, and it contributes to the change of total energy density, $\varepsilon + \frac{1}{2}\dot{\mathbf{x}}^2$ [See (3.4.11)];
2. $\dot{\mathbf{x}}\cdot\operatorname{div}\mathbf{T} =$ a rate of work/vol of the surface forces; however, it contributes only to the change in the kinetic energy [See (3.4.19)]

We can conclude that the difference, $\operatorname{div}\left(\mathbf{T}\dot{\mathbf{x}}\right) - \dot{\mathbf{x}}\cdot\operatorname{div}\mathbf{T}$, is a rate of work of the surface forces that contributes to a change in the internal energy density. Therefore, except in certain trivial cases, there is always a coupling between the rate of work of the surface forces and the internal energy.

If the term $\operatorname{div}\left(\mathbf{T}\dot{\mathbf{x}}\right)$ is expanded, it follows that

$$\operatorname{div}\left(\mathbf{T}\dot{\mathbf{x}}\right) = \operatorname{tr}\left(\mathbf{T}\mathbf{L}\right) + \dot{\mathbf{x}}\cdot\operatorname{div}\mathbf{T} \tag{3.4.21}$$

where (3.3.10) has been used. Equation (3.4.21) is a decomposition of the rate of work per unit volume of the context forces into a part which represents that part of the rate of work per unit volume of the contact forces which changes the kinetic energy and a part that changes the internal energy.

In summary, the balance energy equation (3.4.11) and the balance of linear momentum equation (3.2.20) imply the following balance of thermodynamic energy equation:

$$\rho\dot{\varepsilon} = \operatorname{tr}\left(\mathbf{TL}\right) - \operatorname{div}\mathbf{q} + \rho r \tag{3.4.22}$$

Conversely, (3.4.22) and (3.2.20) imply (3.4.11).

Exercise 3.4.5

Show that (3.4.22) can be written

$$\rho\dot{\varepsilon} = \operatorname{tr}\left(\mathbf{TD}\right) - \operatorname{div}\mathbf{q} + \rho r \tag{3.4.23}$$

Thus, the skew part of $\mathbf{L}$ does not contribute to the thermodynamic energy equation. The term $\operatorname{tr}\mathbf{TL} = \operatorname{tr}\mathbf{TD}$ in (3.4.22) and (3.4.23) is usually called the *stress power.*

Exercise 3.4.6

Show that the material version of (3.4.22) is

$$\rho_R\dot{\varepsilon} = \operatorname{tr}\left(\mathbf{T}_R^T\dot{\mathbf{F}}\right) - \operatorname{Div}\mathbf{q}_R + \rho_R r \tag{3.4.24}$$

The jump statement of balance of energy can be derived by allowing $\chi(\mathscr{P},t)$ to contain a surface of discontinuity as was done in Section 3.1 for balance of mass.

Exercise 3.4.7

Show that

$$[\rho\left(\varepsilon + \frac{1}{2}\dot{\mathbf{x}}^2\right)\left(\dot{\mathbf{x}} - u_n\mathbf{n}\right) - \mathbf{T}\dot{\mathbf{x}} + \mathbf{q}]\cdot\mathbf{n} = 0 \tag{3.4.25}$$

Exercise 3.4.8

Show that

$$\gamma[\left(\varepsilon + \frac{1}{2}\dot{\mathbf{x}}^2\right)] - [\mathbf{T}\dot{\mathbf{x}}]\cdot\mathbf{n} + [\mathbf{q}]\cdot\mathbf{n} = 0 \tag{3.4.26}$$

where γ is defined below equation (3.2.35).

Exercise 3.4.9

If $\mathbf{T} = -\pi\mathbf{I}$ and $[\mathbf{q}]\cdot\mathbf{n} = 0$, show that (3.4.26) can be written

$$[\varepsilon + \frac{\pi}{\rho} + \frac{1}{2}(\dot{x}_n - u_n)^2] = 0 \qquad (3.4.27)$$

Exercise 3.4.10

Show that (3.4.27) can be written

$$[\varepsilon] + \frac{1}{2}(\pi^+ + \pi^-)[1/\rho] = 0 \qquad (3.4.28)$$

Equation (3.4.28) is called the *Hugoniot* relation and is the three-dimensional generalization of (1.4.5).

Exercise 3.4.11

Derive the three-dimensional version of equation (1.4.12).

Several interesting applications require statements of the axiom of balance of energy where the velocities are measured with respect to a translating and rotating coordinate system. The required equations are easily derived if one first recognizes that (3.4.22) or (3.4.23) is a scalar equation which is invariant under a transformation to translating and rotating coordinate systems.

Exercise 3.4.12

Show that with respect to a translating and rotating coordinate system (3.4.11) takes the form.

$$\rho\frac{\delta}{\delta t}\left(\varepsilon + \frac{1}{2}\frac{\delta\mathbf{p}}{\delta t}\cdot\frac{\delta\mathbf{p}}{\delta t}\right) = \operatorname{div}\left(\mathbf{T}\frac{\delta\mathbf{p}}{\delta t}\right) - \operatorname{div}\mathbf{q} + \rho\frac{\delta\mathbf{p}}{\delta t}\cdot\left(\mathbf{b} - \ddot{\mathbf{c}} - \dot{\boldsymbol{\omega}}\times\mathbf{p} - \boldsymbol{\omega}\times\left(\boldsymbol{\omega}\times\mathbf{p}\right)\right) + \rho r \qquad (3.4.29)$$

Exercise 3.4.13

On the assumption that $\dot{\boldsymbol{\omega}} = \mathbf{0}$, show that (3.4.29) can be written

$$\rho\frac{\delta}{\delta t}\left[\varepsilon + \frac{1}{2}\frac{\delta\mathbf{p}}{\delta t}\cdot\frac{\delta\mathbf{p}}{\delta t} - \frac{1}{2}(\boldsymbol{\omega}\times\mathbf{p})\cdot(\boldsymbol{\omega}\times\mathbf{p})\right] = \operatorname{div}\left(\mathbf{T}\frac{\delta\mathbf{p}}{\delta t}\right) - \operatorname{div}\mathbf{q} + \rho\frac{\delta\mathbf{p}}{\delta t}\cdot(\mathbf{b} - \ddot{\mathbf{c}}) + \rho r \qquad (3.4.30)$$

It is possible to formulate an argument which yields the axioms of balance of mass, linear momentum and angular momentum from the axiom of balance of energy and assumptions of invariance [Refs. 8, 9, 10, 11, 12]. Insight into this argument is provided by the following exercises.

Exercise 3.4.14

Without utilizing balance of mass, linear momentum and angular momentum show that the local statement of balance of energy which follows from (3.4.6) is

$$\frac{\partial}{\partial t}\rho\left(\varepsilon+\frac{1}{2}\dot{\mathbf{x}}^2\right)+\operatorname{div}\left(\rho\left(\varepsilon+\frac{1}{2}\dot{\mathbf{x}}^2\right)\dot{\mathbf{x}}\right)=\operatorname{div}\left(\mathbf{T}^{\mathrm{T}}\dot{\mathbf{x}}\right)-\operatorname{div}\mathbf{q}+\rho\dot{\mathbf{x}}\cdot\mathbf{b}+\rho r \tag{3.4.31}$$

at points $\mathbf{x}$ not occupied by a singular surface.

Exercise 3.4.15

Show that (3.4.31) can be written

$$\rho\dot{\varepsilon}-\operatorname{tr}\left(\mathbf{T}^{\mathrm{T}}\mathbf{L}\right)+\operatorname{div}\mathbf{q}-\rho r+\dot{\mathbf{x}}\cdot\left(\rho\ddot{\mathbf{x}}-\operatorname{div}\mathbf{T}-\rho\mathbf{b}\right)+\left(\varepsilon+\frac{1}{2}\dot{\mathbf{x}}^2\right)\left(\dot{\rho}+\rho\operatorname{tr}\mathbf{L}\right)=0 \tag{3.4.32}$$

Exercise 3.4.16

Consider a motion $\chi_{\kappa}^{*}(\mathbf{X},t)$ defined in terms of $\chi_{\kappa}(\mathbf{X},t)$ by [see (A.7.8)]

$$\chi_{\kappa}^{*}(\mathbf{X},t)=\mathbf{c}(t)+\mathbf{Q}(t)\left(\chi_{\kappa}(\mathbf{X},t)-\mathbf{0}\right) \tag{3.4.33}$$

where $\mathbf{c}(t)$ is a point in $\mathscr{E}$ for each t in $(-\infty,\infty)$ and $\mathbf{Q}(t)$ is in $\mathscr{O}(\mathscr{V})$ for each t in $(-\infty,\infty)$. If an asterisk denotes quantities associated with the motion χ_{κ}^{*}, show that

$$\mathbf{F}^{*}=\mathbf{QF} \tag{3.4.34}$$

$$\dot{\mathbf{x}}^{*}=\dot{\mathbf{c}}(t)+\mathbf{Q}\dot{\mathbf{x}}+\dot{\mathbf{Q}}\left(\chi_{\kappa}(\mathbf{X},t)-\mathbf{0}\right) \tag{3.4.35}$$

$$\ddot{\mathbf{x}}^{*}=\ddot{\mathbf{c}}(t)+\mathbf{Q}\ddot{\mathbf{x}}+2\dot{\mathbf{Q}}\dot{\mathbf{x}}+\ddot{\mathbf{Q}}\left(\chi_{\kappa}(\mathbf{X},t)-\mathbf{0}\right) \tag{3.4.36}$$

and

$$\mathbf{L}^{*}=\mathbf{QLQ}^{\mathrm{T}}+\dot{\mathbf{Q}}\mathbf{Q}^{\mathrm{T}} \tag{3.4.37}$$

where $\dot{\mathbf{Q}}\mathbf{Q}^{\mathrm{T}}$ is a skew-symmetric linear transformation.

Exercise 3.4.17

If, in addition to (3.4.34) through (3.4.37), the following transformation rules hold:

$$\rho^{*}=\rho \tag{3.4.38}$$

$$\varepsilon^* = \varepsilon \tag{3.4.39}$$

$$\mathbf{T}^* = \mathbf{Q}\mathbf{T}\mathbf{Q}^{\mathrm{T}} \tag{3.4.40}$$

$$\mathbf{q}^* = \mathbf{Q}\mathbf{q} \tag{3.4.41}$$

$$r^* = r \tag{3.4.42}$$

and

$$\mathbf{b}^* = \mathbf{Q}\mathbf{b} + \ddot{\mathbf{c}} + 2\dot{\mathbf{Q}}\dot{\mathbf{x}} + \ddot{\mathbf{Q}}\left(\chi_\kappa(\mathbf{X},t) - \mathbf{0}\right) \tag{3.4.43}$$

then show that if we postulate

$$\begin{aligned} &\rho^*\dot{\varepsilon}^* - \operatorname{tr}\left(\mathbf{T}^{*\mathrm{T}}\mathbf{L}^*\right) + \operatorname{div}^*\mathbf{q}^* - \rho^* r^* + \dot{\mathbf{x}}^* \cdot \left(\rho^*\ddot{\mathbf{x}}^* - \operatorname{div}^*\mathbf{T}^* - \rho^*\mathbf{b}^*\right) \\ &+ \left(\varepsilon^* + \frac{1}{2}\dot{\mathbf{x}}^{*2}\right)\left(\dot{\rho}^* + \rho^*\operatorname{tr}\mathbf{L}^*\right) = 0 \end{aligned} \tag{3.4.44}$$

for all $\mathbf{c}(\cdot):(-\infty,\infty)\to\mathscr{E}$ which have second derivatives and all $\mathbf{Q}(\cdot):(-\infty,\infty)\to\mathscr{O}(\mathscr{V})$ which have second derivatives, then it is necessary and sufficient that

$$\dot{\rho} + \rho\operatorname{tr}\mathbf{L} = 0 \tag{3.4.45}$$

$$\rho\ddot{\mathbf{x}} = \operatorname{div}\mathbf{T} + \rho\mathbf{b} \tag{3.4.46}$$

and

$$\mathbf{T} = \mathbf{T}^{\mathrm{T}} \tag{3.4.47}$$

3.5. The Entropy Inequality

As in Section 1.6, the mathematical statement of the second axiom of thermodynamics takes the form of an *entropy inequality* or the *Clausius-Duhem* inequality [Ref. 2, 6, 7, 8, 9, 13, 14, 15,16,17].

In order to discuss the second axiom of thermodynamics it is necessary to accept as fundamental the concepts of an *entropy density* η, *body entropy supply density* k and a *contact entropy supply density* h.

Definition. The entropy of a part $\mathscr{P}$ in $\mathscr{B}$ at the time t is defined by

$$S(\mathscr{P},t) = \int_{\chi(\mathscr{P},t)} \rho\eta\, dv \tag{3.5.1}$$

Definition. The entropy flux into $\mathscr{P}$ in $\mathscr{B}$ at the time t is defined by

$$M(\mathscr{P},t) = -\oint_{\partial\chi(\mathscr{P},t)} h(\mathbf{x},\mathscr{P},t)\,ds + \int_{\chi(\mathscr{P},t)} \rho k(\mathbf{x},t)\,dv \tag{3.5.2}$$

Axiom (The Second Axiom of Thermodynamics). For all parts $\mathscr{P}$ of $\mathscr{B}$,

$$\dot{S}(\mathscr{P},t) \geq M(\mathscr{P},t) \tag{3.5.3}$$

By the same kind of argument which produced (3.2.11), (3.2.13) and (3.4.8), it follows that

$$h(\mathbf{x},\mathscr{P},t) = h(\mathbf{x},\mathbf{n},t) = \mathbf{h}(\mathbf{x},t)\cdot\mathbf{n} \tag{3.5.4}$$

[Ref. 6]. It is appropriate to call $\mathbf{h}$ the *entropy flux vector*. By use of (3.5.1), (3.5.2) and (3.5.4),the axiom (3.5.3) can be written

$$\overline{\int_{\chi(\mathscr{P},t)} \rho\eta\,dv}^{\,\cdot} \geq -\oint_{\partial\chi(\mathscr{P},t)} \mathbf{h}(\mathbf{x},t)\cdot d\mathbf{s} + \int_{\chi(\mathscr{P},t)} \rho k(\mathbf{x},t)\,dv \tag{3.5.5}$$

As in Section 1.6, the temperature is introduced with the formal assumption that the ratios q/h and r/k are equal.

Definition. The *temperature* θ at $\mathbf{x}$ in $\chi(\mathscr{P},t)$ at the time t is defined to be the common ratio

$$\theta = \frac{q}{h} = \frac{r}{k} \tag{3.5.6}$$

By assumption, θ shall be regarded as a positive valued function of $(\mathbf{x},t)$, i.e.

$$\theta = \theta(\mathbf{x},t) > 0 \tag{3.5.7}$$

Formulations of the Second Axiom of Thermodynamics do exist for which the definition (3.5.6) is not adopted. These formulations are available in Refs. 13 and 16. Reference 17 contains a slightly different formulation of the entropy inequality from that given here and in Refs. 13 and 16. These alternate statements of the Second Axiom of Thermodynamics are briefly discussed in Section 5.4. The definition (3.5.6) is essentially classical and has been adopted here for that reason. Given (3.5.6), (3.5.4) and (3.4.8), we can replace (3.5.5) by

$$\overline{\int_{\chi(\mathscr{P},t)} \rho\eta\,dv}^{\,\cdot} \geq -\oint_{\partial\chi(\mathscr{P},t)} \frac{1}{\theta}\mathbf{q}\cdot d\mathbf{s} + \int_{\chi(\mathscr{P},t)} \frac{\rho r}{\theta}\,dv \tag{3.5.8}$$

Next, we wish to use (3.5.8) to derive local statements of the second axiom of thermodynamics. It follows from (A.8.33) and (3.1.23) that (3.5.8) can be written

$$\int_{\chi(\mathscr{P},t)} \left(\rho\dot{\eta} + \operatorname{div}(\mathbf{q}/\theta) - \rho r/\theta\right) dv - \int_{\Sigma(t)} \left([\rho\eta(\dot{x}_n - u_n)] + [\mathbf{q}/\theta]\cdot\mathbf{n}\right) d\sigma \geq 0 \tag{3.5.9}$$

where $\Sigma(t)$ is a singular surface. Because $\mathscr{P}$ is arbitrary, (3.5.9) implies that

$$\rho\dot{\eta} + \operatorname{div}(\mathbf{q}/\theta) - \rho r/\theta \geq 0 \tag{3.5.10}$$

for all $\mathbf{x}$ not on a singular surface and

$$[\rho\eta(\dot{x}_n - u_n)] + [\mathbf{q}/\theta]\cdot\mathbf{n} \leq 0 \tag{3.5.11}$$

for all $\mathbf{x}$ on $\Sigma(t)$. Equations (3.5.10) and (3.5.11) generalize (1.6.7) and (1.6.8), respectively.

As in Section 1.6, it is useful to eliminate the term $\operatorname{div}\mathbf{q} - \rho r$ between (3.5.10) and the thermodynamic energy equation (3.4.22). This calculation can be carried out if we write (3.5.10) in the form

$$\rho\dot{\eta} + \frac{1}{\theta}(\operatorname{div}\mathbf{q} - \rho r) - \frac{1}{\theta^2}\mathbf{q}\cdot\operatorname{grad}\theta \geq 0 \tag{3.5.12}$$

It immediately follows from (3.4.22) that (3.5.12) can be written

$$\rho(\theta\dot{\eta} - \dot{\varepsilon}) + \operatorname{tr}\mathbf{TL} - \frac{\mathbf{q}}{\theta}\cdot\operatorname{grad}\theta \geq 0 \tag{3.5.13}$$

which is obviously the three dimensional generalization of (1.6.10). As in Section 1.6, equation (3.5.13) is a combined statement of the entropy inequality (3.5.10) and the energy equation (3.4.22).

The *free energy density* or the *Helmholtz free energy density* is defined as in Section 1.6, by

$$\psi = \varepsilon - \eta\theta \tag{3.5.14}$$

In terms of ψ the inequality (3.5.13) is

$$-\rho\left(\dot{\psi} + \eta\dot{\theta}\right) + \operatorname{tr}\mathbf{TL} - \frac{\mathbf{q}}{\theta}\cdot\operatorname{grad}\theta \geq 0 \tag{3.5.15}$$

Exercise 3.5.1

Show that the material form of the inequality (3.5.15) is

$$-\rho_R\left(\dot{\psi}+\eta\dot{\theta}\right)+\operatorname{tr}\mathbf{T}_R^T\dot{\mathbf{F}}-\frac{\mathbf{q}_R}{\theta}\cdot\operatorname{GRAD}\theta\geq 0 \tag{3.5.16}$$

Exercise 3.5.2

Show that (3.4.9) can be written

$$\begin{aligned}
&\dot{\overline{\int_{\chi(\mathscr{P},t)}\rho\left(\varepsilon-\theta_0\eta+\frac{1}{2}\dot{\mathbf{x}}^2\right)dv}}\\
&\quad=\oint_{\partial\chi(\mathscr{P},t)}\dot{\mathbf{x}}\cdot(\mathbf{T}d\mathbf{s})-\oint_{\partial\chi(\mathscr{P},t)}\left(1-\frac{\theta_0}{\theta}\right)\mathbf{q}\cdot d\mathbf{s}+\int_{\chi(\mathscr{P},t)}\rho\dot{\mathbf{x}}\cdot\mathbf{b}dv\\
&\quad+\int_{\chi(\mathscr{P},t)}\left(1-\frac{\theta_0}{\theta}\right)\rho r dv\\
&\quad-\theta_0\left(\dot{\overline{\int_{\chi(\mathscr{P},t)}\rho\eta dv}}+\oint_{\partial\chi(\mathscr{P},t)}\frac{1}{\theta}\mathbf{q}\cdot d\mathbf{s}-\int_{\chi(\mathscr{P},t)}\frac{1}{\theta}\rho r dv\right)
\end{aligned} \tag{3.5.17}$$

where θ_0 is any constant.

Exercise 3.5.3

If θ_0 is a positive number, show that

$$\begin{aligned}
&\dot{\overline{\int_{\chi(\mathscr{P},t)}\rho\left(\psi+\eta(\theta-\theta_0)+\frac{1}{2}\dot{\mathbf{x}}^2\right)dv}}\\
&\qquad\leq\oint_{\partial\chi(\mathscr{P},t)}\dot{\mathbf{x}}\cdot(\mathbf{T}d\mathbf{s})-\oint_{\partial\chi(\mathscr{P},t)}\left(1-\frac{\theta_0}{\theta}\right)\mathbf{q}\cdot d\mathbf{s}\\
&\qquad\quad+\int_{\chi(\mathscr{P},t)}\rho\dot{\mathbf{x}}\cdot\mathbf{b}dv+\int_{\chi(\mathscr{P},t)}\left(1-\frac{\theta_0}{\theta}\right)\rho r dv
\end{aligned} \tag{3.5.18}$$

Equation (3.5.18) is the three dimensional version of (1.6.13). It is useful in the study of thermodynamic stability [Ref. 18 through 31]. For linearized models of materials, (3.5.18) is useful in the proof of uniqueness theorems. (See Exercises 8, 9, 10, of Section 1.11 and Exercise 6 of Section 4.12.)

Exercise 3.5.4

If the temperature θ is only a function of t, show that

$$\theta\dot{S}(\mathscr{P},t)\geq Q(\mathscr{P},t) \tag{3.5.19}$$

Exercise 3.5.5

Use the definition (3.5.14) and show that the thermodynamic energy equation (3.4.22) can be written

$$\rho\theta\dot{\eta}=-\rho\left(\dot{\psi}+\eta\dot{\theta}\right)+\text{tr}\mathbf{TL}-\text{div}\mathbf{q}+\rho r \tag{3.5.20}$$

Exercise 3.5.6

Show that the material version of (3.5.20) is

$$\rho_R\theta\dot{\eta}=-\rho_R\left(\dot{\psi}+\eta\dot{\theta}\right)+\text{tr}\mathbf{T}_R^{\mathsf{T}}\dot{\mathbf{F}}-\text{Div}\mathbf{q}_R+\rho_R r \tag{3.5.21}$$

3.6. Jump Equations of Balance - Material Versions

Equations (3.1.21), (3.2.33), (3.4.25) and (3.5.11) are the spatial forms of the jump balance equations. In this section the corresponding material forms are derived. The argument is complicated by the fact that $\dot{\mathbf{x}}$ and $\mathbf{F}$ are not necessarily continuous across the singular surface $\Sigma(t)$. Indeed, for a *shock wave*, these quantities have nonzero jumps.

If $\Sigma(t)$ is represented in the form

$$f(\mathbf{x},t)=0 \tag{3.6.1}$$

then it is true that

$$\mathbf{n}=\text{grad}\, f(\mathbf{x},t)\big/\left\|\text{grad}\, f(\mathbf{x},t)\right\| \tag{3.6.2}$$

is the *unit normal* to $\Sigma(t)$, and

$$u_n=-\frac{\partial f(\mathbf{x},t)}{\partial t}\Bigg/\left\|\text{grad}\, f(\mathbf{x},t)\right\| \tag{3.6.3}$$

is the *normal speed* of $\Sigma(t)$.

It is convenient to consider a family of surfaces defined by

$$f(\mathbf{x},t)=\alpha \tag{3.6.4}$$

where $\alpha = 0$ defines the surface $\Sigma(t)$. Without loss of generality we can regard $\mathbf{n}$ to be directed from the surface $\alpha = 0$ towards surfaces with $\alpha > 0$. We assume that (3.6.4) can be inverted to yield

$$t = \hat{t}(\mathbf{x}, \alpha) \tag{3.6.5}$$

Given (3.6.5), a function φ of $(\mathbf{x}, t)$ can be replaced by a function $\hat{\varphi}$ of $(\mathbf{x}, \alpha)$ by the rule

$$\hat{\varphi}(\mathbf{x}, \alpha) = \varphi(\mathbf{x}, \hat{t}(\mathbf{x}, \alpha)) \tag{3.6.6}$$

It follows from (3.6.6) that

$$\operatorname{grad}\hat{\varphi}(\mathbf{x}, \alpha) = \operatorname{grad}\varphi(\mathbf{x}, t) + \frac{\partial \varphi(\mathbf{x}, t)}{\partial t}\operatorname{grad}\hat{t}(\mathbf{x}, \alpha) \tag{3.6.7}$$

Likewise it follows from (3.6.4) and (3.6.5) that

$$\operatorname{grad}\hat{t}(\mathbf{x}, \alpha) = -\operatorname{grad} f(\mathbf{x}, t) \Big/ \frac{\partial f(\mathbf{x}, t)}{\partial t} \tag{3.6.8}$$

By formulas like (3.6.2) and (3.6.3), (3.6.8) can be written

$$\operatorname{grad}\hat{t}(\mathbf{x}, \alpha) = \frac{\mathbf{n}_{(\alpha)}}{u_{n(\alpha)}} \tag{3.6.9}$$

where $\mathbf{n}_{(\alpha)}$ is the unit normal to the surface (3.6.4) and $u_{n(\alpha)}$ is its normal speed. The special surface $\alpha = 0$ has its normal and normal speed denoted by $\mathbf{n}$ and u_n, respectively. Given (3.6.9), equation (3.6.7) becomes

$$\operatorname{grad}\hat{\varphi}(\mathbf{x}, \alpha) = \operatorname{grad}\varphi(\mathbf{x}, t) + \frac{\mathbf{n}_{(\alpha)}}{u_{n(\alpha)}}\frac{\partial \varphi(\mathbf{x}, t)}{\partial t} \tag{3.6.10}$$

The *jump* of φ across the surface $\alpha = 0$ is defined by

$$[\varphi] = \varphi^{-} - \varphi^{+} \tag{3.6.11}$$

where

$$\varphi^{-} = \lim_{\alpha \uparrow 0^{-}} \hat{\varphi}(\mathbf{x}, \alpha) \tag{3.6.12}$$

and

$$\varphi^{+} = \lim_{\alpha \downarrow 0^{+}} \hat{\varphi}(\mathbf{x},\alpha) \tag{3.6.13}$$

Because differentiation at constant α commutes with the jump operation, the jump of (3.6.10) yields

$$\operatorname{grad}[\varphi] = [\operatorname{grad}\varphi(\mathbf{x},t)] + \left[\frac{\partial \varphi}{\partial t}\right]\frac{\mathbf{n}}{u_n} \tag{3.6.14}$$

If, as a special case, the jump of $[\varphi]$ is zero, (3.6.14) yields

$$[\operatorname{grad}\varphi(\mathbf{x},t)] = -\left[\frac{\partial \varphi}{\partial t}\right]\frac{\mathbf{n}}{u_n} \tag{3.6.15}$$

Equation (3.6.15) is known as *Maxwell's theorem* [Ref. 7, Sec. 175].

Given a motion (2.1.5), we can construct a material image of each spatial surface defined by (3.6.4). Each of these surfaces is given by

$$f_{\kappa}(\mathbf{X},t) = \alpha \tag{3.6.16}$$

where f_{κ} is defined by

$$f_{\kappa}(\mathbf{X},t) = f(\chi_{\kappa}(\mathbf{X},t),t) \tag{3.6.17}$$

The material image of $\Sigma(t)$, defined by $\alpha = 0$, is denoted by $\Sigma_{\kappa}(t)$. Differentiation of (3.6.17) yields

$$\dot{f}_{\kappa}(\mathbf{X},t) = \frac{\partial f(\mathbf{x},t)}{\partial t} + (\operatorname{grad} f(\mathbf{x},t))\cdot\dot{\mathbf{x}} \tag{3.6.18}$$

and

$$\operatorname{GRAD} f_{\kappa}(\mathbf{X},t) = \mathbf{F}^{T}\operatorname{grad} f(\mathbf{x},t) \tag{3.6.19}$$

Formulas like (3.6.2) and (3.6.3) allow (3.6.18) and (3.6.19) to be written

$$\dot{f}_{\kappa}(\mathbf{X},t) = -\|\operatorname{grad} f(\mathbf{x},t)\|\left(u_{n(\alpha)} - \mathbf{n}_{(\alpha)}\cdot\dot{\mathbf{x}}\right) \tag{3.6.20}$$

and

$$\text{GRAD}\, f_{\kappa}(\mathbf{X},t) = \left\|\text{grad}\, f(\mathbf{x},t)\right\| \mathbf{F}^{\text{T}} \mathbf{n}_{(\alpha)} \tag{3.6.21}$$

At this point, it is convenient to define a normal speed and a unit normal for each of the surfaces (3.6.16) by the usual definitions

$$U_{N_{(\alpha)}} = -\dot{f}_{\kappa}(\mathbf{X},t) / \left\|\text{GRAD} f_{\kappa}(\mathbf{X},t)\right\| \tag{3.6.22}$$

and

$$\mathbf{N}_{(\alpha)} = \text{GRAD} f_{\kappa}(\mathbf{X},t) / \left\|\text{GRAD} f_{\kappa}(\mathbf{X},t)\right\| \tag{3.6.23}$$

Given these definitions, (3.6.20) and (3.6.21) become

$$U_{N_{(\alpha)}} = \frac{\left\|\text{grad} f(\mathbf{x},t)\right\|}{\left\|\text{GRAD} f_{\kappa}(\mathbf{X},t)\right\|}\left(u_{n(\alpha)} - \mathbf{n}_{(\alpha)} \cdot \dot{\mathbf{x}}\right) \tag{3.6.24}$$

and

$$\mathbf{N}_{(\alpha)} = \frac{\left\|\text{grad} f(\mathbf{x},t)\right\|}{\left\|\text{GRAD} f_{\kappa}(\mathbf{X},t)\right\|} \mathbf{F}^{\text{T}} \mathbf{n}_{(\alpha)} \tag{3.6.25}$$

If the surface $\Sigma(t)$ is a shock wave, then $[\dot{\mathbf{x}}]$ and $[\mathbf{F}]$ are not zero. As a result, (3.6.24) yields two values for the normal speed on $\Sigma_{\kappa}(t)$ and (3.6.25) yields two values of the unit normal. If (3.6.24) and (3.6.25) are evaluated in the limit as each side of $\Sigma(t)$ is approached, these multiple values are given by

$$U_N^{\pm} = \frac{\left\|\text{grad} f(\mathbf{x},t)\right\|}{\left\|\text{GRAD} f_{\kappa}(\mathbf{X},t)\right\|^{\pm}}\left(u_n - \mathbf{n} \cdot \dot{\mathbf{x}}^{\pm}\right) \tag{3.6.26}$$

and

$$\mathbf{N}^{\pm} = \frac{\left\|\text{grad} f(\mathbf{x},t)\right\|}{\left\|\text{GRAD} f_{\kappa}(\mathbf{X},t)\right\|^{\pm}} \mathbf{F}^{\pm^{\text{T}}} \mathbf{n} \tag{3.6.27}$$

Fortunately, we can prove that $\mathbf{N}^{+}$ equals $\mathbf{N}^{-}$ and, in addition, that U_N^{+} equals U_N^{-}. The proof of the first assertion will be given next.

In order to establish that $[\mathbf{N}] = \mathbf{N}^{+} - \mathbf{N}^{-}$ is zero, we utilize (3.6.15) where φ is the jth component of $\boldsymbol{\chi}_{\kappa}^{-1}(\mathbf{x},t)$. The resulting identity can be written

$$[\mathbf{F}^{-1}] = \frac{1}{u_n}\left[\frac{\partial \boldsymbol{\chi}_{\kappa}^{-1}(\mathbf{x},t)}{\partial t}\right] \otimes \mathbf{n} \tag{3.6.28}$$

Because $\mathbf{n}$ is the unit normal to $\Sigma(t)$, if $\mathbf{c}$ is any vector tangent to $\Sigma(t)$ then

$$[\mathbf{F}^{-1}]\mathbf{c} = -\frac{1}{u_n}\left[\frac{\partial \boldsymbol{\chi}_{\kappa}^{-1}(\mathbf{x},t)}{\partial t}\right](\mathbf{n}\cdot\mathbf{c}) = 0 \tag{3.6.29}$$

If $d\mathbf{s}$ is the vector element of area of $\Sigma(t)$, it can be written

$$d\mathbf{s} = d\mathbf{x}_1 \times d\mathbf{x}_2 \tag{3.6.30}$$

for tangent vectors $d\mathbf{x}_1$ and $d\mathbf{x}_2$. The image of $d\mathbf{s}$ under the motion is determined by the transformation rule (2.3.11). It follows from (2.3.11) that

$$d\mathbf{S}^{\pm} = \left|\det\mathbf{F}\right|^{\pm^{-1}} \mathbf{F}^{\pm^{T}} d\mathbf{s} \tag{3.6.31}$$

If (3.6.30) is used, (3.6.31) can be written

$$d\mathbf{S}^{\pm} = \mathbf{F}^{\pm^{-1}} d\mathbf{x}_1 \times \mathbf{F}^{\pm^{-1}} d\mathbf{x}_2 \tag{3.6.32}$$

where (A.5.25) has been used. The result (3.6.29) shows that $[\mathbf{F}^{-1}]d\mathbf{x}_1 = [\mathbf{F}^{-1}]d\mathbf{x}_2 = \mathbf{0}$. Therefore (3.6.32) yields

$$[d\mathbf{S}] = d\mathbf{S}^{-} - d\mathbf{S}^{+} = \mathbf{0} \tag{3.6.33}$$

Because

$$\mathbf{N} = \frac{d\mathbf{S}}{\left\|d\mathbf{S}\right\|} \tag{3.6.34}$$

it follows that

$$[\mathbf{N}] = \mathbf{N}^{-} - \mathbf{N}^{+} = \mathbf{0} \tag{3.6.35}$$

The result $U_N^{+} = U_N^{-}$ will be established next. Given (3.6.33), it follows from (3.6.31) that

$$\|d\mathbf{S}\| = |\det\mathbf{F}|^{\pm^{-1}} \left\|\mathbf{F}^{\pm^{\mathrm{T}}}\mathbf{n}\right\|\|d\mathbf{s}\| \tag{3.6.36}$$

Therefore,

$$\rho_R U_N^{\pm}\|d\mathbf{S}\| = \rho^{\pm}\left(u_n - \mathbf{n}\cdot\dot{\mathbf{x}}^{\pm}\right)\|d\mathbf{s}\| \tag{3.6.37}$$

where (3.1.12) and the result

$$\left\|\mathbf{F}^{\pm^{\mathrm{T}}}\mathbf{n}\right\| = \frac{u_n - \mathbf{n}\cdot\dot{\mathbf{x}}^{\pm}}{U_N^{\pm}} \tag{3.6.38}$$

have been used. Equation (3.6.38) follows from (3.6.26), (3.6.27), (3.6.35) and the fact that $\mathbf{N}$ is a unit vector. The jump balance of mass statement (3.1.21) tells us that

$$\rho^{+}\left(u_n - \mathbf{n}\cdot\dot{\mathbf{x}}^{+}\right) = \rho^{-}\left(u_n - \mathbf{n}\cdot\dot{\mathbf{x}}^{-}\right) \tag{3.6.39}$$

Therefore, (3.6.37) yields the desired result

$$[U_N] = U_N^{-} - U_N^{+} = 0 \tag{3.6.40}$$

As our derivation indicates, the result (3.6.40) is, in effect, the material version of jump balance of mass. The remaining jump balance equations are given below.

Given jump balance of linear momentum in the form (3.2.35), the material version is

$$\rho_R U_N[\dot{\mathbf{x}}] + [\mathbf{T}_R]\mathbf{N} = \mathbf{0} \tag{3.6.41}$$

where (3.6.37), (3.2.28) and (3.6.34) have been used. Likewise, the material version of jump balance of energy (3.4.26) is

$$\rho_R U_N[\varepsilon + \frac{1}{2}\dot{\mathbf{x}}^2] + [\mathbf{T}_R^{\mathrm{T}}\dot{\mathbf{x}}]\cdot\mathbf{N} - [\mathbf{q}_R]\cdot\mathbf{N} = 0 \tag{3.6.42}$$

Exercise 3.6.1

Show that the material version of (3.5.11) is

$$\rho_R U_N[\eta] - [\mathbf{q}_R/\theta]\cdot\mathbf{N} \geq 0 \tag{3.6.43}$$

Exercise 3.6.2

By the same derivation that produced (3.6.28), except for an interchange of $\mathbf{x}$ and $\mathbf{X}$, it follows that

$$[\mathbf{F}] = -\frac{1}{U_N}[\dot{\mathbf{x}}] \otimes \mathbf{N} \tag{3.6.44}$$

Use (3.6.44) and (3.6.41) and derive the *Rankine-Hugoniot* relation

$$\rho_R U_N^2 = \frac{\mathbf{n} \cdot ([\mathbf{T}_R]\mathbf{N})}{\mathbf{n} \cdot ([\mathbf{F}]\mathbf{N})} \tag{3.6.45}$$

Equation (3.6.45) generalizes (1.3.12) and is useful in the study of shock waves in solids.

References

1. TRUESDELL, C., A Program Toward Rediscovering the Rational Mechanics of the Age of Reason, *Arch. History of Exact Sciences*, **1**, 3-36 (1960).
2. TRUESDELL, C., and W. NOLL, *The Nonlinear Field Theories of Mechanics*, Handbuch der Physik, Vol.III/3 (ed. S. Flugge), Springer-Verlag, Berlin (1965).
3. NOLL, W., *The Foundations of Classical Mechanics in the Light of Recent Advances in Continuum Mechanics, The Axiomatic Method*, North Holland Amsterdam (1959).
4. GURTIN, M. E., V. J. MIZEL and W. 0. WILLIAMS, A Note on Cauchy's Stress Theorem, *J. Math. Anal. Applic.*, **22**, 398-401, (1968).
5. GURTIN, M. E., *The Linear Theory of Elasticity*, Handbuch der Physik Vol VIa/2 (ed. C. Truesdell), Springer-Verlag,Berlin (1972).
6. TRUESDELL, C, and C.-C. WANG, *Introduction to Rational Elasticity*, Noordhoff, Leyden (1973).
7. TRUESDELL, C., and R. A. TOUPIN, *The Classical Field Theories*, Handbuch der Physik, Vol. III/l (ed. S. Flugge), Springer-Verlag, Berlin (1960).
8. GURTIN, M. E., *Modern Continuum Mechanics, Mechanics Today*, Vol 1 (ed. S. Nemat-Nasser). Pergamon, New York (1972).
9. GURTIN, M. E., W. NOLL and W. 0. WILLIAMS, On the Foundations of Thermodynamics, Research Report 68-42, Dept. of Math., Carnegie Mellon Univ. (1968).
10. GREEN, A. E. and R. S. RIVLIN, On Cauchy's Equations of Motion, *ZAMP*, **15**, 290-292 (1964).
11. NOLL, W., *La Mecanique Classique, Basee sur un Axiome d' Objectivite, in La Methode Axiomatique dans les Mecaniques Classiques et Nouvelles*, Gauthier-Villars, Paris (1963).
12. BEATTY, M. F., On the Foundation Principles of General Classical Mechanics, *Arch. Rational Mech. Anal.*, **24**, 264-273 (1967).
13. GURTIN, M. E. and W. 0. WILLIAMS, An Axiomatic Foundation for Continuum Thermodynamics, *Arch. Rational Mech. Anal.*, **26**, 83-117 (1967).
14. COLEMAN, B. D., and W. NOLL, The Thermodynamics of Elastic Materials with Heat Conduction and Viscosity, *Arch. Rational Mech. Anal.*, **13**, 245-261 (1963).
15. COLEMAN, B. D., and V. J. MIZEL, Existence of Caloric Equations of State in Thermodynamics, *J. Chem. Phys.*, **40**, 1116-1125 (1964).
16. MÜLLER, I., On the Entropy Inequality, *Arch. Rational Mech. Anal.*, **26**, 118-141 (1967).

17. GREEN, A. E., and N. LAWS, On the Entropy Production Inequality, *Arch. Rational Mech. Anal.*, **45**, 47-53 (1972).
18. ERICKSEN, J. L., A Thermo-Kinetic View of Elastic Stability Theory, *Intl. J. Solids Structures*, **2**, 573-580 (1966).
19. COLEMAN, B. D., and E. H. DILL, On Thermodynamics and the Stability of Motions of Materials with Memory, *Arch. Rational Mech. Analysis*, **51**, 1-53 (1973).
20. GURTIN, M. E., Thermodynamics and the Energy Criterion for Stability, *Arch. Rational Mech. Analysis*, **52**, 93-103 (1973).
21. GURTIN, M. E., Thermodynamics and the Potential Energy of an Elastic Body, *J. Elasticity*, **3**, 23-26 (1973).
22. GURTIN, M E., and S. J. SPECTOR, On the Sign of the Specific Heat, *Quart. J. Mech. Appl. Math.*, **29**, 493-497 (1976).
23. KNOPS, R. J., and E. W. WILKES, *Theory of Elastic Stability*, Handbuch der Physik, Vol. VIa/3 (ed. C. Truesdell), Springer-Verlag, Berlin (1973).
24. DAFERMOS, C. M., On the Existence and the Asymptotic Stability of Solutions to the Equations of Linear Thermoelasticity, *Arch. Rational Mech. Anal.*, **29**, 241-271 (1968).
25. DAFERMOS, C. M., The Second Law and Stability, *Arch. Rational Mech. Anal.*, **70**, 167-179 (1979).
26. COLEMAN, B. D., and E. H. DILL, On the Stability of Certain Motions of Incompressible Materials with Memory, *Arch. Rational Mech. Anal.*, **30**, 197-224 (1968).
27. COLEMAN, B. D., On the Stability of Equilibrium States of General Fluids, *Arch. Rational Mech. Anal.*, **36**, 1-32 (1970).
28. COLEMAN, B. D., On the Dynamical Stability of Fluid Phases, *Proc. IUTAM Symposium on Instability of Continuous Systems* (ed. H. Leipholz). Springer-Verlag, Berlin (1971).
29. COLEMAN, B. D., On the Thermodynamics and the Stability of Motions of Materials with Memory, *Arch. Rational Mech. Anal.*, **51**, 1-53 (1973).
30. GURTIN, M. E., Thermodynamics and Stability, *Arch. Rational Mech. Anal.*, **59**, 63-96 (1975).
31. NAGHDI, P. M., and J. A. TRAPP, On the General Theory of Stability for Elastic Bodies, *Arch. Rational Mech. Anal.*, **51**, 165-191 (1973).

Bibliography

32. TRUESDELL, C., *A First Course in Rational Continuum Mechanics*, Vol. 1, Academic Press, New York (1977).
33. GURTIN, M. E., *An Introduction to Continuum Mechanics*, Academic Press, New York (1981).
34. CHADWICK, P., *Continuum Mechanics*, George Allen & Unwin Ltd., London (1976).
35. SPENCER, A. J. M., *Continuum Mechanics*, Longman, London (1980).
36. MALVERN, L. E., *Introduction to the Mechanics of a Continuous Medium*, Prentice-Hall, Englewood Cliffs (1969).
37. LEIGH, D. C., *Nonlinear Continuum Mechanics*, McGraw-Hill, New York (1968).
38. JAUNZEMIS, W., *Continuum Mechanics*, Macmillan, New York (1967).
39. FUNG, Y. C., *A First Course in Continuum Mechanics*, (2nd Ed.), Prentice-Hall, Englewood Cliffs (1977).

40. ERINGEN, A. C., Continuum Mechanics of Single-Substance Bodies - Part I, in *Continuum Physics*, Vol. II (ed. A. C. Eringen), Academic Press, New York (1975).
41. GURTIN, M. E., and L. C. MARTINS, Cauchy's Theorem in Classical Physics, *Arch. Rational Mech. Anal.*, **66**, 305-324 (1976).
42. GURTIN, M. E., and W. O. WILLIAMS, On the Clauius-Duhem Inequality, *Z. Angew. Math. Physik*, **17**, 626-633 (1966).
43. GURTIN, M. E., and W. O. WILLIAMS, On the First Law of Thermodynamics, *Arch. Rational Mech. Anal.*, **42**, 77-92 (1971).
44. GURTIN, M. E., On the Existence of a Single Temperature in Continuum Thermodynamics, *Z. Angew. Math. Physik*, **27**, 775-779 (1976).

4

MODELS OF MATERIAL BEHAVIOR

This chapter is concerned with the investigation of various models of material behavior. Since materials are defined by constitutive equations, we will be examining various constitutive assumptions and the implication of these assumptions. It is useful, therefore, to list briefly the general requirements that constitutive equations must obey [Ref. 1, Sect. 293].

1. *Consistency*: Constitutive equations must be consistent with the axioms of balance of mass, momentum, energy, and the entropy inequality.
2. *Coordinate invariance*: Constitutive equations must be stated by a rule that is equally valid in all fixed coordinate systems.
3. *Just setting*: Boundary value problems resulting from the constitutive equations must be well posed. In other words, there should exist unique solutions corresponding to appropriate initial and boundary data, and these solutions should depend continuously on this data.
4. *Material frame indifference*: The response of the material as characterized by the constitutive equations must be independent of the frame of reference.
5. *Material symmetry*: If the material possesses any symmetry, the fact must be reflected in the constitutive equations.
6. *Equipresence*: An independent variable present in one constitutive equation of a material should also be present in all constitutive equations unless its presence can be shown to be in contradiction with 1 – 5 above.

The restriction 3 will not be considered in this work. The restriction 2 will be satisfied trivially by writing our equations in either direct notation or in a fixed rectangular Cartesian coordinate system. This chapter will be concerned with proposing constitutive equations for various types of materials and then systematically examining the restrictions imposed by 1, 4, 5 and 6.

4.1. Examples

In order to give direction to the discussion of this chapter, in this section we shall give two examples of constitutive equations which occur in the classical theories of continuous materials. The material in this section, in some aspects, parallels the material in Section 1.7.

In order to model a particular material we need constitutive equations for ψ, η, $\mathbf{T}$, and $\mathbf{q}$. The first example constitutive equations are those of a *compressible conducting viscous fluid*. In this case, the material is defined by the following constitutive equations:

$$\psi = \psi(\theta, \rho) \tag{4.1.1}$$

$$\eta = \eta(\theta, \rho) \tag{4.1.2}$$

$$\mathbf{T} = -\pi(\theta, \rho)\mathbf{I} + \lambda(\theta, \rho)(\text{tr}\mathbf{D})\mathbf{I} + 2\mu(\theta, \rho)\mathbf{D} \tag{4.1.3}$$

and

$$\mathbf{q} = -\kappa(\theta, \rho)\,\text{grad}\,\theta \tag{4.1.4}$$

where

$$\eta(\theta, \rho) = -\frac{\partial \psi(\theta, \rho)}{\partial \theta} \tag{4.1.5}$$

$$\pi(\theta, \rho) = \rho^2 \frac{\partial \psi(\theta, \rho)}{\partial \rho} \tag{4.1.6}$$

$$\lambda(\theta, \rho) + \tfrac{2}{3}\mu(\theta, \rho) \geq 0 \tag{4.1.7}$$

$$\mu(\theta, \rho) \geq 0 \tag{4.1.8}$$

and

$$\kappa(\theta, \rho) \geq 0 \tag{4.1.9}$$

The quantity π is the *pressure*. The coefficients λ and μ are the *viscosities*, while κ is the *thermal conductivity*. The quantity $\lambda + \frac{2}{3}\mu$ is called the *bulk viscosity* and μ is called the *shear viscosity*. In many applications the bulk viscosity is taken to be zero. Given (4.1.1) and (4.1.2), (3.5.14) shows that

$$\varepsilon = \varepsilon(\theta, \rho) \tag{4.1.10}$$

An example expression for ψ as a function of (θ, ρ) is [see (1.7.6)]

$$\psi(\theta, \rho) = c_v\theta - c_v\theta \ln\theta + R\theta \ln\rho + \varepsilon^+ - \theta\eta^+ \tag{4.1.11}$$

where c_v and R are positive constants and ε^+ and η^+ are constants. It follows from (4.1.5), (4.1.6), and (4.1.11) that

$$\eta(\theta, \rho) = c_v \ln\theta - R\ln\rho + \eta^+ \tag{4.1.12}$$

and

$$\pi(\theta,\rho)=\rho R\theta \tag{4.1.13}$$

The reader will recognize (4.1.11) as defining a *perfect gas* with constant *specific heat* c_v. The constant R is the *gas constant*. Equations (4.1.11) and (4.1.12) combine with (3.5.14) to yield

$$\varepsilon(\theta,\rho)=c_v\theta+\varepsilon^{+} \tag{4.1.14}$$

Thus, for a perfect gas the internal energy density is independent of density.

Returning to the more general case (4.1.1), we see that

$$\dot{\psi}=\frac{\partial\psi}{\partial\theta}\dot{\theta}+\frac{\partial\psi}{\partial\rho}\dot{\rho} \tag{4.1.15}$$

By (4.1.5) and (4.1.6), we can write (4.1.15) in the form

$$\dot{\psi}=-\eta\dot{\theta}+\frac{\pi}{\rho^2}\dot{\rho} \tag{4.1.16}$$

Equation (4.1.16) is called the *Gibbs relation* for the material being discussed. By use of (3.5.14), it follows from (4.1.16) that

$$\dot{\varepsilon}=\theta\dot{\eta}+\frac{\pi}{\rho^2}\dot{\rho} \tag{4.1.17}$$

Next we wish to present the special forms taken by the axioms of balance of linear momentum and energy for the material defined by (4.1.1) through (4.1.4). Note that the axiom of balance of angular momentum is automatically satisfied because $\mathbf{T}$ in (4.1.3) is symmetric. First we shall consider the axiom of balance of linear momentum. If (4.1.3) is substituted into (3.2.20), we can use (2.2.22), (2.2.25) and (2.4.29) to obtain

$$\rho\ddot{\mathbf{x}}=-\operatorname{grad}\pi+\operatorname{grad}(\lambda\operatorname{div}\dot{\mathbf{x}})+\operatorname{div}(\mu\operatorname{grad}\dot{\mathbf{x}})+\operatorname{div}\left(\mu(\operatorname{grad}\dot{\mathbf{x}})^T\right)+\rho\mathbf{b} \tag{4.1.18}$$

In the special case where the viscosities are constant, (4.1.18) can be written

$$\rho\ddot{\mathbf{x}}=-\operatorname{grad}\pi+(\lambda+\mu)\operatorname{grad}(\operatorname{div}\dot{\mathbf{x}})+\mu\Delta\dot{\mathbf{x}}+\rho\mathbf{b} \tag{4.1.19}$$

where $\Delta\dot{\mathbf{x}}$ denotes the *Laplacian* of $\dot{\mathbf{x}}$. This operator is defined by equation (A.8.25). Equations (4.1.19) are called the *Navier-Stokes equations*.

The special form taken by the thermodynamic energy equation in this case is

$$\rho\theta\dot{\eta} = \operatorname{div}\left(\kappa \operatorname{grad}\theta\right) + \Phi + \rho r \tag{4.1.20}$$

where Φ is called the *viscous dissipation* and is defined by

$$\Phi = \lambda\left(\operatorname{tr}\mathbf{D}\right)^2 + 2\mu \operatorname{tr}\mathbf{D}^2 \tag{4.1.21}$$

If κ is a constant, (4.1.20) becomes

$$\rho\theta\dot{\eta} = \kappa\Delta\theta + \Phi + \rho r \tag{4.1.22}$$

Exercise 4.1.1

Derive (4.1.20).

Exercise 4.1.2

Show that (4.1.21) can be written

$$\Phi = \left(\lambda + \frac{2}{3}\mu\right)\left(\operatorname{tr}\mathbf{D}\right)^2 + 2\mu \operatorname{tr}\left(\mathbf{D} - \frac{1}{3}\left(\operatorname{tr}\mathbf{D}\right)\mathbf{I}\right)^2 \tag{4.1.23}$$

Exercise 4.1.3

Show that (4.1.21) can be written

$$\Phi = \left(\lambda + 2\mu\right)\mathrm{I}_{\mathbf{D}}^2 - 4\mu \mathrm{II}_{\mathbf{D}} \tag{4.1.24}$$

where $\mathrm{I}_{\mathbf{D}}$ and $\mathrm{II}_{\mathbf{D}}$ are the first two fundamental invariants of $\mathbf{D}$. In those cases where λ, μ, and κ are constant, equations (3.1.15), (4.1.19) and (4.1.22) represent the field equations appropriate to the viscous compressible fluid with heat conduction.

The next example we wish to discuss is that of an *isotropic linear thermoelastic solid* with *heat conduction*. Unlike the previous example, this one is approximate in the sense that it only holds when the displacement gradients are small. Thus, it corresponds to a thermomechanical version of classical linear elasticity.

The defining constitutive equations are

$$\rho_R\psi = -\frac{\rho_R c_v}{2\theta_0}\left(\theta - \theta_0\right)^2 - \beta\left(\theta - \theta_0\right)\operatorname{tr}\tilde{\mathbf{E}} + \tfrac{1}{2}\lambda\left(\operatorname{tr}\tilde{\mathbf{E}}\right)^2 + \mu \operatorname{tr}\tilde{\mathbf{E}}^2 \tag{4.1.25}$$

$$\rho_R \eta = \rho_R c_v \left(\frac{\theta - \theta_0}{\theta_0} \right) + \beta \operatorname{tr} \tilde{\mathbf{E}} \tag{4.1.26}$$

$$\mathbf{T}_R = \lambda \left(\operatorname{tr} \tilde{\mathbf{E}} \right) \mathbf{I} + 2\mu \tilde{\mathbf{E}} - \beta \left(\theta - \theta_0 \right) \mathbf{I} \tag{4.1.27}$$

and

$$\mathbf{q}_R = -\kappa \, \mathrm{GRAD}\, \theta \tag{4.1.28}$$

where c_v, θ_0, β, λ, μ, and κ are constants. The coefficient c_v is a positive number called the *specific heat*, θ_0 is a reference temperature, λ and μ are the *isothermal Lame parameters* and κ is the *thermal conductivity*. The conductivity must obey the inequality

$$\kappa \geq 0 \tag{4.1.29}$$

The coefficient μ is also called the *isothermal shear modulus* while $\lambda + \frac{2}{3}\mu$ is the *isothermal bulk modulus*. The constant α, defined by

$$\alpha = \frac{\beta}{3\left(\lambda + \frac{2}{3}\mu \right)} \tag{4.1.30}$$

is called the *coefficient of thermal expansion*. The constant ν, defined by

$$\nu = \frac{\lambda}{2\left(\lambda + \mu \right)} \tag{4.1.31}$$

is the *isothermal Poisson's ratio*. Finally, the constant E ,defined by

$$E = \frac{\mu\left(3\lambda + 2\mu \right)}{\lambda + \mu} \tag{4.1.32}$$

is the *isothermal Young's modulus*.

Equations (4.1.26) and (4.1.27) can be written

$$\eta = -\frac{\partial \psi}{\partial \theta} \tag{4.1.33}$$

and

$$\mathbf{T}_R = \rho_R \frac{\partial \psi}{\partial \tilde{\mathbf{E}}} \tag{4.1.34}$$

It easily follows from (4.1.27) that

$$\mathbf{T}_R = \mathbf{T}_R^T \tag{4.1.35}$$

In the small displacement approximation being explained here, it is possible to show that (3.3.10) and (4.1.35) are equivalent. Thus, balance of angular momentum is satisfied in this theory. It is possible to show that the axioms of balance of linear momentum and energy in this model become

$$\rho_R \ddot{\mathbf{w}} = (\lambda+\mu)\operatorname{GRAD}(\operatorname{Div}\mathbf{w}) + \mu \operatorname{Div}(\operatorname{GRAD}\mathbf{w}) - \beta \operatorname{GRAD}\theta \tag{4.1.36}$$

and

$$\frac{\theta}{\theta_0}\left(\rho_R c_v \dot{\theta} + \beta\theta_0 \operatorname{Div}\dot{\mathbf{w}}\right) = \kappa \operatorname{Div}(\operatorname{GRAD}\theta) \tag{4.1.37}$$

respectively. In the classical theory the left side of (4.1.37) is linearized by approximating the ratio θ/θ_0 by unity. The result is a linear equation of the form

$$\rho_R c_v \dot{\theta} + \beta\theta_0 \operatorname{Div}\dot{\mathbf{w}} = \kappa \operatorname{Div}(\operatorname{GRAD}\theta) \tag{4.1.38}$$

Equations (4.1.36) and (4.1.38) are the field equations of classical linear thermoelasticity. They represent the three dimensional generalizations of of (1.11.7) and (1.11.8).

Exercise 4.1.4

Show that the Gibbs relation for the material defined by (4.1.25) through (4.1.38) is

$$\dot{\psi} = -\eta\dot{\theta} + \frac{1}{\rho_R}\operatorname{tr}\mathbf{T}_R\dot{\tilde{\mathbf{E}}} \tag{4.1.39}$$

Exercise 4.1.5

Show that $\mathbf{T}_R$ and $\tilde{\mathbf{E}}$ have the same eigenvectors.

Exercise 4.1.6

Show that (4.1.27) can be solved for $\tilde{\mathbf{E}}$ and that the result is

$$\tilde{\mathbf{E}} = \frac{1+\nu}{E}\mathbf{T}_R - \frac{\nu}{E}(\operatorname{tr}\mathbf{T}_R)\mathbf{I} + \alpha(\theta-\theta_0)\mathbf{I} \tag{4.1.40}$$

Exercise 4.1.7

Eliminate $\theta-\theta_0$ from (4.1.27) by use of (4.1.26) and show that

$$\mathbf{T}_R=\left(\lambda+\frac{\theta_0}{\rho_R c_v}\beta^2\right)\left(\operatorname{tr}\tilde{\mathbf{E}}\right)\mathbf{I}+2\mu\tilde{\mathbf{E}}-\frac{\beta\theta_0}{c_v}\eta\mathbf{I} \tag{4.1.41}$$

The quantity

$$\lambda+\frac{\theta_0}{\rho_R c_v}\beta^2+\frac{2}{3}\mu$$

is the *isentropic bulk modulus*.

Exercise 4.1.8

Solve (4.1.41) for $\tilde{\mathbf{E}}$ and show that

$$\tilde{\mathbf{E}}=\frac{1+\nu_\eta}{E_\eta}\mathbf{T}_R-\frac{\nu_\eta}{E_\eta}\left(tr\,\mathbf{T}_R\right)\mathbf{I}+\frac{\alpha_\eta\theta_0}{c_v}\eta\mathbf{I} \tag{4.1.42}$$

where

$$\nu_\eta=\frac{\lambda+\dfrac{\theta_0}{\rho_R c_v}\beta^2}{2\left(\lambda+\dfrac{\theta_0}{\rho_R c_v}\beta^2+\mu\right)} \tag{4.1.43}$$

$$\alpha_\eta=\frac{\beta}{3\left(\lambda+\dfrac{\theta_0}{\rho_R c_v}\beta^2+\dfrac{2}{3}\mu\right)} \tag{4.1.44}$$

and

$$E_\eta=\frac{\mu\left[3\left(\lambda+\dfrac{\theta_0}{\rho_R c_v}\beta^2\right)+2\mu\right]}{\lambda+\dfrac{\theta_0}{\rho_R c_v}\beta^2+\mu} \tag{4.1.45}$$

The quantity ν_η is the *isentropic Poisson's ratio* and E_η is the *isentropic Young's modulus.*

Exercise 4.1.9

Adopt the governing equations of linear thermoelasticity given in this section and derive the three dimensional version of (1.11.28).

In closing this section it is useful to mention a few additional examples of constitutive equations which occur in the applications. The first is one which has been proven to be useful in rubber elasticity. It is a *Mooney-Rivlin* material. This model is one where thermodynamic influences and compressibility are taken to be unimportant. As a result, the only constitutive equation of importance is the one for the stress. In this case, this constitutive equation takes the form

$$\mathbf{T} = -p\mathbf{I} + \mu\left(\tfrac{1}{2} + \beta\right)\mathbf{B} - \mu\left(\tfrac{1}{2} - \beta\right)\mathbf{B}^{-1} \tag{4.1.46}$$

where μ and β are constants, p is the hydrostatic stress, and $\mathbf{B}$ is the left Cauchey-Green tensor defined by (2.4.2). In Section 4.5, we shall formulate a model of an *incompressible isothermal elastic material*. Equation (4.1.46) is a special case of the class of incompressible materials discussed in this section. Another example is that of a *Rivlin-Ericksen* fluid. This model is also valid in the isothermal case and is defined by

$$\mathbf{T} = -\pi(\rho)\mathbf{I} + \mathbf{G}(\rho, \mathbf{A}_1, \mathbf{A}_2, \ldots, \mathbf{A}_n) \tag{4.1.47}$$

where $\mathbf{A}_n$ is the n th Rivlin-Ericksen tensor defined by (2.4.37). More precisely, (4.1.47) defines a Rivlin-Ericksen fluid of grade n. In the special case where $n = 1$, (4.1.47) defines what is called a *Reiner-Rivlin* fluid. In the incompressible case, (4.1.47) is altered by replacing the pressure $\pi(\rho)$ by an indeterminate pressure p and omitting the dependence of $\mathbf{G}$ on the density ρ. A fluid of the *second grade* is a special incompressible Rivlin-Ericksen fluid of grade 2 defined by

$$\mathbf{T} = -p\mathbf{I} + \mu\mathbf{A}_1 + \alpha_1\mathbf{A}_2 + \alpha_2\mathbf{A}_1^2 \tag{4.1.48}$$

where μ, α_1, and α_2 are material constants.

4.2. Isothermal Elasticity – Thermodynamic Restrictions

As our first model of a material, we shall consider a purely mechanical theory of an *elastic material*. First we shall investigate the implications of the entropy inequality on the constitutive equations which define this type of material. In the following sections we shall examine the additional restrictions imposed by material frame indifference and by material symmetry.

Because we are interested in an isothermal model, the entropy inequality (3.5.15) reduces to

$$-\rho\dot{\psi} + \operatorname{tr}(\mathbf{TL}) \geq 0 \tag{4.2.1}$$

Therefore, as far as the entropy inequality is concerned, η and $\mathbf{q}$ are indeterminate. This indeterminacy indicates that we are formulating a theory of a *constrained material*. Rather than presenting a detailed discussion of the implication of various types of constraints, it will simply be stated that the isothermal constraint causes the energy equation (3.4.22) to be identically satisfied. The necessary field equations for our isothermal formulation are balance of mass and balance of momentum.

The constitutive equations which *define* an elastic material are

$$\psi(\mathbf{X},t)=u_{\kappa}\left(\mathbf{F}(\mathbf{X},t),\mathbf{X}\right) \tag{4.2.2}$$

and

$$\mathbf{T}(\mathbf{X},t)=\mathbf{G}_{\kappa}\left(\mathbf{F}(\mathbf{X},t),\mathbf{X}\right) \tag{4.2.3}$$

As the notation indicates, the *response functions*, whose values are ψ and $\mathbf{T}$, depend upon the reference configuration $\boldsymbol{\kappa}$. This dependence is necessary because $\mathbf{F}$ is calculated from (2.1.5), and the deformation function $\boldsymbol{\chi}_{\kappa}$ depends upon the reference configuration $\boldsymbol{\kappa}$. The response function u_{κ} and $\mathbf{G}_{\kappa}$ are allowed to depend upon the particle X through the position $\mathbf{X}=\boldsymbol{\kappa}(X)$. As a result, if two different particles are subjected to the same deformation gradient, they need not have the same free energy and the same stress. Therefore, the elastic material defined by (4.2.2) and (4.2.3) is allowed to be *inhomogeneous*.

A comment is in order regarding the domain of the response functions u_{κ} and $\mathbf{G}_{\kappa}$. The deformation gradient $\mathbf{F}$ must obey (2.2.16), which rejects a deformation gradient with zero determinant. Therefore, the first argument of the response functions must lie in a subset of $\mathscr{GL}(\mathscr{V})$, the set of automorphisms in $\mathscr{L}(\mathscr{V};\mathscr{V})$, such that all elements have the same sign for their determinants. Without loss of generality, we can take this subset of $\mathscr{GL}(\mathscr{V})$ to contain elements with *positive* determinant.

If we regard the density ρ_R to be a known function of $\mathbf{X}$, a *thermodynamic process* for an isothermal model is the following set of five functions of $(\mathbf{X},t)$: $\boldsymbol{\chi}_{\kappa}(\mathbf{X},t)$, $\psi(\mathbf{X},t)$, $\mathbf{T}(\mathbf{X},t)$, $\rho(\mathbf{X},t)$, and $\mathbf{b}(\mathbf{X},t)$, which obey balance of mass,

$$\rho\left|\det\mathbf{F}\right|=\rho_R \tag{4.2.4}$$

and balance of linear momentum,

$$\rho\ddot{\mathbf{x}}=\operatorname{div}\mathbf{T}+\rho\mathbf{b} \tag{4.2.5}$$

An *admissible thermodynamic process* is a thermodynamic process which satisfies (4.2.2) and (4.2.3). Note that for every choice of the function χ_κ there exists an admissible thermodynamic process. The proof of this assertion is essentially the same as the corresponding assertion given in Section 1.8. Given χ_κ one can calculate $\rho(\mathbf{X},t)$ from (4.2.4), $\psi(\mathbf{X},t)$ from (4.2.2) and $\mathbf{T}(\mathbf{X},t)$ from (4.2.3). This information can then be used to calculate $\mathbf{b}(\mathbf{X},t)$ from (4.2.5).

As in Section 1.8, we require that (4.2.1) hold *for every* admissible thermodynamic process. In addition, balance of moment of momentum, in the form (3.3.10), is required to hold for every admissible thermodynamic process. Thus (4.2.1) and (3.3.10) are restrictions on the response functions u_κ and $\mathbf{G}_\kappa$. The restriction implied by (3.3.10) is simply that the function $\mathbf{G}_\kappa$ has *symmetric* values. We shall assume this restriction is satisfied and next investigate the restrictions implied by (4.2.1). It follows from (4.2.2) that

$$\dot{\psi} = \operatorname{tr}\left(\left(\frac{\partial u_\kappa(\mathbf{F},\mathbf{X})}{\partial \mathbf{F}}\right)^T \dot{\mathbf{F}}\right) = \operatorname{tr}\left(\left(\mathbf{F}\frac{\partial u_\kappa(\mathbf{F},\mathbf{X})^T}{\partial \mathbf{F}}\right)\mathbf{L}\right) \tag{4.2.6}$$

where (2.2.24) and (A.5.39) have been used. The partial derivative of u_κ with respect to the linear transformation $\mathbf{F}$ which appears in (4.2.6) requires a careful definition. For each $\mathbf{X}$, u_κ is defined on a certain open subset of $\mathscr{GL}(\mathscr{V})$. The gradient of u_κ with respect to $\mathbf{F}$ is defined by the definition given in Section 8 of Appendix A. In the notation used in (4.2.6), the formula (A.8.6) takes the special form

$$\operatorname{tr}\left(\left(\frac{\partial u_\kappa(\mathbf{F},\mathbf{X})}{\partial \mathbf{F}}\right)^T \mathbf{A}\right) = \frac{d}{d\tau}u_\kappa(\mathbf{F}+\tau\mathbf{A},\mathbf{X})\Big|_{\tau=0} \tag{4.2.7}$$

where $\mathbf{A}$ is an arbitrary linear transformation in $\mathscr{L}(\mathscr{V};\mathscr{V})$. Equation (4.2.7) defines the partial derivative $\partial u_\kappa(\mathbf{F},\mathbf{X})/\partial\mathbf{F}$. If (4.2.6) and (4.2.3) are substituted into (4.2.1), the result is

$$\operatorname{tr}\left(\left(\mathbf{G}_\kappa(\mathbf{F},\mathbf{X}) - \rho\mathbf{F}\frac{\partial u_\kappa(\mathbf{F},\mathbf{X})^T}{\partial \mathbf{F}}\right)\mathbf{L}\right) \geq 0 \tag{4.2.8}$$

Equation (4.2.8) must hold for every admissible thermodynamic process. Thus, it must hold for every deformation function. Next, we shall prove that $\mathbf{L}$ in (4.2.8) can be assigned arbitrarily. Let $\mathbf{F}$ be a fixed but arbitrary element in the domain of u_κ and $\mathbf{G}_\kappa$, and let $\mathbf{X}$ and t be fixed. Given these quantities, we can define a deformation χ_κ^* by

$$\chi_\kappa^*(\mathbf{X}^*,t^*) = \chi_\kappa(\mathbf{X},t) + \left\{\left[\mathbf{I}+\mathbf{A}(t^*-t)\right]\mathbf{F}\right\}(\mathbf{X}^*-\mathbf{X}) \tag{4.2.9}$$

where $\mathbf{A}$ is an arbitrary linear transformation. The deformation gradient of the deformation function χ_κ^* is

$$\mathbf{F}^*\left(\mathbf{X}^*,t^*\right)=\left[\mathbf{I}+\mathbf{A}\left(t^*-t\right)\right]\mathbf{F} \tag{4.2.10}$$

and its rate is

$$\dot{\mathbf{F}}^*\left(\mathbf{X}^*,t^*\right)=\mathbf{AF} \tag{4.2.11}$$

Therefore, at $t^*=t$ and $\mathbf{X}^*=\mathbf{X}$,

$$\mathbf{F}^*\left(\mathbf{X},t\right)=\mathbf{F} \tag{4.2.12}$$

and

$$\mathbf{L}^*=\dot{\mathbf{F}}^*\left(\mathbf{X},t\right)\mathbf{F}^{*-1}=\mathbf{A} \tag{4.2.13}$$

If (4.2.8) is evaluated on the deformation function χ_κ^*, and we then let $\mathbf{X}^*\rightarrow\mathbf{X}$ and $t^*\rightarrow t$, the result is

$$\operatorname{tr}\left(\left(\mathbf{G}_\kappa\left(\mathbf{F},\mathbf{X}\right)-\rho\mathbf{F}\frac{\partial u_\kappa\left(\mathbf{F},\mathbf{X}\right)^T}{\partial\mathbf{F}}\right)\mathbf{A}\right)\geq 0 \tag{4.2.14}$$

Because (4.2.14) must hold for every linear transformation $\mathbf{A}$, it immediately follows that the response function $\mathbf{G}_\kappa$ is determined by u_κ through the formula

$$\mathbf{G}_\kappa\left(\mathbf{F},\mathbf{X}\right)=\rho\mathbf{F}\frac{\partial u_\kappa\left(\mathbf{F},\mathbf{X}\right)^T}{\partial\mathbf{F}} \tag{4.2.15}$$

Conversely, given (4.2.15), the entropy inequality (4.2.8) is trivially satisfied. Thus, (4.2.15) is necessary and sufficient for the entropy inequality (4.2.1) to hold for every admissible thermodynamic process. Equation (4.2.15) is the three dimensional generalization of the isothermal version of (1.8.33).

As indicated above, moment of momentum requires that $\mathbf{G}_\kappa$ have symmetric values and, as a result, (4.2.15) yields

$$\mathbf{F}\frac{\partial u_\kappa\left(\mathbf{F},\mathbf{X}\right)^T}{\partial\mathbf{F}}=\frac{\partial u_\kappa\left(\mathbf{F},\mathbf{X}\right)}{\partial\mathbf{F}}\mathbf{F}^T \tag{4.2.16}$$

Equation (4.2.16) is a restriction which must be obeyed by the function u_κ.

Exercise 4.2.1

Show that (4.2.3) and (4.2.15) combine to yield

$$\mathbf{T}_R = \rho_R \frac{\partial u_\kappa(\mathbf{F},\mathbf{X})}{\partial \mathbf{F}} \tag{4.2.17}$$

Exercise 4.2.2

Show that the material defined by (4.2.2) and (4.2.3) obeys the following Gibb's relation

$$\dot{\psi} = \frac{1}{\rho_R}\operatorname{tr}\mathbf{T}_R^T\dot{\mathbf{F}} = \frac{1}{\rho}\operatorname{tr}\mathbf{TL} = \frac{1}{\rho}\operatorname{tr}\mathbf{TD} \tag{4.2.18}$$

Because of the simplicity of (4.2.17), it is convenient to replace the equation of motion in the form (4.2.5) by its material version (3.2.29). In components, (3.2.29) is

$$\rho_R \ddot{x}_j = \frac{\partial T_{R_{jJ}}}{\partial X_J} + \rho_R b_j \tag{4.2.19}$$

If (4.2.17) is substituted into (4.2.19) the result is

$$\rho_R \ddot{x}_j = A_{jJkK}\frac{\partial^2 x_k}{\partial X_J \partial X_K} + q_j + \rho_R b_j \tag{4.2.20}$$

where the fourth-order *elasticity tensor* $\mathbf{A}$ is defined by

$$A_{jJkK}(\mathbf{F},\mathbf{X}) = \rho_R \frac{\partial^2 u_\kappa(\mathbf{F},\mathbf{X})}{\partial F_{jJ}\partial F_{kK}} \tag{4.2.21}$$

where $F_{jJ} = \partial x_j / \partial X_J$, and

$$q_j(\mathbf{F},\mathbf{X}) = \frac{\partial}{\partial X_J}\left(\rho_R \frac{\partial u_\kappa(\mathbf{F},\mathbf{X})}{\partial F_{jJ}}\right)\Bigg|_{\mathbf{F}=\text{const}} \tag{4.2.22}$$

Equation (4.2.20) is the governing differential equation of finite elasticity. The term q_j arises from the explicit dependence of u_κ on $\mathbf{X}$ and, thus, represents a body force resulting from the inhomogeneous nature of the material. From (4.2.21) it is readily seen that the elasticity tensor $\mathbf{A}$ is symmetric in the following sense:

$$A_{jJkK} = A_{kKjJ} \tag{4.2.23}$$

Exercise 4.2.3

Show that for the isothermal elastic material defined in this section, (3.5.17) reduces to

$$\overline{\int_{\kappa(\mathscr{P})} \rho_R \left(\psi + \tfrac{1}{2}\dot{\mathbf{x}}^2\right) dV}^{\bullet} = \oint_{\partial\kappa(\mathscr{P})} \dot{\mathbf{x}} \cdot \left(\mathbf{T}_R\, d\mathbf{S}\right) + \int_{\kappa(\mathscr{P})} \rho_R \dot{\mathbf{x}} \cdot \mathbf{b}\, dV \tag{4.2.24}$$

Equation (4.2.24) is in the form of an energy equation for our isothermal elastic material. However, because of the isothermal assumption, it is an *identity* derivable from the equations in this section.

4.3. Isothermal Elasticity-Material Frame Indifference

In this section we shall introduce the axiom of material frame indifference and show how it restricts the constitutive equations for an isothermal elastic material.

By definition a *frame of reference* is the set $\mathscr{E} \times \mathscr{R}$. In words, a frame of reference is a set that provides information about position in $\mathscr{E}$ and the time t. We are interested in mappings of $\mathscr{E} \times \mathscr{R} \to \mathscr{E} \times \mathscr{R}$ with the properties that distances, time intervals, and temporal order are preserved. Such mappings are called *changes of frame*. However, they are mappings of $\mathscr{E} \times \mathscr{R}$ into itself. The formal definition is as follows:

Definition. A *change of frame* is a one-to-one mapping $\mathscr{E} \times \mathscr{R} \to \mathscr{E} \times \mathscr{R}$ defined by

$$\mathbf{x}^* = \mathbf{c}(t) + \mathbf{Q}(t)(\mathbf{x} - \mathbf{0}) \tag{4.3.1}$$

and

$$t^* = t - a \tag{4.3.2}$$

where $\mathbf{c}(t)$ is a time-dependent element of $\mathscr{E}$, $\mathbf{Q}(t)$ is a time-dependent of $\mathscr{O}(\mathscr{V})$, and a is in $\mathscr{R}$. The geometric interpretation of (4.3.1) has been discussed in Section A.7 of Appendix A.

Because of the special nature of the constitutive equations of an elastic material, we actually will not need to capitalize on the time dependence of $\mathbf{c}$ and $\mathbf{Q}$. However, for the more general materials to be investigated later, this time dependence will be important. Also, in this section, we can always take $a = 0$ without any loss of generality. Given a motion $\boldsymbol{\chi}_\kappa$, we construct a new motion χ^*_κ by use of (4.3.1). The result is

$$\chi_{\kappa}^{*}(\mathbf{X},t)=\mathbf{c}(t)+\mathbf{Q}(t)\left(\chi_{\kappa}(\mathbf{X},t)-\mathbf{0}\right) \tag{4.3.3}$$

The deformation gradient associated with the new motion is by definition

$$\mathbf{F}^{*}(\mathbf{X},t)=\mathrm{GRAD}\,\chi_{\kappa}^{*}(\mathbf{X},t) \tag{4.3.4}$$

It follows from (4.3.3) that

$$\mathbf{F}^{*}=\mathbf{QF} \tag{4.3.5}$$

Exercise 4.3.1

Show that

$$\mathbf{C}^{*}=\mathbf{C} \tag{4.3.6}$$

$$\mathbf{U}^{*}=\mathbf{U} \tag{4.3.7}$$

$$\mathbf{R}^{*}=\mathbf{QR} \tag{4.3.8}$$

$$\mathbf{B}^{*}=\mathbf{QBQ}^{T} \tag{4.3.9}$$

and

$$\rho^{*}=\rho \tag{4.3.10}$$

Equation (4.3.5) indicates how the *independent* variables in (4.2.2) and (4.2.3) transform under the transformation (4.3.3). The *dependent* variables are *required* to transform according to the rules

$$\psi^{*}=\psi \tag{4.3.11}$$

and

$$\mathbf{T}^{*}=\mathbf{QTQ}^{T} \tag{4.3.12}$$

The derivation of (4.3.12) is as follows. Given (3.2.13), we require that the stress vector $\mathbf{t}$ and the normal $\mathbf{n}$ transform by the following rules:

$$\mathbf{t}^{*}=\mathbf{Qt} \tag{4.3.13}$$

and

$$\mathbf{n}^* = \mathbf{Q}\mathbf{n} \tag{4.3.14}$$

It follows from (3.2.13) and (4.3.14) that (4.3.13) can be written

$$\mathbf{t}^* = \mathbf{Q}\mathbf{T}\mathbf{Q}^T\mathbf{n}^* \tag{4.3.15}$$

Equation (4.3.15) implies (4.3.12) because $\mathbf{t}^*$ and $\mathbf{n}^*$ must be related by

$$\mathbf{t}^* = \mathbf{T}^*\mathbf{n}^* \tag{4.3.16}$$

for every normal vector $\mathbf{n}^*$.

The axiom of material frame indifference for our isothermal elastic material is the following statement [Ref. 3, Sect. 19].

Axiom. The constitutive equations (4.2.2) and (4.2.3) must be invariant under a change of frame. In particular, if

$$\psi(\mathbf{X},t) = u_\kappa\left(\mathbf{F}(\mathbf{X},t),\mathbf{X}\right) \tag{4.3.17}$$

and

$$\mathbf{T}(\mathbf{X},t) = \mathbf{G}_\kappa\left(\mathbf{F}(\mathbf{X},t),\mathbf{X}\right) \tag{4.3.18}$$

then

$$\psi^*(\mathbf{X},t) = u_\kappa\left(\mathbf{F}^*(\mathbf{X},t),\mathbf{X}\right) \tag{4.3.19}$$

and

$$\mathbf{T}^*(\mathbf{X},t) = \mathbf{G}_\kappa\left(\mathbf{F}^*(\mathbf{X},t),\mathbf{X}\right) \tag{4.3.20}$$

where

$$\psi^*(\mathbf{X},t) = \psi(\mathbf{X},t) \tag{4.3.21}$$

$$\mathbf{T}^* = \mathbf{Q}\mathbf{T}(\mathbf{X},t)\mathbf{Q}^T \tag{4.3.22}$$

and

$$\mathbf{F}^*(\mathbf{X},t) = \mathbf{Q}\mathbf{F}(\mathbf{X},t) \tag{4.3.23}$$

for all orthogonal linear transformations $\mathbf{Q}$, such that $\mathbf{F}^*(\mathbf{X},t)$ is in the domain of u_κ and $\mathbf{G}_\kappa$.

It is important to note that the reference configuration is fixed in the statement of the above axiom. Equations (4.3.17) through (4.3.23) are summarized by the following two equations:

$$u_\kappa(\mathbf{F},\mathbf{X}) = u_\kappa(\mathbf{QF},\mathbf{X}) \tag{4.3.24}$$

and

$$\mathbf{QG}_\kappa(\mathbf{F},\mathbf{X})\mathbf{Q}^T = \mathbf{G}_\kappa(\mathbf{QF},\mathbf{X}) \tag{4.3.25}$$

Equations (4.3.24) and (4.3.25) must hold for every orthogonal linear transformation $\mathbf{Q}$ such that $\mathbf{QF}$ is in the domain of u_κ and $\mathbf{G}_\kappa$. As explained in Section 4.2, the domains of u_κ and $\mathbf{G}_\kappa$ contains linear transformations $\mathbf{F}$ which obey $\det\mathbf{F} > 0$. Therefore, because $\det\mathbf{QF} = \det\mathbf{Q}\det\mathbf{F}$, the orthogonal linear transformations $\mathbf{Q}$ in (4.3.24) and (4.3.25) must have *positive* determinants.

Because of the thermodynamic restriction (4.2.15), it is reasonable to suspect that (4.3.24) implies (4.3.25). In fact, this is the case, and we shall now present the proof. The key formula is

$$\frac{\partial u_\kappa(\mathbf{F},\mathbf{X})}{\partial \mathbf{F}}^T = \frac{\partial u_\kappa(\mathbf{QF},\mathbf{X})}{\partial \mathbf{F}}^T \mathbf{Q} \tag{4.3.26}$$

This equation results from differentiation of (4.3.24) with respect to $\mathbf{F}$ according to the definition (4.2.7). It follows from (4.3.26) that

$$\mathbf{Q}\left(\rho\mathbf{F}\frac{\partial u_\kappa(\mathbf{F},\mathbf{X})}{\partial \mathbf{F}}^T\right)\mathbf{Q}^T = \rho\mathbf{QF}\frac{\partial u_\kappa(\mathbf{QF},\mathbf{X})}{\partial \mathbf{F}}^T \tag{4.3.27}$$

Because of (4.2.15), equation (4.3.27) implies (4.3.25).

Exercise 4.3.2

Derive (4.3.26).

Next, we shall deduce the solution of (4.3.24). Because (4.3.24) must hold for every orthogonal $\mathbf{Q}$ with positive determinant, we can obtain *necessary* conditions by making various choices for $\mathbf{Q}$. Following Noll, if we take $\mathbf{Q} = \mathbf{R}^T$, (2.4.1*a*) allows (4.3.24) to be written

$$u_\kappa(\mathbf{F},\mathbf{X}) = u_\kappa(\mathbf{U},\mathbf{X}) \tag{4.3.28}$$

[Ref. 4, Sect. 15]. Therefore, (4.3.24) yields the necessary condition that u_{κ} evaluated on $\mathbf{F}$ equals u_{κ} evaluated on $\mathbf{U}$. Thus, the rotation part of $\mathbf{F}=\mathbf{RU}$ does not effect the value of u_{κ}. If we can show that (4.3.28) is sufficient for (4.3.24) to be valid for all $\mathbf{Q}$, then we have established that (4.3.28) is a solution of (4.3.24). If we assume (4.3.28), then

$$\begin{aligned} u_{\kappa}(\mathbf{QF},\mathbf{X}) &= u_{\kappa}\left(\left((\mathbf{QF})^T\mathbf{QF}\right)^{1/2},\mathbf{X}\right) = u_{\kappa}\left(\left(\mathbf{F}^T\mathbf{Q}^T\mathbf{QF}\right)^{1/2},\mathbf{X}\right) \\ &= u_{\kappa}\left(\left(\mathbf{F}^T\mathbf{F}\right)^{1/2},\mathbf{X}\right) = u_{\kappa}(\mathbf{U},\mathbf{X}) \end{aligned} \tag{4.3.29}$$

Therefore,

$$u_{\kappa}(\mathbf{QF},\mathbf{X}) = u_{\kappa}(\mathbf{F},\mathbf{X}) \tag{4.3.30}$$

for all orthogonal linear transformations $\mathbf{Q}$. Thus, (4.3.28) reflects all of the restrictions implied by the axiom of material frame indifference for an isothermal elastic material.

Given (4.3.28), we can define a function $\hat{u}_{\kappa}$ of $\mathbf{C}$ and $\mathbf{X}$ by

$$\hat{u}_{\kappa}(\mathbf{C},\mathbf{X}) = u_{\kappa}(\mathbf{U},\mathbf{X}) \tag{4.3.31}$$

The choice of $\mathbf{C}$ as an independent variable is convenient because, from (2.4.3), $\mathbf{C}$ is a rational function of the deformation gradient. In summary, if

$$\psi(\mathbf{X},t) = \hat{u}_{\kappa}(\mathbf{C}(\mathbf{X},t),\mathbf{X}) \tag{4.3.32}$$

then the axiom of material frame indifference is satisfied.

The stress is calculated from (4.2.15). It follows from (4.3.28), (4.3.31) and (2.4.3) that

$$\frac{\partial u_{\kappa}(\mathbf{F},\mathbf{X})^T}{\partial \mathbf{F}} = 2\frac{\partial \hat{u}_{\kappa}(\mathbf{C},\mathbf{X})}{\partial \mathbf{C}}\mathbf{F}^T \tag{4.3.33}$$

Therefore,

$$\mathbf{G}_{\kappa}(\mathbf{F},\mathbf{X}) = 2\rho\mathbf{F}\frac{\partial \hat{u}_{\kappa}(\mathbf{C},\mathbf{X})}{\partial \mathbf{C}}\mathbf{F}^T \tag{4.3.34}$$

and, from (4.2.3),

$$\mathbf{T}(\mathbf{F},\mathbf{X}) = 2\rho\mathbf{F}\frac{\partial \hat{u}_{\kappa}(\mathbf{C},\mathbf{X})}{\partial \mathbf{C}}\mathbf{F}^T \tag{4.3.35}$$

As a final observation of this section, note that the symmetry condition (4.2.16) is satisfied. This fact follows from (4.3.33) or, more directly, from (4.3.35) which clearly yields a symmetric stress tensor.

Exercise 4.3.3

Derive (4.3.33).

Exercise 4.3.4

Show that

$$\mathbf{T}_R = 2\rho_R \mathbf{F} \frac{\partial \hat{u}_\kappa (\mathbf{C}, \mathbf{X})}{\partial \mathbf{C}} \tag{4.3.36}$$

and

$$\tilde{\mathbf{T}} = 2\rho_R \frac{\partial \hat{u}_\kappa (\mathbf{C}, \mathbf{X})}{\partial \mathbf{C}} \tag{4.3.37}$$

Exercise 4.3.5

Take $\mathbf{Q} = \mathbf{R}$ in (4.3.24) and show that the resulting necessary condition is *not* sufficient to satisfy (4.3.24) for all orthogonal $\mathbf{Q}$.

Exercise 4.3.6

Given (4.2.15), it was shown in the text that (4.3.24) implies (4.3.25). Show that (4.3.25) implies (4.3.24). Therefore, given (4.2.15), (4.3.24) and (4.3.25) are equivalent.

Exercise 4.3.7

Given (4.2.15), it was shown in the text that (4.3.24) implies (4.2.16). Show that (4.2.16) implies (4.3.24). Therefore, given (4.2.15), (4.2.16) and (4.3.25) are equivalent.

Exercises 4.3.6 and 4.3.7 combine to yield Noll's result that frame indifference of the free energy, frame indifference of the stress and symmetry of the stress are equivalent for the material defined by (4.2.2) and (4.2.3) [Ref. 4, Sect. 4].

Exercise 4.3.8

Define a stress tensor $\mathbf{T}^{(1)}$ by the formula

$$\mathbf{T}^{(1)} = \frac{1}{2}\left(\mathbf{T}_R^T \mathbf{R} + \mathbf{R}^T \mathbf{T}_R\right) \tag{4.3.38}$$

and show that

$$\mathbf{T}^{(1)} = \rho_R \frac{\partial u_\kappa(\mathbf{U},\mathbf{X})}{\partial \mathbf{U}} \tag{4.3.39}$$

The stress tensor $\mathbf{T}^{(1)}$ is sometimes called the Biot stress tensor or the Jaumann stress tensor. [Ref. 6]

4.4. Isothermal Elasticity-Material Symmetry

In this section we shall present the concept of material symmetry and show how the response function $\hat{u}_\kappa$ is restricted for special types of material symmetry. The presentation will be brief. Material Symmetry will be discussed in greater detail later in Section 4.10.

Roughly speaking, the concept of material symmetry arises when one attempts to determine in what fashion the response function $\hat{u}_\kappa$ depends upon the choice of reference configuration. Recall from (2.1.4), a reference configuration is a mapping $\boldsymbol{\kappa}:\mathscr{B}\rightarrow\mathscr{E}$, and we wrote

$$\mathbf{X} = \boldsymbol{\kappa}(X) \tag{4.4.1}$$

to indicate the position occupied by X in the reference configuration $\boldsymbol{\kappa}$. If $\hat{\boldsymbol{\kappa}}$ is a different reference configuration, then

$$\hat{\mathbf{X}} = \hat{\boldsymbol{\kappa}}(X) \tag{4.4.2}$$

is the position occupied by X in the reference configuration $\hat{\boldsymbol{\kappa}}$. It follows from (4.4.1) and (4.4.2) that the positions $\hat{\mathbf{X}}$ and $\mathbf{X}$ are related by

$$\hat{\mathbf{X}} = \hat{\boldsymbol{\kappa}}\left(\boldsymbol{\kappa}^{-1}(\mathbf{X})\right) \equiv \boldsymbol{\lambda}(\mathbf{X}) \tag{4.4.3}$$

Figure 4.4.1 reflects the construction of the function $\boldsymbol{\lambda}$. Equation (4.4.3) is a *change of reference configuration.* It maps particles which occupy points in $\boldsymbol{\kappa}(\mathscr{B})$ into the region $\hat{\boldsymbol{\kappa}}(\mathscr{B})$. In general such a mapping will deform the body $\mathscr{B}$. We shall denote by $\mathbf{H}$ the gradient defined by

$$\mathbf{H} = \text{GRAD}\,\boldsymbol{\lambda}(\mathbf{X}) \tag{4.4.4}$$

There should be no confusion between this quantity and the displacement gradient defined by (2.2.20). $\mathbf{H}$ is the deformation gradient associated with the deformation (4.4.3).

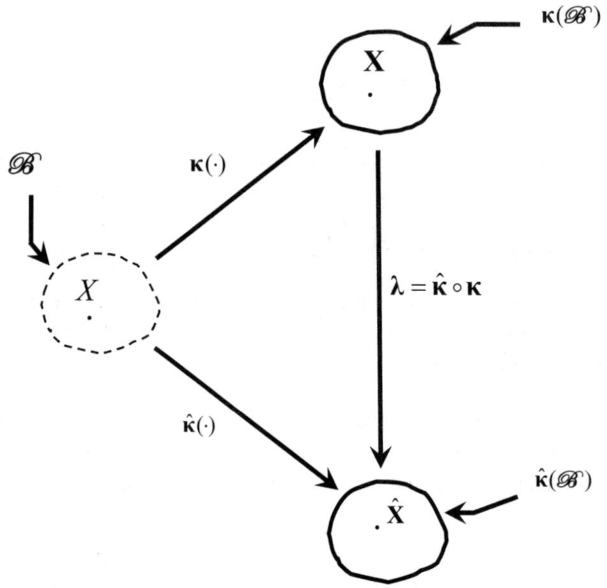

Figure 4.4.1

The deformation function χ_κ is defined by (2.1.5). Likewise, we can define a deformation function $\hat{\chi}_{\hat{\kappa}}$ by

$$\hat{\chi}_{\hat{\kappa}}\left(\hat{\mathbf{X}},t\right)=\hat{\chi}\left(\hat{\boldsymbol{\kappa}}^{-1}\left(\hat{\mathbf{X}}\right),t\right) \tag{4.4.5}$$

where $\hat{\chi}$ is the motion for $\mathscr{B}$. Figure 4.4.2 illustrates the construction of χ_κ and $\hat{\chi}_{\hat{\kappa}}$. Of course, if the motion $\hat{\chi}$ is the same as χ, the spatial regions $\hat{\chi}\left(\mathscr{B},t\right)$ and $\chi\left(\mathscr{B},t\right)$ are the same.

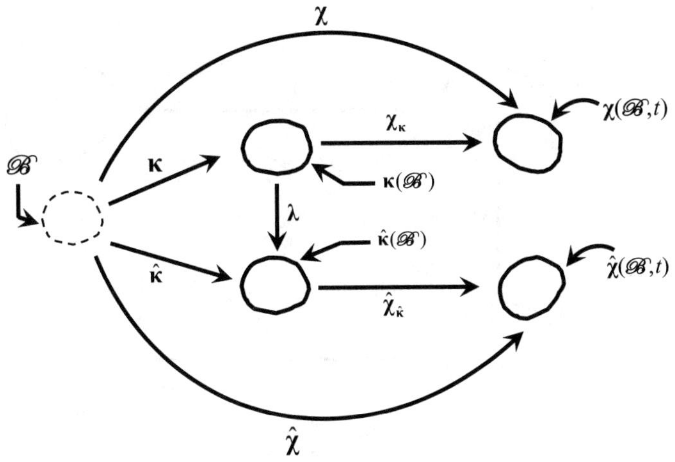

Figure 4.4.2

The deformation function $\hat{\chi}_{\hat{\kappa}}$ can be expressed in terms of positions in $\boldsymbol{\kappa}$ by use of (4.4.3). The result is

$$\hat{\mathbf{x}} = \hat{\chi}_{\hat{\kappa}}\left(\hat{\mathbf{X}},t\right) = \hat{\chi}_{\hat{\kappa}}\left(\boldsymbol{\lambda}\left(\mathbf{X}\right),t\right) \tag{4.4.6}$$

The deformation gradient associated with the deformation function $\hat{\chi}_{\hat{\kappa}}$ is denoted by $\hat{\mathbf{F}}$ and is defined by

$$\hat{\mathbf{F}} = \text{GRAD}\,\hat{\chi}_{\hat{\kappa}}\left(\hat{\mathbf{X}},t\right) \tag{4.4.7}$$

If we use (4.4.6) to define a deformation function $\hat{\chi}_{\kappa}$ by

$$\hat{\chi}_{\kappa}\left(\mathbf{X},t\right) = \hat{\chi}_{\hat{\kappa}}\left(\boldsymbol{\lambda}\left(\mathbf{X}\right),t\right) \tag{4.4.8}$$

then $\hat{\chi}_{\kappa}$ is a deformation function which, for each t, describes deformations from $\boldsymbol{\kappa}\left(\mathscr{B}\right)$ to $\hat{\chi}_{\kappa}\left(\mathscr{B},t\right)$. If we differentiate (4.4.8) and use the definitions (4.4.7) and (4.4.4), it follows that

$$\text{GRAD}\,\hat{\chi}_{\kappa}\left(\mathbf{X},t\right) = \hat{\mathbf{F}}\mathbf{H} \tag{4.4.9}$$

This equation relates the deformation gradients constructed from viewing the same motion from two different reference configurations. Figure 4.43 illustrates the construction of these deformation gradients.

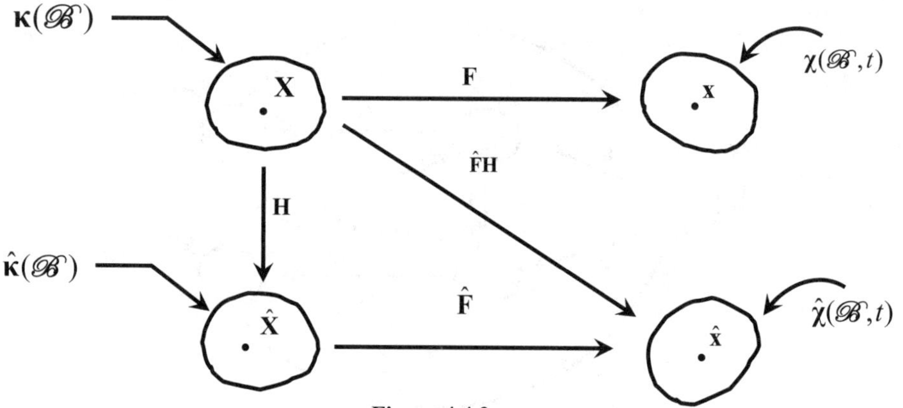

Figure 4.4.3

The concept of material symmetry arises when one tries to characterize those changes of reference configuration which, in some sense, do not affect the response of the isothermal elastic material. If the material is subjected to a deformation gradient $\mathbf{F}$, then (2.4.3) and (4.3.32) yield

$$\psi = \hat{u}_{\kappa}\left(\mathbf{C},\mathbf{X}\right) \tag{4.4.10}$$

The value of ψ is the free energy density one obtains when the particle at $\mathbf{X}$ in the reference configuration $\boldsymbol{\kappa}$ is subjected to a deformation $\mathbf{F}$. If the same particle is subjected to the deformation $\mathbf{FH}$, then the resulting value of the free energy density is

$$\psi^{*} = \hat{u}_{\kappa}\left(\mathbf{H}^{T}\mathbf{CH},\mathbf{X}\right) \tag{4.4.11}$$

Conceptually, (4.4.11) arises when one deforms $\mathbf{X}$ by the amount $\mathbf{H}$ followed by a deformation of the amount $\mathbf{F}$. The second deformation is the same as that which appeared in (4.4.10) except that it is applied from $\hat{\boldsymbol{\kappa}}\left(\mathscr{B}\right)$ to $\hat{\chi}\left(\mathscr{B},t\right)$. Figure 4.4.4 should be useful.

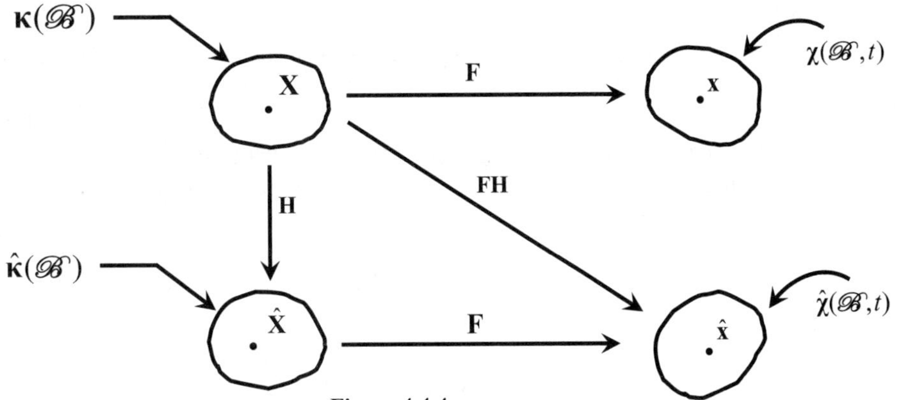

Figure 4.4.4

In rough terms, we are interested in characterizing those linear transformations $\mathbf{H}$ which produce the *same* value of the free energy from (4.4.11) as obtained from (4.4.10). In order that the mass density $\hat{\rho} = \rho_R \left|\det \mathbf{FH}\right|^{-1}$ equals the mass density $\rho = \rho_R \left|\det \mathbf{F}\right|^{-1}$, we shall only consider changes of reference configuration which obey

$$\left|\det \mathbf{H}\right| = 1 \tag{4.4.12}$$

Because $\mathbf{F}$ and $\mathbf{H}$ are nonsingular, they are members of the general linear group $\mathscr{GL}(\mathscr{V})$, defined in Section A.2. The set of linear transformations $\mathbf{H}$ in $\mathscr{GL}(\mathscr{V})$ which obey (4.4.12) form a subgroup of $\mathscr{GL}(\mathscr{V})$ called the *unimodular group*. It is denoted by the symbol $\mathscr{U}(\mathscr{V})$. Roughly speaking, a subgroup is a subset of a group which obeys all of the group axioms. The orthogonal group $\mathscr{O}(\mathscr{V})$, defined in Section A.3, is a subgroup of $\mathscr{U}(\mathscr{V})$ because all of its elements obey (4.4.12). However, one can construct examples of unimodular linear transformations which are not orthogonal. Thus, $\mathscr{O}(\mathscr{V})$ is a *proper* subgroup of $\mathscr{U}(\mathscr{V})$. The groups $\mathscr{GL}(\mathscr{V})$, $\mathscr{U}(\mathscr{V})$ and $\mathscr{O}(\mathscr{V})$ have *positive* components. These positive components are defined to be the subsets, of each group, which have positive determinants. For example, $\mathscr{GL}(\mathscr{V})^+$ is the set of nonsingular linear transformations in $\mathscr{L}(\mathscr{V};\mathscr{V})$ which have positive determinant. The sets $\mathscr{U}(\mathscr{V})^+$ and $\mathscr{O}(\mathscr{V})^+$ are defined accordingly. Because of (A.5.22), $\mathscr{GL}(\mathscr{V})^+$ is actually a subgroup of the group $\mathscr{GL}(\mathscr{V})$. Identical observations yield that $\mathscr{U}(\mathscr{V})^+$ is a subgroup of $\mathscr{U}(\mathscr{V})$ and $\mathscr{O}(\mathscr{V})^+$ is a subgroup of $\mathscr{O}(\mathscr{V})$. It should also be clear that $\mathscr{O}(\mathscr{V})^+$ is a subgroup of $\mathscr{U}(\mathscr{V})^+$, which, in turn, is a subgroup of $\mathscr{GL}(\mathscr{V})^+$. The group $\mathscr{O}(\mathscr{V})^{+}$ is often called the *proper orthogonal group* because its elements represent proper rotations. An improper rotation is illustrated by the inversion $-\mathbf{I}$ in $\mathscr{O}(\mathscr{V})$. Likewise, one could call $\mathscr{U}(\mathscr{V})^{+}$ the proper unimodular group. More often, it is referred to as the *special linear group*.

Definition. The *symmetry group* $\mathscr{G}_{\kappa}(X)$ (isotropy group) of the isothermal elastic particle X in the reference configuration $\boldsymbol{\kappa}$ is the set of linear transformations $\mathbf{H}$ which obey the following two conditions:

1. $$\det \mathbf{H} = 1 \tag{4.4.13}$$

and

2. $$\hat{u}_{\kappa}(\mathbf{C},\mathbf{X}) = \hat{u}_{\kappa}(\mathbf{H}^{T}\mathbf{C}\mathbf{H},\mathbf{X}) \tag{4.4.14}$$

for all symmetric linear transformations $\mathbf{C}$.

It is important to note that, as defined, the symmetry group depends upon the particle X and the reference configuration $\boldsymbol{\kappa}$. This dependence arises because the response function $\hat{u}_{\kappa}$ depends upon $\boldsymbol{\kappa}$ and $\mathbf{X}$. If $\hat{u}_{\kappa}$ is independent of $\mathbf{X}$, then accordingly $\mathscr{G}_{\kappa}(X)$ is independent of X. In this case we write $\mathscr{G}_{\kappa}$ for the symmetry group of the body. The dependence of $\mathscr{G}_{\kappa}(X)$ on the reference configuration will be examined briefly later in this section.

Exercise 4.4.1

Prove that $\mathscr{G}_{\kappa}(X)$ is in fact a group. In particular, prove that it is a subgroup of the general linear group $\mathscr{U}(\mathscr{V})^{+}$.

Note that (4.4.14) is unchanged if $\mathbf{H}$ is replaced by $-\mathbf{H}$. For this reason, one could have defined the symmetry group to be the set of linear transformations $\mathbf{H}$ which obey (4.4.14) and $|\det \mathbf{H}| = 1$ rather than (4.4.13). In this case, $\mathscr{G}_{\kappa}(X)$ would be a subgroup of $\mathscr{U}(\mathscr{V})$ and could contain elements with negative determinant. Insofar as the results of this section are concerned, these two possible definitions are equivalent. For more general materials, like the one discussed in Section 4.6, the condition which replaces (4.4.14) is altered if $\mathbf{H}$ is replaced by $-\mathbf{H}$.

The dependence of $\mathscr{G}_{\kappa}(X)$ on $\boldsymbol{\kappa}$ can be characterized by investigating how it is altered under a change of reference configuration. Such a change is caused by a mapping

$$\hat{\mathbf{X}} = \boldsymbol{\alpha}(\mathbf{X}) \tag{4.4.15}$$

whose gradient is written

$$\mathbf{P} = \mathrm{GRAD}\,\boldsymbol{\alpha}(\mathbf{X}) \tag{4.4.16}$$

By the same type of calculation that yielded (4.4.9), if $\mathbf{F}$ is the deformation gradient from $\boldsymbol{\kappa}(\mathscr{B}) \to \chi(\mathscr{B},t)$, then $\mathbf{FP}^{-1}$ is the deformation gradient from $\hat{\boldsymbol{\kappa}}(\mathscr{B})$ to $\chi(\mathscr{B},t)$. These relationships are illustrated in Figure 4.4.5.

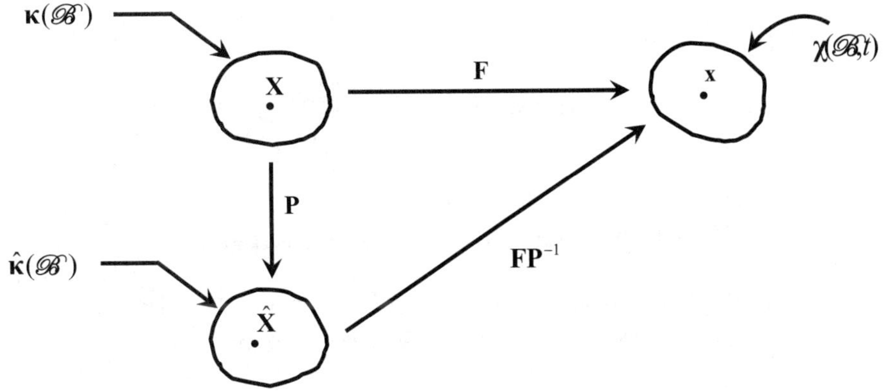

Figure 4.4.5

Because the free energy density of X at the time t must be independent of the choice of reference configuration, we define $\hat{u}_{\hat{\kappa}}$, the response function relative to $\hat{\boldsymbol{\kappa}}$, by

$$\hat{u}_{\hat{\kappa}}\left(\hat{\mathbf{C}},\hat{\mathbf{X}}\right) = \hat{u}_{\kappa}\left(\mathbf{C},\mathbf{X}\right) \tag{4.4.17}$$

where $\hat{\mathbf{X}} = \boldsymbol{\alpha}(\mathbf{X})$ and, from (2.4.3),

$$\hat{\mathbf{C}} = \left(\mathbf{FP}^{-1}\right)^T\left(\mathbf{FP}^{-1}\right) = \mathbf{P}^{-1^T}\mathbf{F}^T\mathbf{FP}^{-1} = \mathbf{P}^{-1^T}\mathbf{CP}^{-1} \tag{4.4.18}$$

is the right Cauchy-Green tensor relative to the reference configuration $\hat{\boldsymbol{\kappa}}$. Equation (4.4.17) is required to hold for every particle X (or position $\mathbf{X}$) and every symmetric linear transformation $\mathbf{C}$. Equation (4.4.17) characterizes to what extent $\hat{u}_{\kappa}$ depends on the reference configuration $\boldsymbol{\kappa}$. Given $\mathbf{C}$ and $\mathbf{X}$, one must know the *value* of $\boldsymbol{\alpha}$ at $\mathbf{X}$ and the value of the *first gradient* of $\boldsymbol{\alpha}$ at $\mathbf{X}$. One does not really need to know any more about the function $\boldsymbol{\alpha}$ in order to construct $\hat{u}_{\hat{\kappa}}$ for the particle at $\mathbf{X}$. This observation shows that $\hat{u}_{\kappa}$ really only depends on $\boldsymbol{\kappa}$ in a neighborhood of $\mathbf{X}$. Such a neighborhood is usually called a *local reference configuration.*

Given the function $\hat{u}_{\hat{\kappa}}$ we can define the symmetry group $\mathcal{G}_{\hat{\kappa}}(X)$ by a definition identical to (4.4.13) and (4.4.14). Because $\hat{u}_{\hat{\kappa}}$ is determined by $\hat{u}_{\kappa}$, it follows that $\mathcal{G}_{\hat{\kappa}}(X)$ is determined by $\mathcal{G}_{\kappa}(X)$. The explicit relationship between these groups follows by writing (4.4.14) in terms of the function $\hat{u}_{\hat{\kappa}}$, which is defined by (4.4.17). It follows from (4.4.17) and (4.4.18) that (4.4.14) can be written

$$\hat{u}_{\hat{\kappa}}\left(\hat{\mathbf{C}},\hat{\mathbf{X}}\right)=\hat{u}_{\hat{\kappa}}\left(\mathbf{P}^{-1^T}\mathbf{H}^T\mathbf{CHP}^{-1},\hat{\mathbf{X}}\right) \tag{4.4.19}$$

or, by (4.4.18),

$$\hat{u}_{\hat{\kappa}}\left(\hat{\mathbf{C}},\hat{\mathbf{X}}\right)=\hat{u}_{\hat{\kappa}}\left(\mathbf{P}^{-1^T}\mathbf{H}^T\mathbf{P}^T\hat{\mathbf{C}}\mathbf{PHP}^{-1},\hat{\mathbf{X}}\right)=\hat{u}_{\hat{\kappa}}\left(\left(\mathbf{PHP}^{-1}\right)^T\hat{\mathbf{C}}\left(\mathbf{PHP}^{-1}\right),\hat{\mathbf{X}}\right) \tag{4.4.20}$$

Because

$$\det\mathbf{PHP}^{-1}=\det\mathbf{P}\det\mathbf{H}\det\mathbf{P}^{-1}=\det\mathbf{PP}^{-1}\det\mathbf{H}=\det\mathbf{H}$$

(4.4.20) shows that if $\mathbf{H}$ is an element of $\mathcal{G}_{\kappa}(X)$, then $\mathbf{PHP}^{-1}$ is an element of $\mathcal{G}_{\hat{\kappa}}(X)$. The result, which was first obtained by Noll, can be written symbolically as

$$\mathcal{G}_{\hat{\kappa}}(X)=\mathbf{P}\mathcal{G}_{\kappa}(X)\mathbf{P}^{-1} \tag{4.4.21}$$

[Ref. 5].

A reference configuration $\boldsymbol{\kappa}$ is said to be *undistorted* for the elastic particle X if $\mathcal{G}_{\kappa}(X)$ contains the proper orthogonal group $\mathcal{O}(\mathscr{V})^+$ or if $\mathcal{G}_{\kappa}(X)$ is contained in the proper orthogonal group $\mathcal{O}(\mathscr{V})^+$. The elastic particle X is *isotropic* if there exists an undistorted reference configuration for X such that $\mathcal{O}(\mathscr{V})^+$ is in $\mathcal{G}_{\kappa}(X)$. The elastic particle X is an elastic *solid particle* if there exists an undistorted reference configuration for X such that $\mathcal{G}_{\kappa}(X)$ is in $\mathcal{O}(\mathscr{V})^+$. These two definitions show that X is an elastic isotropic solid particle if there exists an undistorted reference configuration for X such that $\mathcal{G}_{\kappa}(X)=\mathcal{O}(\mathscr{V})^+$. It is important in these definitions of an isotropic particle and of a solid particle to require the existence of an undistorted reference configuration. If $\boldsymbol{\kappa}$ is such a configuration for X and $\mathbf{H}$ is an orthogonal element of $\mathcal{G}_{\kappa}(X)$, then a simple manipulation shows that the corresponding element in $\mathcal{G}_{\hat{\kappa}}(X)$, $\mathbf{PHP}^{-1}$, is not orthogonal for arbitrary $\mathbf{P}$. Therefore, for example, if X is an elastic isotropic solid particle in the configuration $\boldsymbol{\kappa}$ its symmetry group $\mathcal{G}_{\hat{\kappa}}(X)$ is not necessarily equal to $\mathcal{O}(\mathscr{V})^+$. Our definition requires that $\mathcal{G}_{\kappa}(X)=\mathcal{O}(\mathscr{V})^+$ for *some* reference configuration $\boldsymbol{\kappa}$.

Exercise 4.4.2

Verify the assertion that if $\mathbf{H}$ is in $\mathcal{O}(\mathscr{V})^+$ then $\mathbf{PHP}^{-1}$ is not in $\mathcal{O}(\mathscr{V})^+$ for arbitrary $\mathbf{P}$.

Following Noll, an elastic particle X is an elastic *fluid* particle if $\mathscr{G}_\kappa(X) = \mathscr{U}(\mathscr{V})^+$ [Ref. 3, Sect 32; Ref. 5]. Note that if $\mathscr{G}_\kappa(X)$ equals $\mathscr{U}(\mathscr{V})^+$, then from (4.4.21) $\mathscr{G}_{\hat{\kappa}}(X)$ also equals $\mathscr{U}(\mathscr{V})^+$. Thus, we need not make reference to a particular reference configuration when asserting that X is a fluid particle. From our definitions, it follows that a fluid particle is an isotropic particle. It is possible to show that isotropic solids and fluids are the only isotropic particles [Refs. 6, 7].

If X is an elastic *fluid particle*, then (4.4.14) must hold identically for every linear transformation $\mathbf{H}$ with determinant equal to $+1$. In order to display how $\hat{u}_\kappa$ is restricted in this case we first derive a necessary condition. Given a deformation gradient $\mathbf{F}$, it readily follows from (A.5.32) that

$$\mathbf{H} = \mathbf{F}^{-1}(\det \mathbf{F})^{1/3} \tag{4.4.22}$$

is in $\mathscr{U}(\mathscr{V})^+$. If (4.4.22) is substituted into (4.4.14), it follows that

$$\hat{u}_\kappa(\mathbf{C}, \mathbf{X}) = \hat{u}_\kappa\left((\det \mathbf{F})^{2/3}\mathbf{I}, \mathbf{X}\right) \tag{4.4.23}$$

Therefore, it is necessary that $\hat{u}_\kappa$ depend on $\mathbf{C}$ only through the determinant of $\mathbf{F}$. From Section (2.3), we know that $\det \mathbf{F}$ measures the deformation of volume elements. Equation (4.4.23) simply states that for an elastic fluid particle it is only the volume deformations which affect the free energy.

Exercise 4.4.3

Show that (4.4.23) is sufficient, as well as necessary, for (4.4.14) to hold *for every* $\mathbf{H}$ in $\mathscr{U}(\mathscr{V})^+$.

Given the result (4.4.23), we can use balance of mass in the form (3.1.12) to define a function $\bar{u}_\kappa$ of ρ and $\mathbf{X}$ by

$$\bar{u}_\kappa(\rho, \mathbf{X}) = \hat{u}_\kappa\left((\rho_R/\rho)^{2/3}\mathbf{I}, \mathbf{X}\right) \tag{4.4.24}$$

Notice that the dependence of $\bar{u}_\kappa$ on the particle position $\mathbf{X}$ arises from the dependence of $\hat{u}_\kappa$ on $\mathbf{X}$ and the possible dependence of ρ_R on $\mathbf{X}$. Given (4.4.23) and (4.4.24), we can use (4.3.35) to calculate the stress on our fluid particle. Because

$$\frac{\partial \hat{u}_\kappa(\mathbf{C},\mathbf{X})}{\partial \mathbf{C}} = \frac{\partial \bar{u}_\kappa(\rho,\mathbf{X})}{\partial \rho}\frac{\partial \rho}{\partial \mathbf{C}} \tag{4.4.25}$$

and

$$2\frac{\partial \rho}{\partial \mathbf{C}} = -\rho \mathbf{F}^{-1}\mathbf{F}^{-1^T} = -\rho \mathbf{C}^{-1} \tag{4.4.26}$$

It follows that (4.3.35) reduces to

$$\mathbf{T}(\mathbf{X},t) = -\rho^2 \frac{\partial \bar{u}_\kappa(\rho,\mathbf{X})}{\partial \rho}\mathbf{I} \tag{4.4.27}$$

Exercise 4.4.4

Use (3.1.12), (2.4.3) and (2.2.29) and derive the identity (4.4.26).

We see from (4.4.27) that our definition of an elastic fluid particle implies that the stress on our isothermal elastic fluid particle is necessarily hydrostatic, i. e. the elastic fluid particle will not support a shear stress. Equation (4.4.27) is usually written

$$\mathbf{T}(\mathbf{X},t) = -\pi \mathbf{I} \tag{4.4.28}$$

where π is the pressure defined by

$$\pi = \rho^2 \frac{\partial \bar{u}_\kappa(\rho,\mathbf{X})}{\partial \rho} \tag{4.4.29}$$

If every particle of $\mathscr{B}$ is a fluid particle, then (4.4.23) and (4.4.29) hold for every X in $\mathscr{B}$. In this case our constitutive equations define an *isothermal elastic fluid* body.

Exercise 4.4.5

Show that for an isothermal elastic fluid body that the equation of motion (3.2.20) can be written

$$\ddot{\mathbf{x}} = -\frac{1}{\rho}\operatorname{grad}\pi + \mathbf{b} \tag{4.4.30}$$

Exercise 4.4.6

If the fluid body is such that $\bar{u}_\kappa$ is independent of $\mathbf{X}$, use the result in Exercise 2.3.3 and show that

$$\dot{\overline{\Gamma\left(\mathscr{Q}(t)\right)}} = 0 \tag{4.4.31}$$

if you assume $\mathbf{b} = -\operatorname{grad} v(\mathbf{x})$, i. e., $\mathbf{b}$ is conservative.

Exercise 4.4.7

If $\mathbf{b}$ is a conservative body force as in Exercise 4.4.6, then according to (4.4.30), in *equilibrium*, the pressure π must obey

$$\operatorname{grad} \pi(\mathbf{x}) + \rho(\mathbf{x}) \operatorname{grad} v(\mathbf{x}) = \mathbf{0} \tag{4.4.32}$$

Continue to assume the fluid is homogeneous and show that (4.4.32) integrates to yield

$$\bar{u}_\kappa\left(\rho(\mathbf{x})\right) + \frac{\pi\left(\rho(\mathbf{x})\right)}{\rho(\mathbf{x})} + v(\mathbf{x}) = \text{const} \tag{4.4.33}$$

Equation (4.4.33) determines the equilibrium density distribution in the presence of the conservative body force $\mathbf{b} = -\operatorname{grad} v(\mathbf{x})$.

Next we shall record the restrictions on the function $\hat{u}_\kappa$ in the case where $\mathbf{X}$ is an elastic isotropic solid particle in its undistorted reference configuration. In this case (4.4.14) must hold for every orthogonal linear transformation in $\mathcal{O}(\mathscr{V})^+$. Thus, we must determine how $\hat{u}_\kappa$ is restricted if it is required to obey

$$\hat{u}_\kappa(\mathbf{C},\mathbf{X}) - \hat{u}_\kappa\left(\mathbf{Q}^T\mathbf{C}\mathbf{Q},\mathbf{X}\right) \tag{4.4.34}$$

for every orthogonal linear transformation in $\mathcal{O}(\mathscr{V})^+$. An immediate necessary condition is obtained if one takes $\mathbf{Q} = \mathbf{R}^T$ in (4.4.34) and utilizes (2.4.6). The result is

$$\hat{u}_\kappa(\mathbf{C},\mathbf{X}) = \hat{u}_\kappa(\mathbf{B},\mathbf{X}) \tag{4.4.35}$$

Therefore, in this case, the value of $\hat{u}_\kappa$ on $\mathbf{C}$ equals its value on $\mathbf{B} = \mathbf{R}\mathbf{C}\mathbf{R}^T$. Equation (4.4.35) is *not* sufficient to satisfy (4.4.34) for all $\mathbf{Q}$ in $\mathcal{O}(\mathscr{V})^+$.

Exercise 4.4.8

Given (4.4.35) show that equation (4.4.34) is obeyed if

$$\hat{u}_\kappa(\mathbf{B},\mathbf{X}) = \hat{u}_\kappa(\mathbf{Q}^T\mathbf{B}\mathbf{Q},\mathbf{X}) \tag{4.4.36}$$

for all orthogonal $\mathbf{Q}$ in $\mathcal{O}(\mathscr{V})^+$.

Exercise 4.4.9

Use (4.4.35) and show that

$$\mathbf{T} = 2\rho\mathbf{B}\frac{\partial\hat{u}_\kappa(\mathbf{B},\mathbf{X})}{\partial\mathbf{B}} = 2\rho\frac{\partial\hat{u}_\kappa(\mathbf{B},\mathbf{X})}{\partial\mathbf{B}}\mathbf{B} \tag{4.4.37}$$

Equation (4.4.34) or, equivalently, (4.4.36) defines what is called a scalar valued *isotropic function* of a symmetric linear transformation. It is a classical result that the solution of (4.4.34) is that $\hat{u}_\kappa$ depends upon $\mathbf{C}$ through the three fundamental invariants of $\mathbf{C}$ [Ref. 3, Sect. 10]. Because $\mathbf{B}$ and $\mathbf{C}$ have the same fundamental invariants (see Section A.5), it follows that (4.4.36) is satisfied if and only if

$$\hat{u}_\kappa(\mathbf{C},\mathbf{X}) = u^*_\kappa(I_\mathbf{B}, II_\mathbf{B}, III_\mathbf{B},\mathbf{X}) \tag{4.4.38}$$

For the sake of completeness, a proof of this representation theorem is given in Appendix B.

Exercise 4.4.10

Show that (4.4.38) is sufficient for (4.4.34) to hold for all orthogonal $\mathbf{Q}$ in $\mathcal{O}(\mathscr{V})^+$.

Given (4.4.38), we can calculate the stress by use of (4.4.37). Clearly,

$$\frac{\partial\hat{u}_\kappa(\mathbf{B},\mathbf{X})}{\partial\mathbf{B}} = \frac{\partial u^*_\kappa}{\partial I_\mathbf{B}}\frac{\partial I_\mathbf{B}}{\partial\mathbf{B}} + \frac{\partial u^*_\kappa}{\partial II_\mathbf{B}}\frac{\partial II_\mathbf{B}}{\partial\mathbf{B}} + \frac{\partial u^*_\kappa}{\partial III_\mathbf{B}}\frac{\partial III_\mathbf{B}}{\partial\mathbf{B}} \tag{4.4.39}$$

From (A.5.44), (A.5.49) and (A.5.46), the fundamental invariants of $\mathbf{B}$ are defined by

$$I_\mathbf{B} = \operatorname{tr}\mathbf{B} \tag{4.4.40}$$

$$II_\mathbf{B} = \tfrac{1}{2}\left((\operatorname{tr}\mathbf{B})^2 - \operatorname{tr}\mathbf{B}^2\right) \tag{4.4.41}$$

and

$$III_\mathbf{B} = \det\mathbf{B} \tag{4.4.42}$$

The derivatives which appear in (4.4.39) are thus given by

$$\frac{\partial I_{\mathbf{B}}}{\partial \mathbf{B}} = \mathbf{I} \tag{4.4.43}$$

$$\frac{\partial II_{\mathbf{B}}}{\partial \mathbf{B}} = I_{\mathbf{B}}\mathbf{I} - \mathbf{B} \tag{4.4.44}$$

and

$$\frac{\partial III_{\mathbf{B}}}{\partial \mathbf{B}} = III_{\mathbf{B}}\mathbf{B}^{-1} \tag{4.4.45}$$

Exercise 4.4.11

Derive the results (4.4.43) through (4.4.45).

If (4.4.43), (4.4.44) and (4.4.45) are substituted into (4.4.39) and the result is substituted into (4.4.37), the following formula is obtained:

$$\mathbf{T} = \alpha_0\left(I_{\mathbf{B}}, II_{\mathbf{B}}, III_{\mathbf{B}}, \mathbf{X}\right)\mathbf{I} + \alpha_1\left(I_{\mathbf{B}}, II_{\mathbf{B}}, III_{\mathbf{B}}, \mathbf{X}\right)\mathbf{B} + \alpha_2\left(I_{\mathbf{B}}, II_{\mathbf{B}}, III_{\mathbf{B}}, \mathbf{X}\right)\mathbf{B}^2 \tag{4.4.46}$$

where

$$\alpha_0\left(I_{\mathbf{B}}, II_{\mathbf{B}}, III_{\mathbf{B}}, \mathbf{X}\right) = 2\rho III_{\mathbf{B}} \frac{\partial u_{\kappa}^*\left(I_{\mathbf{B}}, II_{\mathbf{B}}, III_{\mathbf{B}}, \mathbf{X}\right)}{\partial III_{\mathbf{B}}} \tag{4.4.47}$$

$$\alpha_1\left(I_{\mathbf{B}}, II_{\mathbf{B}}, III_{\mathbf{B}}, \mathbf{X}\right) = 2\rho \frac{\partial u_{\kappa}^*\left(I_{\mathbf{B}}, II_{\mathbf{B}}, III_{\mathbf{B}}, \mathbf{X}\right)}{\partial I_{\mathbf{B}}} + 2\rho I_{\mathbf{B}} \frac{\partial u_{\kappa}^*\left(I_{\mathbf{B}}, II_{\mathbf{B}}, III_{\mathbf{B}}, \mathbf{X}\right)}{\partial II_{\mathbf{B}}} \tag{4.4.48}$$

and

$$\alpha_2\left(I_{\mathbf{B}}, II_{\mathbf{B}}, III_{\mathbf{B}}, \mathbf{X}\right) = -2\rho \frac{\partial u_{\kappa}^*\left(I_{\mathbf{B}}, II_{\mathbf{B}}, III_{\mathbf{B}}, \mathbf{X}\right)}{\partial II_{\mathbf{B}}} \tag{4.4.49}$$

Equation (4.4.46) is the representation for the stress of an isothermal isotropic elastic solid particle in its undistorted reference configuration. In the state $\mathbf{B} = \mathbf{I}$, (4.4.46) reduces to

$$\mathbf{T} = -p\mathbf{I} \tag{4.4.50}$$

where p is the residual pressure defined by

$$p = -\alpha_0\left(3,3,1,\mathbf{X}\right) - \alpha_1\left(3,3,1,\mathbf{X}\right) - \alpha_2\left(3,3,1,\mathbf{X}\right) \tag{4.4.51}$$

Equation (4.4.50) shows that the stress in the state $\mathbf{B}=\mathbf{I}$, the *residual stress*, is necessarily hydrostatic for the isotropic particle in its undistorted reference configuration.

Exercise 4.4.12

It is desired to approximate (4.4.46) for the case where $\mathbf{B}-\mathbf{I}$ is small in some sense. Expand (4.4.46) about the state $\mathbf{B}=\mathbf{I}$ and show that the result is

$$\mathbf{T}=-p\mathbf{I}+\tfrac{1}{2}(\lambda+p)\left(\operatorname{tr}(\mathbf{B}-\mathbf{I})\right)\mathbf{I}+(\mu-p)(\mathbf{B}-\mathbf{I}) \tag{4.4.52}$$

where second-order terms have been dropped and the coefficients λ and μ are defined by

$$\lambda+p=2\left(\frac{\partial}{\partial I_{\mathbf{B}}}+2\frac{\partial}{\partial II_{\mathbf{B}}}+\frac{\partial}{\partial III_{\mathbf{B}}}\right)(\alpha_0+\alpha_1+\alpha_2)\Big|_{\substack{I_{\mathbf{B}}=II_{\mathbf{B}}=3\\ III_{\mathbf{B}}=1}} \tag{4.4.53}$$

and

$$\mu-p=\alpha_1(3,3,1,\mathbf{X})+2\alpha_2(3,3,1,\mathbf{X}) \tag{4.4.54}$$

Of course, p is defined by (4.4.51). Equation (4.4.52) defines what is known as a *finite-linear* elastic particle. As a function of strain $\mathbf{B}$, $\mathbf{T}$ is linear. As a function of displacement gradient, $\mathbf{B}$ is *not linear* (see (2.2.14) and (2.2.2)). Thus, (4.4.52) is a possible model for large (finite) displacement elasticity.

Exercise 4.4.13

If the model developed in Exercise 4.4.12 is specialized by the assumptions of no residual stress $(p=0)$ and infinitesimal strains, show that (4.4.52) reduces to

$$\mathbf{T}=\lambda\left(\operatorname{tr}\tilde{\mathbf{E}}\right)\mathbf{I}+2\mu\tilde{\mathbf{E}} \tag{4.4.55}$$

Exercise 4.4.14

Equation (4.4.55) is almost the isothermal version of (4.1.27). The difference is the appearance of the Cauchy stress in (4.4.55) and the first Piola-Kirchhoff stress in (4.1.27). Show that to the order of approximation used in Exercises 4.4.12 and 4.4.13 $\mathbf{T}$ and $\mathbf{T}_R$ are equal and, thus, (4.4.55) reduces to

$$\mathbf{T}_R=\lambda\left(\operatorname{tr}\tilde{\mathbf{E}}\right)\mathbf{I}+2\mu\tilde{\mathbf{E}} \tag{4.4.56}$$

Clearly, the coefficients λ and μ are the isothermal Lame parameters introduced in Section 4.1.

Exercise 4.4.15

Given (4.4.56), use (3.2.29) and show that the displacement $\mathbf{w}(\mathbf{X},t)$ must be a solution of the *Navier* equation of motion

$$\rho_R \ddot{\mathbf{w}} = (\lambda+\mu)\,\mathrm{GRAD}(\mathrm{Div}\,\mathbf{w}) + \mathrm{Div}(\mathrm{GRAD}\,\mathbf{w}) \tag{4.4.57}$$

Notice that the body force $\mathbf{b}$ in (3.2.29) has been taken to be zero. This choice reflected our assumption that $\mathbf{w}=\mathbf{0}$ is a solution to the equation of motion. Equation (4.4.57) is the isothermal version of (4.1.36). During the derivation of (4.4.57), the reader will need to assume λ and μ are constants, independent of $\mathbf{X}$. Such a dependence is still allowed in (4.4.52) and, thus, (4.4.56).

Exercise 4.4.16

Equation (4.4.57) is a vector partial differential equation for the displacement vector $\mathbf{w}$. Viewed as a system of three partial differential equations, the system is coupled. Decouple the system (4.4.57) and show that the displacement, $\mathbf{w}$, is a solution of

$$\left(\frac{\partial^2}{\partial t^2} - \frac{\lambda+2\mu}{\rho_R}\Delta\right)\left(\frac{\partial^2}{\partial t^2} - \frac{\mu}{\rho_R}\Delta\right)\mathbf{w} = \mathbf{0} \tag{4.4.58}$$

Each factor in the operator in (4.4.58) is a wave operator. The squared speeds are $(\lambda+2\mu)/\rho_R$ and μ/ρ_R. These speeds are called the *longitudinal* and *transverse* speeds, respectively. The result (4.4.58) was first obtained by *Cauchy* in 1840 [Ref. 9].

Exercise 4.4.17

Consider the group $\hat{\mathscr{G}}_\kappa(X)$ defined by the set of linear transformations $\mathbf{H}$ that obey

$$\det \mathbf{H} = 1 \tag{4.4.59}$$

and

$$\mathbf{G}_\kappa(\mathbf{F},\mathbf{X}) = \mathbf{G}_\kappa(\mathbf{FH},\mathbf{X}) \tag{4.4.60}$$

for all linear transformations $\mathbf{F}$ in the domain of $\mathbf{G}_\kappa$. Given (4.2.15), (4.3.28) and (4.3.31), show that $\hat{\mathscr{G}}_\kappa(X)$ can be defined by the set of linear transformations $\mathbf{H}$ that obey (4.4.59) and

$$\hat{u}_\kappa(\mathbf{C},\mathbf{X}) = \hat{u}_\kappa(\mathbf{H}^T\mathbf{CH},\mathbf{X}) + \hat{u}_\kappa(\mathbf{I},\mathbf{X}) - \hat{u}_\kappa(\mathbf{H}^T\mathbf{H},\mathbf{X}) \tag{4.4.61}$$

for all $\mathbf{C}$ in the domain of $\hat{u}_\kappa$. Equation (4.4.61) shows that $\mathscr{G}_\kappa(X)$ is a subgroup of $\hat{\mathscr{G}}_\kappa(X)$ [Ref. 10]. In other words, if an $\mathbf{H}$ in $\mathscr{U}(\mathscr{V})^+$ obeys (4.4.14), then (4.4.61) and, thus, (4.4.60) is satisfied.

4.5. Incompressible Isothermal Elasticity

As explained in Section 4.2, the isothermal assumption defines a constrained material. If, in addition, we assume the deformations are constrained such that

$$\rho(\mathbf{X},t) = \rho_R(\mathbf{X}) \tag{4.5.1}$$

then the material point $\mathbf{X}$ is *incompressible*. From (3.1.12) and (3.1.15), we see that (4.5.1) implies that

$$|\det \mathbf{F}| = 1 \tag{4.5.2}$$

and

$$\operatorname{div} \dot{\mathbf{x}} = \operatorname{tr} \mathbf{L} = 0 \tag{4.5.3}$$

where (2.2.25) has been used. Notice that if the body $\mathscr{B}$ is incompressible, it is not necessarily true that $\rho(\mathbf{X},t)$ is uniform in $\mathbf{X}$. Our definition allows for the possibility that $\operatorname{GRAD} \rho(\mathbf{X},t) = \operatorname{GRAD} \rho_R(\mathbf{X}) \neq \mathbf{0}$. In any case, given (4.5.3) we see that the isothermal entropy inequality (4.2.1) allows for $\mathbf{T}$ to have an *indeterminate* hydrostatic part. This assertion becomes clear if one observes that

$$\operatorname{tr}\left((\mathbf{T} + p\mathbf{I})\mathbf{L}\right) = \operatorname{tr}(\mathbf{TL}) + p \operatorname{tr} \mathbf{L} = \operatorname{tr}(\mathbf{TL}) \tag{4.5.4}$$

for *arbitrary* real number p. This indeterminacy suggests that our constitutive equations (4.2.2) and (4.2.3) be replaced by

$$\psi(\mathbf{X},t) = u_\kappa\left(\mathbf{F}(\mathbf{X},t),\mathbf{X}\right) \tag{4.5.5}$$

and

$$\mathbf{T}(\mathbf{X},t) + p\mathbf{I} = \mathbf{G}_\kappa\left(\mathbf{F}(\mathbf{X},t),\mathbf{X}\right) \tag{4.5.6}$$

where, at this point, p is simply an arbitrary multiplier.

Our objective in this section is to deduce the thermodynamic restrictions on the constitutive equations (4.5.5) and (4.5.6). The formalism is roughly the same as that in Section 4.2 except that we must utilize the constraint (4.5.3) in some fashion. Our first technical assumption concerns the response functions u_κ and $\mathbf{G}_\kappa$. In Section 4.2, the response functions were defined on an *open subset* of $\mathscr{GL}(\mathscr{V})^+$. As a result, if $\mathbf{F}$ is in this domain, $\mathbf{F}+\tau\mathbf{A}$ is in the domain for arbitrary $\mathbf{A}$ in $\mathscr{L}(\mathscr{V};\mathscr{V})$ for some nonzero real numbers τ. This feature of the domain of u_κ and $\mathbf{G}_\kappa$ make it meaningful to require the response functions to be differentiable. Without this feature, (4.2.7) is meaningless. In this section, the elements in the domain of u_κ and $\mathbf{G}_\kappa$ must obey (4.5.2). Without loss of generality, we can take these elements to have determinants equal to $+1$. Therefore, the domain of u_κ and $\mathbf{G}_\kappa$ is a subset of $\mathscr{U}(\mathscr{V})^+$, the special linear group. Unfortunately, this domain is not an open set. As a result, without further assumptions, it is meaningless to regard u_κ and $\mathbf{G}_\kappa$ as being differentiable. The technical assumption we shall make is to assume the domains u_κ and $\mathbf{G}_\kappa$ have differentiable extensions to an open set. This purely formal assumption allows us to write, from (4.5.5),

$$\dot{\psi} = \operatorname{tr}\left(\left(\frac{\partial u_\kappa(\mathbf{F},\mathbf{X})}{\partial \mathbf{F}}\right)^T \dot{\mathbf{F}}\right) = \operatorname{tr}\left(\left(\mathbf{F}\frac{\partial u_\kappa(\mathbf{F},\mathbf{X})^T}{\partial \mathbf{F}}\right)\mathbf{L}\right) \tag{4.5.7}$$

where the partial derivative $\partial u_\kappa(\mathbf{F},\mathbf{X})/\partial \mathbf{F}$ is again defined by (4.2.7).

If (4.5.7) and (4.5.6) are substituted into (4.2.1), it follows that

$$\operatorname{tr}\left(\left(\mathbf{G}_\kappa(\mathbf{F},\mathbf{X}) - \rho\mathbf{F}\frac{\partial u_\kappa(\mathbf{F},\mathbf{X})^T}{\partial \mathbf{F}}\right)\mathbf{L}\right) \geq 0 \tag{4.5.8}$$

where (4.5.4) has been used. Equations (4.5.8) and (4.2.8) are formally identical. Like (4.2.8), (4.5.8) must hold for every admissible thermodynamic process. However, for an incompressible particle, the definition of an admissible thermodynamic process must be such that the constraint (4.5.2) must be obeyed. By the same argument that produced (4.2.14), we can again obtain

$$\operatorname{tr}\left(\left(\mathbf{G}_\kappa(\mathbf{F},\mathbf{X}) - \rho\mathbf{F}\frac{\partial u_\kappa(\mathbf{F},\mathbf{X})^T}{\partial \mathbf{F}}\right)\mathbf{A}\right) \geq 0 \tag{4.5.9}$$

where $\mathbf{A}$ is the velocity gradient for the motion χ_κ^*. Because of the constraint (4.5.3), at this point in the discussion (4.5.9) holds for all $\mathbf{A}$ such that

$$\operatorname{tr}\mathbf{A} = 0 \tag{4.5.10}$$

Our next formal step will be familiar to readers proficient in analytical mechanics or the calculus of variations [Refs. 10, 11]. We shall allow p to take on whatever value is necessary as a consequence of taking $\mathbf{A}$ in (4.5.9) to be arbitrary. When the constraint (4.5.10) is freed, then (4.5.9) yields

$$\mathbf{G}_\kappa(\mathbf{F},\mathbf{X}) = \rho\mathbf{F}\frac{\partial u_\kappa(\mathbf{F},\mathbf{X})^T}{\partial \mathbf{F}} \tag{4.5.11}$$

just as in Section 4.2. The difference is that $\mathbf{T}$ is *not* determined by u_κ. From (4.5.11) and (4.5.6), we see that

$$\mathbf{T} = -p\mathbf{I} + \rho\mathbf{F}\frac{\partial u_\kappa(\mathbf{F},\mathbf{X})^T}{\partial \mathbf{F}} \tag{4.5.12}$$

In physical terms, p is the *extra hydrostatic pressure* necessary to constrain the motion to obey (4.5.2). It becomes one of the unknowns in the problem. Unlike in Section 4.2, where the appropriate field equations are (3.1.12) and (3.2.20), for an incompressible body one has the extra field equation given by the constraint (4.5.2) to aid in the computation of $\chi_\kappa(\mathbf{X},t)$ *and* p.

The restrictions implied by material frame indifference and material symmetry explained in Sections 4.3 and 4.4 carry over to the incompressible case with minor modification. One point worthy of mention here is that for an incompressible isothermal isotropic elastic solid particle, in its undistorted reference configuration the representation (4.4.46) is replaced by

$$\mathbf{T} = -p\mathbf{I} + \alpha_1(I_\mathbf{B}, II_\mathbf{B}, \mathbf{X})\mathbf{B} + \alpha_2(I_\mathbf{B}, II_\mathbf{B}, \mathbf{X})\mathbf{B}^2 \tag{4.5.13}$$

because $III_\mathbf{B} = 1$ and the hydrostatic term proportional to α_0 is simply incorporated into the indeterminate term. Equation (4.1.46), which defines a Mooney-Rivlin material, is a special case of (4.5.13).

Exercise 4.5.1

There is an interesting argument which produces the results of this section as a limit of the compressible case. In this exercise the argument is developed for an isothermal infinitesimal elastic material. The isothermal version of the inverse of (4.4.56) is a special case of (4.1.40) and is

$$\tilde{\mathbf{E}} = \frac{1+\nu}{E}\mathbf{T}_R - \frac{\nu}{E}(\operatorname{tr}\mathbf{T}_R)\mathbf{I} \tag{4.5.14}$$

where ν and E are defined by (4.1.31) and (4.1.32). If the material is incompressible, (4.5.14) must be made consistent with the constraint

$$\operatorname{tr}\tilde{\mathbf{E}} = 0 \tag{4.5.15}$$

Equation (4.5.15) follows from (4.5.2) and (2.5.18*b*). Show that (4.5.14) implies that

$$\operatorname{tr}\tilde{\mathbf{E}} = \frac{1-2\nu}{E}\left(\operatorname{tr}\mathbf{T}_R\right) \tag{4.5.16}$$

Therefore, (4.5.16) tells us that

$$\nu = \frac{1}{2} \tag{4.5.17}$$

for the incompressible limit. Show that in this limit

$$1/\lambda = 0 \tag{4.5.18}$$

$$E = 3\mu \tag{4.5.19}$$

and

$$\mathbf{T}_R = -p\mathbf{I} + 2\mu\tilde{\mathbf{E}} \tag{4.5.20}$$

where p is arbitrary. Equation (4.5.20) is the infinitesimal elasticity version of (4.5.13).

The reader interested in a carefully constructed formulation of the constitutive theory of elastic materials with internal constraints should consult Cohen and Wang [Refs. 12 and 13].

4.6. Thermoelastic Material with Heat Conduction and Viscous Dissipation – Constitutive Assumptions

Sections 4.2 through 4.4 serve to illustrate how one uses the entropy inequality, material frame indifference and material symmetry for a rather simple material model. In the remainder of this chapter we shall, essentially, retrace the material in Sections 4.2 through 4.4 except that the material model is more complicated. This model is one which has presented the effects of heat conduction and viscous dissipation in addition to the nonlinear elasticity effects of the previous model. Many of the fundamental definitions and concepts are repeats of topics presented in Sections 4.2 through 4.4. In some cases these ideas are refined and stated with more precision than originally. It is hoped that the reader will find this repetition helpful in understanding the ideas being presented. The material in Sections 1.8, 1.9 and 1.10 also relates to the contents of the remainder of this chapter. The constitutive assumptions adopted in Section 1.8 are one-dimensional versions of constitutive equations which will be studied in detail in this and the following sections.

Before we present the explicit special constitutive assumptions, it is useful to characterize our constitutive equations in rather general terms. We are interested in constitutive equations *for each particle* X. *The general class of materials of interest are those for which* ψ, η, $\mathbf{T}$, and $\mathbf{q}$

for the particle X are determined by the motion χ and the temperature θ. Given a motion χ, we can define a function $\Theta : \mathscr{B} \times \mathscr{R} \to \mathscr{R}^+$ by

$$\Theta(X,t) = \theta(\chi(X,t),t) \tag{4.6.1}$$

for all (X,t) in $\mathscr{B} \times \mathscr{R}$. With the definition (4.6.1), both χ and Θ are functions defined on $\mathscr{B} \times \mathscr{R}$. Like the definition (2.1.5), if we are given a reference configuration $\boldsymbol{\kappa}$ we can define a function $\Theta_\kappa : \boldsymbol{\kappa}(\mathscr{B}) \times \mathscr{R} \to \mathscr{R}^+$ by

$$\Theta_\kappa(\mathbf{X},t) = \Theta(\boldsymbol{\kappa}^{-1}(\mathbf{X}),t) = \theta(\chi_\kappa(\mathbf{X},t),t) \tag{4.6.2}$$

Therefore, the two fields which are going to determine ψ, η, $\mathbf{T}$, and $\mathbf{q}$ are Θ and χ or Θ_κ and χ_κ, when a reference configuration is given. In the following discussion, we shall assume an arbitrary but fixed reference configuration $\boldsymbol{\kappa}$ is given. In addition, as in Section 4.2, we shall regard the density ρ_R in $\boldsymbol{\kappa}$ to be a prescribed function of $\mathbf{X}$ in $\boldsymbol{\kappa}(\mathscr{B})$.

Definition. A *thermodynamic process* is a set consisting of the two functions Θ_κ and χ_κ and the seven functions of $(\mathbf{X},t)$ whose values are ψ, η, $\mathbf{T}$, $\mathbf{q}$, ρ, r and $\mathbf{b}$ which satisfy

1. Balance of Mass (3.1.12):

$$\rho\left|\det \mathbf{F}\right| = \rho_R \tag{4.6.3}$$

2. Balance of Linear Momentum (3.2.20):

$$\rho\ddot{\mathbf{x}} = \operatorname{div}\mathbf{T} + \rho\mathbf{b} \tag{4.6.4}$$

3. Balance of Thermodynamic Energy (3.5.20):

$$\rho\theta\dot{\eta} = -\rho\left(\dot{\psi} + \eta\dot{\theta}\right) + \operatorname{tr}\mathbf{TL} - \operatorname{div}\mathbf{q} + \rho r \tag{4.6.5}$$

As we have stated, we are interested in a theory of material behavior for which ψ, η, $\mathbf{T}$, and $\mathbf{q}$ are determined by Θ and χ. More precisely, we shall require that ψ, η, $\mathbf{T}$, and $\mathbf{q}$ for each particle X at the time t be determined by the *histories* of Θ_κ and χ_κ up to the time t defined by

$$\Theta_\kappa^{(t)}(\mathbf{X},s) = \Theta_\kappa(\mathbf{X},t-s) \tag{4.6.6}$$

and

$$\chi_\kappa^{(t)}(\mathbf{X},s)=\chi_\kappa(\mathbf{X},t-s) \tag{4.6.7}$$

for all $(\mathbf{X},s)$ in $\boldsymbol{\kappa}(\mathscr{B})\times[0,\infty)$. Symbolically, we can write

$$\left(\psi(\mathbf{X},t),\eta(\mathbf{X},t),\mathbf{T}(\mathbf{X},t),\mathbf{q}(\mathbf{X},t)\right)=f_\kappa\left(\Theta_\kappa^{(t)}(\cdot,\cdot),\chi_\kappa^{(t)}(\cdot,\cdot),X\right) \tag{4.6.8}$$

Equation (4.6.8) shows that we have assumed that ψ, η, $\mathbf{T}$, and $\mathbf{q}$ are *not* influenced by future values of Θ_κ and χ_κ. Equation (4.6.8) does allow these quantities to be influenced by all *past* values of Θ_κ and χ_κ as well as their values over the region $\boldsymbol{\kappa}(\mathscr{B})$. A dependence on the particle X is included in order to allow for the possibility that the values ψ, η, $\mathbf{T}$, and $\mathbf{q}$ could be different for two particles X_1 and X_2 subjected to the *same* $\Theta_\kappa^{(t)}$ and $\chi_\kappa^{(t)}$. Since a reference configuration $\boldsymbol{\kappa}$ is presumed to be given, we can always use (2.1.4) to express the particle X in terms of its position $\mathbf{X}$ in $\boldsymbol{\kappa}$. However, it is convenient at this point to take f_κ to be an explicit function of X. The function f_κ is called the *response function* relative to $\boldsymbol{\kappa}$. It is important to stress that the *values* of f_κ for a particle X are *independent* of $\boldsymbol{\kappa}$. If this were not the case, one could alter ψ, η, $\mathbf{T}$, and $\mathbf{q}$ by simply selecting a different reference configuration. It is also important to note that (4.6.8) obeys the *equipresence* condition mention in the Introduction to this chapter.

As in Section 4.2, an *admissible thermodynamic process* is a thermodynamic process which is consistent with the constitutive assumption (4.6.8). Also, as in Section 4.2, for every choice of Θ_κ and χ_κ there exists and admissible thermodynamic process [Refs. 2, 14]. The proof of this assertion is as follows. Given Θ_κ and χ_κ, we can compute ψ, η, $\mathbf{T}$, and $\mathbf{q}$ from (4.6.8). Given χ_κ and the known value of ρ_r, we can calculate ρ from (4.6.3). This collection of information can then be used to calculate $\mathbf{b}$ from (4.6.4) and r from (4.6.5).

Notice that the definition of a thermodynamic process made no mention of balance of angular momentum (3.3.10) or the entropy inequality (3.5.15). We shall utilize these axioms by requiring that they be *satisfied for all admissible thermodynamic processes*. In this way (3.3.10) and (3.5.15) *restrict* the constitutive function f_κ. For example, if (3.3.10) is satisfied for every admissible thermodynamic process it must be satisfied for all choices of the functions Θ_κ and χ_κ. Immediately, we see from (4.6.8) that f_κ must be restricted such that it yields a symmetric stress tensor. In the remaining sections we shall assume that this restriction is obeyed and focus our attention on determining the restrictions on the response function forced by the entropy inequality. When this has been done, the requirement of *consistency*, discussed in the Introduction to this chapter, will be satisfied.

The special case of (4.6.8) which we wish to investigate in detail is the one with the following properties:

1. For each $\mathbf{X}$ in $\mathscr{B}$, the dependence on $\Theta_\kappa^{(t)}(\cdot,\cdot)$ is only through the *present value* at $\mathbf{X}$ and the *present gradient* at $\mathbf{X}$.
2. For each $\mathbf{X}$ in $\mathscr{B}$, the dependence on $\boldsymbol{\chi}_\kappa^{(t)}(\cdot,\cdot)$ is only through the *present deformation gradient* at $\mathbf{X}$ and the *present rate of deformation gradient* at $\mathbf{X}$.

It will be clear in Section 4.9 that a dependence on the value of $\boldsymbol{\chi}_\kappa^{(t)}$ at $(\mathbf{X},0)$ would violate material frame indifference. For simplicity, we have omitted such a dependence at this point in the discussion. As a result of our specialization, the constitutive assumption which defines each particle in the body is

$$\begin{aligned}&\left(\psi(\mathbf{X},t),\eta(\mathbf{X},t),\mathbf{T}(\mathbf{X},t),\mathbf{q}(\mathbf{X},t)\right)\\&\qquad=\bar{f}_\kappa\left(\Theta_\kappa(\mathbf{X},t),\operatorname{GRAD}\Theta_\kappa(\mathbf{X},t),\mathbf{F}(\mathbf{X},t),\dot{\mathbf{F}}(\mathbf{X},t),\mathbf{X}\right)\end{aligned} \tag{4.6.9}$$

where

$$\Theta_\kappa(\mathbf{X},t)=\Theta_\kappa^{(t)}(\mathbf{X},0) \tag{4.6.10}$$

$$\operatorname{GRAD}\Theta_\kappa(\mathbf{X},t)=\operatorname{GRAD}\Theta_\kappa^{(t)}(\mathbf{X},0) \tag{4.6.11}$$

$$\mathbf{F}(\mathbf{X},t)=\operatorname{GRAD}\boldsymbol{\chi}_\kappa^{(t)}(\mathbf{X},0)=\mathbf{F}^{(t)}(\mathbf{X},0) \tag{4.6.12}$$

and

$$\dot{\mathbf{F}}(\mathbf{X},t)=-\left.\frac{d\mathbf{F}^{(t)}(\mathbf{X},s)}{ds}\right|_{s=0}=\left.\frac{d\mathbf{F}(\mathbf{X},\tau)}{d\tau}\right|_{\tau=t} \tag{4.6.13}$$

Since (4.6.2) shows that

$$\operatorname{GRAD}\Theta_\kappa(\mathbf{X},t)=\mathbf{F}^T(\mathbf{X},t)\operatorname{grad}\theta(\mathbf{x},t) \tag{4.6.14}$$

we can use $\operatorname{grad}\theta(\mathbf{x},t)$ as an independent variable if we chose. It is convenient to define a vector $\mathbf{g}(\mathbf{x},t)$ by

$$\mathbf{g}(\mathbf{x},t)=\operatorname{grad}\theta(\mathbf{x},t) \tag{4.6.15}$$

and rewrite the defining constitutive assumption in the simplified form

$$(\psi,\eta,\mathbf{T},\mathbf{q})=k_\kappa\left(\theta,\mathbf{g},\mathbf{F},\dot{\mathbf{F}},\mathbf{X}\right) \tag{4.6.16}$$

where k_κ is defined by

$$k_\kappa\left(\theta,\mathbf{g},\mathbf{F},\dot{\mathbf{F}},\mathbf{X}\right)=\bar{f}_\kappa\left(\theta,\mathbf{F}^T\mathbf{g},\mathbf{F},\dot{\mathbf{F}},\mathbf{X}\right) \tag{4.6.17}$$

Equation (4.6.16) does not contain all of the important classical theories of continuous materials. It does contain, as special cases, the two examples summarized in Section 4.1. The most obvious case excluded by (4.6.16) is the case of a viscoelastic material. Viscoelastic materials have among their independent variables the quantity $\mathbf{F}^{(t)}(\mathbf{X},\cdot)$ [Refs. 3, 14, 15, 16, 17]. It is perhaps helpful to think of the dependence of k_κ on $\left(\mathbf{F},\dot{\mathbf{F}}\right)$ as arising as an approximation of a dependence on $\mathbf{F}^{(t)}(\mathbf{X},\cdot)$. This approximation is suggested by the formal series expansion of $\mathbf{F}^{(t)}(\mathbf{X},\cdot)$ about $s=0$ in the form

$$\mathbf{F}^{(t)}(\mathbf{X},s)=\mathbf{F}^{(t)}(\mathbf{X},0)+\left.\frac{d\mathbf{F}^{(t)}(\mathbf{X},s)}{ds}\right|_{s=0}s+\cdots=\mathbf{F}-\dot{\mathbf{F}}s+\cdots \tag{4.6.18}$$

Clearly, more general types of materials would allow for a dependence on more derivatives of $\mathbf{F}^{(t)}(\mathbf{X},\cdot)$ at $s=0$.

The constitutive assumption (4.6.16) reflects another special feature. The point made in the last paragraph above can be explained by saying that ψ, η, $\mathbf{T}$, and $\mathbf{q}$ at $(\mathbf{X},t)$ depend upon $\Theta_\kappa^{(t)}(\mathbf{X},\cdot)$ and $\chi_\kappa^{(t)}(\mathbf{X},\cdot)$ *local* to the point $s=0$. Likewise, ψ, η, $\mathbf{T}$, and $\mathbf{q}$ at $(\mathbf{X},t)$ depend upon $\Theta_\kappa^{(t)}(\cdot,s)$ and $\chi_\kappa^{(t)}(\cdot,s)$ *local* to the point $\mathbf{X}$. By including a dependence on higher gradients of $\Theta_\kappa^{(t)}(\cdot,s)$ and $\chi_\kappa^{(t)}(\cdot,s)$ than their first, one can formulate a theory of material behavior for which *long-range spatial effects* are allowed [Refs. 18, 19, 20, 21].

Some comments are in order concerning the domain of the response function k_κ in (4.6.16). The temperature θ must obey the natural restriction

$$\theta>0 \tag{4.6.19}$$

In addition, the deformation gradient $\mathbf{F}$ must obey the condition (2.2.16),

$$\det\mathbf{F}\neq 0 \tag{4.6.20}$$

Since $\mathscr{R}^+$ denotes the subset of $\mathscr{R}$ consisting of the *positive* real numbers, then θ is in $\mathscr{R}^+$. Equation (4.6.20) shows that $\mathbf{F}(\mathbf{X},t)$, as a nonsingular linear transformation, must either have a positive or negative determinant. Recall from Appendix A that $\mathscr{GL}(\mathscr{V})$ denotes the set of automorphisms of $\mathscr{V}$, i.e. the set of regular or *nonsingular* linear transformations in $\mathscr{L}(\mathscr{V};\mathscr{V})$. We write $\mathscr{GL}(\mathscr{V})^{+}$ and $\mathscr{GL}(\mathscr{V})^{-}$ for the subsets of $\mathscr{GL}(\mathscr{V})$ defined by

$$\mathscr{GL}(\mathscr{V})^{+} = \left\{\mathbf{A} \middle| \mathbf{A} \text{ in } \mathscr{L}(\mathscr{V};\mathscr{V}) \text{ and } \det \mathbf{A} > 0\right\} \tag{4.6.21}$$

and

$$\mathscr{GL}(\mathscr{V})^{-} = \left\{\mathbf{A} \middle| \mathbf{A} \text{ in } \mathscr{L}(\mathscr{V};\mathscr{V}) \text{ and } \det \mathbf{A} < 0\right\} \tag{4.6.22}$$

The subset $\mathscr{GL}(\mathscr{V})^{+}$ was introduced in Section 4.4. The condition (4.6.20) shows that $\mathbf{F}(\mathbf{X},t)$ must lie in only one of the components of $\mathscr{GL}(\mathscr{V})$ defined by (4.6.21) and (4.6.22). As in Section 4.2, without loss of generality, we can assume that $\mathbf{F}(\mathbf{X},t)$ lies in $\mathscr{GL}(\mathscr{V})^{+}$. Therefore, for each $\mathbf{X}$, the domain of $k_{\boldsymbol{\kappa}}$ is a *subset* of $\mathscr{R}^{+} \times \mathscr{V} \times \mathscr{GL}(\mathscr{V})^{+} \times \mathscr{L}(\mathscr{V};\mathscr{V})$. It has values in the set $\mathscr{R} \times \mathscr{R} \times \mathscr{S}(\mathscr{V};\mathscr{V}) \times \mathscr{V}$, where $\mathscr{S}(\mathscr{V};\mathscr{V})$ denotes the *symmetric* members of $\mathscr{L}(\mathscr{V};\mathscr{V})$. We assume that, for each $\mathbf{X}$, $k_{\boldsymbol{\kappa}}$ is at least of class C^2 on an open subset of $\mathscr{R}^{+} \times \mathscr{V} \times \mathscr{GL}(\mathscr{V})^{+} \times \mathscr{L}(\mathscr{V};\mathscr{V})$.

Next we wish to characterize the dependence of $k_{\boldsymbol{\kappa}}$ on the reference configuration $\boldsymbol{\kappa}$. By (2.1.5), we can define a deformation function relative to a reference configuration $\hat{\boldsymbol{\kappa}}$ by

$$\chi_{\hat{\boldsymbol{\kappa}}}\left(\hat{\mathbf{X}},t\right) = \chi\left(\hat{\boldsymbol{\kappa}}^{-1}\left(\hat{\mathbf{X}}\right),t\right) \tag{4.6.23}$$

It easily follows from (2.1.5) and (4.6.23) that

$$\chi_{\boldsymbol{\kappa}}(\mathbf{X},t) = \chi_{\hat{\boldsymbol{\kappa}}}(\boldsymbol{\lambda}(\mathbf{X}),t) \tag{4.6.24}$$

where $\boldsymbol{\lambda} : \boldsymbol{\kappa}(\mathscr{B}) \to \hat{\boldsymbol{\kappa}}(\mathscr{B})$ is defined by

$$\boldsymbol{\lambda} = \hat{\boldsymbol{\kappa}} \circ \boldsymbol{\kappa}^{-1} \tag{4.6.25}$$

By differentiation of (4.6.24), it follows that

$$\mathbf{F} = \hat{\mathbf{F}}\mathbf{P} \tag{4.6.26}$$

where

$$\hat{\mathbf{F}} = \operatorname{GRAD} \chi_{\hat{\boldsymbol{\kappa}}}\left(\hat{\mathbf{X}},t\right) = \hat{\mathbf{F}}\left(\hat{\mathbf{X}},t\right) \tag{4.6.27}$$

and

$$\mathbf{P} = \mathrm{GRAD}\boldsymbol{\lambda}(\mathbf{X}) = \mathbf{P}(\mathbf{X}) \tag{4.6.28}$$

In understanding (4.6.23) through (4.6.28), it is helpful to consider the Figure 4.6.1.

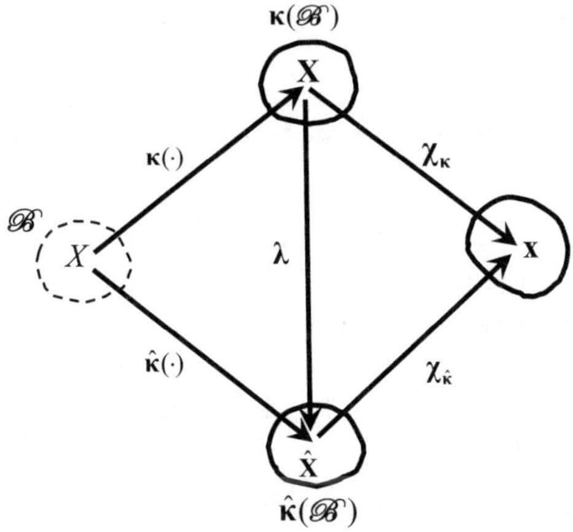

Figure 4.6.1

Because *values* of k_{κ} must be independent of $\boldsymbol{\kappa}$, it follows from (4.6.16) that one can define a response function with respect to $\hat{\boldsymbol{\kappa}}$ by

$$k_{\hat{\kappa}}\left(\theta, \mathbf{g}, \hat{\mathbf{F}}, \dot{\hat{\mathbf{F}}}, \mathbf{X}\right) = k_{\kappa}\left(\theta, \mathbf{g}, \mathbf{F}, \dot{\mathbf{F}}, \mathbf{X}\right) \tag{4.6.29}$$

or, with (4.6.26) and (4.6.28),

$$k_{\hat{\kappa}}\left(\theta, \mathbf{g}, \mathbf{F}\mathbf{P}^{-1}, \dot{\mathbf{F}}\mathbf{P}^{-1}, \mathbf{X}\right) = k_{\kappa}\left(\theta, \mathbf{g}, \mathbf{F}, \dot{\mathbf{F}}, \mathbf{X}\right) \tag{4.6.30}$$

Equation (4.6.30) must hold for all $\left(\theta, \mathbf{g}, \mathbf{F}, \dot{\mathbf{F}}, \mathbf{X}\right)$ in the domain of k_{κ}. The important message in (4.6.30) is that the dependence of the response function on the reference configuration is such that only the gradient of $\boldsymbol{\lambda}$ at $\mathbf{X}$ is needed in order to characterize the change of reference configuration. This result reflects again the *local* nature of the constitutive assumption (4.6.16).

In order to make the local dependence of k_{κ} on $\boldsymbol{\kappa}$ explicit, we shall introduce the concept of a *local reference configuration* [Ref. 3, Sect. 22; Refs. 5, 22, 23].

Definition. Two reference configurations $\boldsymbol{\kappa}_1$ and $\boldsymbol{\kappa}_2$ are *equivalent* at X in $\mathscr{B}$ if

$$\text{GRAD}\left(\boldsymbol{\kappa}_1 \circ \boldsymbol{\kappa}_2^{-1}\right) = \mathbf{I} \tag{4.6.31}$$

Exercise 4.6.1

If $\boldsymbol{\kappa}_1$ and $\boldsymbol{\kappa}_2$ and $\boldsymbol{\kappa}_1$ and $\boldsymbol{\kappa}_3$ are equivalent at X in $\mathscr{B}$, then show that $\boldsymbol{\kappa}_2$ and $\boldsymbol{\kappa}_3$ are equivalent.

Definition. The *local reference configuration* at X in $\mathscr{B}$ is the set $\mathbf{K}_X$ of all reference configurations equivalent at X.

It follows from (4.6.31) that, when $\boldsymbol{\kappa}$ and $\hat{\boldsymbol{\kappa}}$ are equivalent at X, $\mathbf{P} = \mathbf{I}$ and from (4.6.30) that

$$k_{\hat{\boldsymbol{\kappa}}} = k_{\boldsymbol{\kappa}} \tag{4.6.32}$$

Therefore, at X in $\mathscr{B}$, $k_{\boldsymbol{\kappa}}$ depends on the configuration $\boldsymbol{\kappa}$ through the local reference configuration $\mathbf{K}_X$. For this reason, we define a function $k_{\mathbf{K}_X}$ by

$$k_{\mathbf{K}_X}\left(\theta, \mathbf{g}, \mathbf{F}, \dot{\mathbf{F}}\right) = k_{\boldsymbol{\kappa}}\left(\theta, \mathbf{g}, \mathbf{F}, \dot{\mathbf{F}}, \mathbf{X}\right) \tag{4.6.33}$$

for all $\left(\theta, \mathbf{g}, \mathbf{F}, \dot{\mathbf{F}}\right)$ in the domain of $k_{\boldsymbol{\kappa}}$ and for all $\boldsymbol{\kappa}$ in $\mathbf{K}_X$. With (4.6.33), our constitutive assumption (4.6.16) becomes

$$\left(\psi, \eta, \mathbf{T}, \mathbf{q}\right) = k_{\mathbf{K}_X}\left(\theta, \mathbf{g}, \mathbf{F}, \dot{\mathbf{F}}\right) \tag{4.6.34}$$

Many of our later manipulations require that we introduce symbols for the component functions of $k_{\mathbf{K}_X}$. These functions are defined by

$$\left(u_{\mathbf{K}_X}, h_{\mathbf{K}_X}, \mathbf{G}_{\mathbf{K}_X}, \mathbf{l}_{\mathbf{K}_X}\right) = k_{\mathbf{K}_X} \tag{4.6.35}$$

For example, (4.6.35) shows that $\mathbf{T}$ is given by

$$\mathbf{T} = \mathbf{G}_{\mathbf{K}_X}\left(\theta, \mathbf{g}, \mathbf{F}, \dot{\mathbf{F}}\right) \tag{4.6.36}$$

4.7. Thermoelastic Material with Heat Conduction and Viscous Dissipation – General Thermodynamic Restrictions

As indicated in the last section, we require that (3.5.15) be satisfied for all admissible thermodynamic processes. In this section we shall investigate the restrictions on the response function $k_{\mathbf{K}_X}$ in (4.6.34) implied by this requirement [Ref. 15].

From (4.6.34) and (4.6.35), we have

$$\psi = u_{\mathbf{K}_X}\left(\theta, \mathbf{g}, \mathbf{F}, \dot{\mathbf{F}}\right) \tag{4.7.1}$$

For sufficient smooth Θ_κ and χ_κ, the material derivative of ψ can be expressed in terms of the material derivatives of θ, $\mathbf{g}$, $\mathbf{F}$ and $\dot{\mathbf{F}}$ by the chain rule. The result is

$$\dot{\psi} = \frac{\partial u_{\mathbf{K}_X}}{\partial \theta}\dot{\theta} + \frac{\partial u_{\mathbf{K}_X}}{\partial \mathbf{g}} \cdot \dot{\mathbf{g}} + \operatorname{tr}\left(\frac{\partial u_{\mathbf{K}_X}}{\partial \mathbf{F}}^T \dot{\mathbf{F}}\right) + \operatorname{tr}\left(\frac{\partial u_{\mathbf{K}_X}}{\partial \dot{\mathbf{F}}}^T \ddot{\mathbf{F}}\right) \tag{4.7.2}$$

If (4.7.2) and (4.6.34) are substituted into (3.5.15), the result can be written

$$\begin{aligned} &-\rho\left(\frac{\partial u_{\mathbf{K}_X}\left(\theta, \mathbf{g}, \mathbf{F}, \dot{\mathbf{F}}\right)}{\partial \theta} + h_{\mathbf{K}_X}\left(\theta, \mathbf{g}, \mathbf{F}, \dot{\mathbf{F}}\right)\right)\dot{\theta} - \rho\frac{u_{\mathbf{K}_X}\left(\theta, \mathbf{g}, \mathbf{F}, \dot{\mathbf{F}}\right)}{\partial \mathbf{g}} \cdot \dot{\mathbf{g}} - \rho \operatorname{tr}\left(\frac{\partial u_{\mathbf{K}_X}\left(\theta, \mathbf{g}, \mathbf{F}, \dot{\mathbf{F}}\right)^T}{\partial \dot{\mathbf{F}}}\ddot{\mathbf{F}}\right) \\ &+ \operatorname{tr}\left(\mathbf{F}^{-1}\left(\mathbf{G}_{\mathbf{K}_X}\left(\theta, \mathbf{g}, \mathbf{F}, \dot{\mathbf{F}}\right) - \rho\mathbf{F}\frac{\partial u_{\mathbf{K}_X}\left(\theta, \mathbf{g}, \mathbf{F}, \dot{\mathbf{F}}\right)^T}{\partial \mathbf{F}}\right)\dot{\mathbf{F}}\right) - \mathbf{I}_{\mathbf{K}_X}\left(\theta, \mathbf{g}, \mathbf{F}, \dot{\mathbf{F}}\right) \cdot \mathbf{g}/\theta \geq 0 \end{aligned} \tag{4.7.3}$$

Equation (4.7.3) must hold *for every* choice of the functions Θ_κ and χ_κ.

Let $\left(\theta, \mathbf{g}, \mathbf{F}, \dot{\mathbf{F}}\right)$ be an *arbitrary* element of the domain of $k_{\mathbf{K}_X}$ and let $(\mathbf{X}, t)$ be fixed. Given these quantities, we can define functions Θ_κ^* and χ_κ^* by

$$\Theta_\kappa^*(\mathbf{Y}, \tau) = \theta + (\tau - t)c + \left[\mathbf{g} + (\tau - t)\mathbf{a}\right] \cdot \left[\mathbf{F}(\mathbf{Y} - \mathbf{X})\right] \tag{4.7.4}$$

and

$$\chi_\kappa^*(\mathbf{Y}, \tau) = \chi_\kappa(\mathbf{X}, t) + \left[\mathbf{F} + (\tau - t)\dot{\mathbf{F}} + \tfrac{1}{2}(\tau - t)^2 \mathbf{A}\right](\mathbf{Y} - \mathbf{X}) \tag{4.7.5}$$

where $(c, \mathbf{a}, \mathbf{A})$ is *any* element in $\mathscr{R} \times \mathscr{V} \times \mathscr{L}(\mathscr{V}; \mathscr{V})$. It follows from (4.7.4) and (4.7.5) that at $\mathbf{Y} = \mathbf{X}$

$$\theta^*(\tau) = \Theta_\kappa^*(\mathbf{X}, \tau) = \theta + (\tau - t)c \tag{4.7.6}$$

$$\mathbf{g}^*(\tau) = \mathbf{F}^{-1^T} \operatorname{GRAD} \Theta_\kappa^*(\mathbf{X}, \tau) = \mathbf{g} + (\tau - t)\mathbf{a} \tag{4.7.7}$$

$$\mathbf{F}^*(\tau) = \operatorname{GRAD} \chi_\kappa^*(\mathbf{X}, \tau) = \mathbf{F} + (\tau - t)\dot{\mathbf{F}} + \tfrac{1}{2}(\tau - t)^2 \mathbf{A} \tag{4.7.8}$$

and

$$\dot{\mathbf{F}}^*(\tau)=\dot{\mathbf{F}}+(\tau-t)\mathbf{A} \tag{4.7.9}$$

The time τ is required to lie in the internal $[t,t+\delta]$, where δ is a positive number chosen so that $\left(\theta^*(\tau),\mathbf{g}^*(\tau),\mathbf{F}^*(\tau),\dot{\mathbf{F}}^*(\tau)\right)$ is in the domain of $k_{\mathbf{K}_X}$ for all τ in $[t,t+\delta]$. Since (4.7.3) must hold for all Θ_κ and $\boldsymbol{\chi}_\kappa$, we can evaluate it on Θ_κ^* and $\boldsymbol{\chi}_\kappa^*$. If this evaluation is done and we then consider the limit of the result as $\tau\to t$, the result is

$$\begin{aligned}
&-\rho\left(\frac{\partial u_{\mathbf{K}_X}\left(\theta,\mathbf{g},\mathbf{F},\dot{\mathbf{F}}\right)}{\partial\theta}+h_{\mathbf{K}_X}\left(\theta,\mathbf{g},\mathbf{F},\dot{\mathbf{F}}\right)\right)c-\rho\frac{\partial u_{\mathbf{K}_X}\left(\theta,\mathbf{g},\mathbf{F},\dot{\mathbf{F}}\right)}{\partial\mathbf{g}}\cdot\mathbf{a}-\rho\operatorname{tr}\left(\frac{\partial u_{\mathbf{K}_X}\left(\theta,\mathbf{g},\mathbf{F},\dot{\mathbf{F}}\right)^T}{\partial\dot{\mathbf{F}}}\mathbf{A}\right)\\
&\quad+\operatorname{tr}\left(\mathbf{F}^{-1}\left(\mathbf{G}_{\mathbf{K}_X}\left(\theta,\mathbf{g},\mathbf{F},\dot{\mathbf{F}}\right)-\rho\mathbf{F}\frac{\partial u_{\mathbf{K}_X}\left(\theta,\mathbf{g},\mathbf{F},\dot{\mathbf{F}}\right)^T}{\partial\mathbf{F}}\right)\dot{\mathbf{F}}\right)-\mathbf{l}_{\mathbf{K}_X}\left(\theta,\mathbf{g},\mathbf{F},\dot{\mathbf{F}}\right)\cdot\mathbf{g}/\theta\geq 0
\end{aligned} \tag{4.7.10}$$

In (4.7.10), the quantities c, $\mathbf{a}$ and $\mathbf{A}$ are arbitrary. By selecting different values of these quantities for fixed $\left(\theta,\mathbf{g},\mathbf{F},\dot{\mathbf{F}}\right)$ we can clearly violate (4.7.10) unless their coefficients are *zero*. Consequently, (4.7.10) yields

$$\frac{\partial u_{\mathbf{K}_X}\left(\theta,\mathbf{g},\mathbf{F},\dot{\mathbf{F}}\right)}{\partial\theta}+h_{\mathbf{K}_X}\left(\theta,\mathbf{g},\mathbf{F},\dot{\mathbf{F}}\right)=0 \tag{4.7.11}$$

$$\frac{\partial u_{\mathbf{K}_X}\left(\theta,\mathbf{g},\mathbf{F},\dot{\mathbf{F}}\right)}{\partial\mathbf{g}}=\mathbf{0} \tag{4.7.12}$$

and

$$\frac{\partial u_{\mathbf{K}_X}\left(\theta,\mathbf{g},\mathbf{F},\dot{\mathbf{F}}\right)}{\partial\dot{\mathbf{F}}}=\mathbf{0} \tag{4.7.13}$$

Because $\left(\theta,\mathbf{g},\mathbf{F},\dot{\mathbf{F}}\right)$ is an arbitrary element of the domain of $k_{\mathbf{K}_X}$, (4.7.12) and (4.7.13) yield the important result that the free energy density of the material defined by (4.6.34) is *independent* of $\mathbf{g}$ and $\dot{\mathbf{F}}$. Thus,

$$\psi=u_{\mathbf{K}_X}\left(\theta,\mathbf{F}\right) \tag{4.7.14}$$

and, from (4.7.11), (4.6.34) and (4.6.35),

$$\eta = h_{\mathbf{K}_X}(\theta,\mathbf{F}) = -\frac{\partial u_{\mathbf{K}_X}(\theta,\mathbf{F})}{\partial \theta} \tag{4.7.15}$$

Equation (4.7.15) shows that $h_{\mathbf{K}_X}$ is *determined* by $u_{\mathbf{K}_X}$.

Given (4.7.14) and (4.7.15), the inequality (4.7.3) [or (4.7.10)] reduces to

$$\operatorname{tr}\left(\mathbf{F}^{-1}\left(\mathbf{G}_{\mathbf{K}_X}(\theta,\mathbf{g},\mathbf{F},\dot{\mathbf{F}}) - \rho\mathbf{F}\frac{\partial u_{\mathbf{K}_X}(\theta,\mathbf{F})^T}{\partial \mathbf{F}}\right)\dot{\mathbf{F}}\right) - \mathbf{l}_{\mathbf{K}_X}(\theta,\mathbf{g},\mathbf{F},\dot{\mathbf{F}})\cdot\mathbf{g}/\theta \geq 0 \tag{4.7.16}$$

Equation (4.7.16) is called the *residual entropy inequality*. Equations (4.7.14), (4.7.15), and (4.7.16) represent *necessary* conditions which follow from requiring that (4.7.3) hold for all admissible thermodynamic processes. They are also *sufficient* conditions because if we assume (4.7.14), (4.7.15), and (4.7.16) are valid, then (4.7.3) is necessarily true.

If we set $\dot{\mathbf{F}} = \mathbf{0}$ in (4.7.16) we obtain the *heat conduction inequality*

$$l_{\mathbf{K}_X}(\theta,\mathbf{g},\mathbf{F},\mathbf{0})\cdot\mathbf{g}/\theta \geq 0 \tag{4.7.17}$$

Equation (4.7.17) shows that when $\dot{\mathbf{F}} = \mathbf{0}$, the angle between a nonzero temperature gradient and a nonzero heat flux vector must be greater than or equal to ninety degrees. If we set $\mathbf{g} = \mathbf{0}$ in (4.7.16) we obtain the *mechanical dissipation inequality*

$$\operatorname{tr}\left(\mathbf{F}^{-1}\left(\mathbf{G}_{\mathbf{K}_X}(\theta,\mathbf{0},\mathbf{F},\dot{\mathbf{F}}) - \rho\mathbf{F}\frac{\partial u_{\mathbf{K}_X}(\theta,\mathbf{F})^T}{\partial \mathbf{F}}\right)\dot{\mathbf{F}}\right) \geq 0 \tag{4.7.18}$$

Exercise 4.7.1

Specialize (4.6.34) by omitting the dependence on $\dot{\mathbf{F}}$ and show that (4.7.14) and (4.7.15) are again obtained and that (4.7.16) is replaced by

$$\mathbf{T} = \mathbf{G}_{\mathbf{K}_X}(\theta,\mathbf{F}) = \rho\mathbf{F}\frac{\partial u_{\mathbf{K}_X}(\theta,\mathbf{F})^T}{\partial \mathbf{F}} \tag{4.7.19}$$

$$\mathbf{F}\frac{\partial u_{\mathbf{K}_X}(\theta,\mathbf{F})^T}{\partial \mathbf{F}} = \frac{\partial u_{\mathbf{K}_X}(\theta,\mathbf{F})}{\partial \mathbf{F}}\mathbf{F}^T \tag{4.7.20}$$

and

$$-\mathbf{l}_{\mathbf{K}_X}(\theta,\mathbf{g},\mathbf{F})\cdot\mathbf{g}/\theta \geq 0 \tag{4.7.21}$$

This example defines a *thermoelastic material* with *heat conduction.*

Exercise 4.7.2

If in Exercise 4.7.1 a dependence on $\mathbf{g}$ is omitted, show that (4.7.14), (4.7.15), (4.7.19) and (4.7.20) remain valid but that (4.7.21) is replaced by

$$\mathbf{q} = \mathbf{l}_{\mathbf{K}_X}(\theta, \mathbf{F}) = \mathbf{0} \tag{4.7.22}$$

This example defines a *thermoelastic nonconductor*. Note that it is not necessarily true that θ is constant for this model. Thus, it does *not* correspond to the one formulated in Section 4.2.

Exercise 4.7.3

In an effort to produce a theory of heat conduction for which thermal disturbances propagate with finite, rather than infinite, speed, it is reasonable to investigate a material for which

$$(\psi, \eta, \mathbf{T}, \mathbf{q}) = k_{\mathbf{K}_X}(\theta, \dot{\theta}, \mathbf{g}, \mathbf{F}, \dot{\mathbf{F}}) \tag{4.7.23}$$

Derive the thermodynamic restrictions in this case. Reference 25 provides information useful for the solution of this exercise.

Exercise 4.7.4

In order to include additional memory effects, the discussion in Section 4.6 suggests that higher time derivatives of $\mathbf{F}$ be included in the response function. Assume that

$$(\psi, \eta, \mathbf{T}, \mathbf{q}) = k_{\mathbf{K}_X}(\theta, \mathbf{g}, \mathbf{F}, \dot{\mathbf{F}}, \ddot{\mathbf{F}}) \tag{4.7.24}$$

Derive the thermodynamic restrictions in this case.

Exercise 4.7.5

In order to include additional spatial effects, the discussion in Section 4.6 suggests that higher gradients of Θ_κ and χ_κ be included in the response function. As an example assume that

$$(\psi, \eta, \mathbf{T}, \mathbf{q}) = k_{\mathbf{K}_X}(\theta, \mathbf{g}, \mathbf{F}, \mathrm{GRAD}\,\mathbf{F}) \tag{4.7.25}$$

and show that the resulting thermodynamic restrictions are such that (4.7.25) reduces to the case of a thermoelastic material with heat conduction studied in Exercise 4.7.1 above except that $\mathbf{l}_{\mathbf{K}_X}$ in this case will depend upon $\mathrm{GRAD}\,\mathbf{F}$. Additional discussion of spatial effects can be found in Refs. 16, 17, 18, and 19.

It is convenient for manipulations which will be carried out in Section 4.8 to define a function $\mathbf{G}^e_{\mathbf{K}_X}$ of $(\theta,\mathbf{g},\mathbf{F},\dot{\mathbf{F}})$ as follows

$$\mathbf{G}^e_{\mathbf{K}_X}(\theta,\mathbf{g},\mathbf{F},\dot{\mathbf{F}}) = \mathbf{G}_{\mathbf{K}_X}(\theta,\mathbf{g},\mathbf{F},\dot{\mathbf{F}}) - \rho\mathbf{F}\frac{\partial u_{\mathbf{K}_X}(\theta,\mathbf{F})^T}{\partial\mathbf{F}} \tag{4.7.26}$$

If we denote the value of $\mathbf{G}^e_{\mathbf{K}_X}$ by $\mathbf{T}^e$, we can rewrite (4.7.16) in the more simple form

$$\operatorname{tr}\mathbf{T}^e\mathbf{L} - \mathbf{q}\cdot\mathbf{g}/\theta \geq 0 \tag{4.7.27}$$

where (2.2.24) and (4.6.34) have been used.

4.8. Thermoelastic Material with Heat Conduction and Viscous Dissipation – Equilibrium Thermodynamic Restrictions

In this section, we shall investigate certain results implied by the residual entropy inequality. They correspond to special conditions which hold at *thermodynamic equilibrium*.

Definition. The material defined by (4.6.34) is in *thermodynamic equilibrium* at $(\mathbf{X},t)$ if $\mathbf{g}=\mathbf{0}$ and $\dot{\mathbf{F}}=\mathbf{0}$.

Of course, we have made the obvious assumption that elements of the form $(\theta,\mathbf{0},\mathbf{F},\mathbf{0})$ are in the domain of the response function $k_{\mathbf{K}_X}$. If we define a function of $(\theta,\mathbf{g},\mathbf{F},\dot{\mathbf{F}})$ by

$$\Phi(\theta,\mathbf{g},\mathbf{F},\dot{\mathbf{F}}) = \operatorname{tr}\mathbf{T}^e\mathbf{L} - \mathbf{q}\cdot\mathbf{g}/\theta \tag{4.8.1}$$

it follows from (4.7.27) that

$$\Phi(\theta,\mathbf{g},\mathbf{F},\dot{\mathbf{F}}) \geq 0 \tag{4.8.2}$$

and

$$\Phi(\theta,\mathbf{0},\mathbf{F},\mathbf{0}) = 0 \tag{4.8.3}$$

Thus, the function Φ is a *minimum* at a thermodynamic equilibrium state. Consequently, it must be true that

$$\left.\frac{d\Phi(\theta,\lambda\mathbf{a},\mathbf{F},\lambda\mathbf{A})}{d\lambda}\right|_{\lambda=0} = 0 \tag{4.8.4}$$

and

$$\left.\frac{d^2\Phi(\theta,\lambda\mathbf{a},\mathbf{F},\lambda\mathbf{A})}{d\lambda^2}\right|_{\lambda=0}\geq 0 \tag{4.8.5}$$

for all vectors $\mathbf{a}$ and linear transformations $\mathbf{A}$.

It follows from (4.8.1), (4.7.26), (4.6.34) and (4.6.35) that

$$\Phi(\theta,\lambda\mathbf{a},\mathbf{F},\lambda\mathbf{A})=\lambda\operatorname{tr}\left(\mathbf{F}^{-1}\mathbf{G}^{e}_{\mathbf{K}_X}(\theta,\lambda\mathbf{a},\mathbf{F},\lambda\mathbf{A})\mathbf{A}\right)-\lambda\mathbf{l}_{\mathbf{K}_X}(\theta,\lambda\mathbf{a},\mathbf{F},\lambda\mathbf{A})\cdot\mathbf{a}/\theta \tag{4.8.6}$$

A simple calculation yields

$$\left.\frac{d\Phi(\theta,\lambda\mathbf{a},\mathbf{F},\lambda\mathbf{A})}{d\lambda}\right|_{\lambda=0}=\operatorname{tr}\left(\mathbf{F}^{-1}\mathbf{G}^{e}_{\mathbf{K}_X}(\theta,\mathbf{0},\mathbf{F},\mathbf{0})\mathbf{A}\right)-\mathbf{l}_{\mathbf{K}_X}(\theta,\mathbf{0},\mathbf{F},\mathbf{0})\cdot\mathbf{a}/\theta \tag{4.8.7}$$

and

$$\begin{aligned}\left.\frac{d^2\Phi(\theta,\lambda\mathbf{a},\mathbf{F},\lambda\mathbf{A})}{d\lambda^2}\right|_{\lambda=0}&=2\operatorname{tr}\left(\mathbf{F}^{-1}\left.\frac{d\mathbf{G}^{e}_{\mathbf{K}_X}(\theta,\lambda\mathbf{a},\mathbf{F},\lambda\mathbf{A})}{d\lambda}\right|_{\lambda=0}\mathbf{A}\right)\\&\quad-\frac{2}{\theta}\left.\frac{d\mathbf{l}_{\mathbf{K}_X}(\theta,\lambda\mathbf{a},\mathbf{F},\lambda\mathbf{A})}{d\lambda}\right|_{\lambda=0}\cdot\mathbf{a}\end{aligned} \tag{4.8.8}$$

where

$$\left.\frac{d\mathbf{G}^{e}_{\mathbf{K}_X}(\theta,\lambda\mathbf{a},\mathbf{F},\lambda\mathbf{A})}{d\lambda}\right|_{\lambda=0}=\frac{\partial\mathbf{G}^{e}_{\mathbf{K}_X}(\theta,\mathbf{0},\mathbf{F},\mathbf{0})}{\partial\dot{\mathbf{F}}}[\mathbf{A}]+\frac{\partial\mathbf{G}^{e}_{\mathbf{K}_X}(\theta,\mathbf{0},\mathbf{F},\mathbf{0})}{\partial\mathbf{g}}[\mathbf{a}] \tag{4.8.9}$$

and

$$\left.\frac{d\mathbf{l}_{\mathbf{K}_X}(\theta,\lambda\mathbf{a},\mathbf{F},\lambda\mathbf{A})}{d\lambda}\right|_{\lambda=0}=\frac{\partial\mathbf{l}_{\mathbf{K}_X}(\theta,\mathbf{0},\mathbf{F},\mathbf{0})}{\partial\dot{\mathbf{F}}}[\mathbf{A}]+\frac{\partial\mathbf{l}_{\mathbf{K}_X}(\theta,\mathbf{0},\mathbf{F},\mathbf{0})}{\partial\mathbf{g}}\mathbf{a} \tag{4.8.10}$$

Equations (4.8.9) and (4.8.10) introduce three algebraic operations which we have not encountered previously. In components, these operations are defined by

$$\frac{\partial\mathbf{G}^{e}_{\mathbf{K}_X}(\theta,\mathbf{0},\mathbf{F},\mathbf{0})}{\partial\dot{\mathbf{F}}}[\mathbf{A}]=\frac{\partial\mathbf{G}^{e}_{\mathbf{K}_X}(\theta,\mathbf{0},\mathbf{F},\mathbf{0})}{\partial\dot{F}_{jJ}}A_{jJ} \tag{4.8.11}$$

$$\frac{\partial \mathbf{G}^e_{\mathbf{K}_X}\left(\theta,\mathbf{0},\mathbf{F},\mathbf{0}\right)}{\partial \mathbf{g}}[\mathbf{a}] = \frac{\partial \mathbf{G}^e_{\mathbf{K}_X}\left(\theta,\mathbf{0},\mathbf{F},\mathbf{0}\right)}{\partial g_j}a_j \tag{4.8.12}$$

and

$$\frac{\partial \mathbf{l}_{\mathbf{K}_X}\left(\theta,\mathbf{0},\mathbf{F},\mathbf{0}\right)}{\partial \dot{\mathbf{F}}}[\mathbf{A}] = \frac{\partial \mathbf{l}_{\mathbf{K}_X}\left(\theta,\mathbf{0},\mathbf{F},\mathbf{0}\right)}{\partial \dot{F}_{jJ}}A_{jJ} \tag{4.8.13}$$

Because $\mathbf{a}$ and $\mathbf{A}$ are arbitrary in (4.8.7), it follows from (4.8.4) that

$$\mathbf{l}_{\mathbf{K}_X}\left(\theta,\mathbf{0},\mathbf{F},\mathbf{0}\right) = \mathbf{0} \tag{4.8.14}$$

and

$$\mathbf{G}^e_{\mathbf{K}_X}\left(\theta,\mathbf{0},\mathbf{F},\mathbf{0}\right) = \mathbf{0} \tag{4.8.15}$$

for all $\left(\theta,\mathbf{F}\right)$ in the domain of $k_{\mathbf{K}_X}$. Equation (4.8.14) shows that the heat flux vector must vanish in equilibrium. Equations (4.8.15) and (4.7.26) show that in equilibrium the stress is determined by the free energy by the formula

$$\mathbf{G}_{\mathbf{K}_X}\left(\theta,\mathbf{0},\mathbf{F},\mathbf{0}\right) = \rho\mathbf{F}\frac{\partial u_{\mathbf{K}_X}\left(\theta,\mathbf{F}\right)^T}{\partial \mathbf{F}} \tag{4.8.16}$$

Since $\mathbf{G}_{\mathbf{K}_X}$ has symmetric values, it follows from (4.8.16) that the condition

$$\mathbf{F}\frac{\partial u_{\mathbf{K}_X}\left(\theta,\mathbf{F}\right)^T}{\partial \mathbf{F}} = \frac{\partial u_{\mathbf{K}_X}\left(\theta,\mathbf{F}\right)}{\partial \mathbf{F}}\mathbf{F}^T \tag{4.8.17}$$

must hold. The result (4.8.16) shows that the value of the function $\mathbf{G}^e_{\mathbf{K}_X}$ defined by (4.7.26) is a symmetric linear transformation which represents the *extra* or *nonequilibrium part of the stress*. If we denote the value of $\mathbf{G}_{\mathbf{K}_X}$ at $\left(\theta,\mathbf{0},\mathbf{F},\mathbf{0}\right)$ by $\mathbf{T}^0$, it follows from (4.7.26) that

$$\mathbf{T} = \mathbf{T}^0 + \mathbf{T}^e \tag{4.8.18}$$

where

$$\mathbf{T}^0 = \rho\mathbf{F}\frac{\partial u_{\mathbf{K}_X}\left(\theta,\mathbf{F}\right)^T}{\partial \mathbf{F}} \tag{4.8.19}$$

and

$$\mathbf{T}^e = \mathbf{G}^e_{\mathbf{K}_X}\left(\theta, \mathbf{g}, \mathbf{F}, \dot{\mathbf{F}}\right) \tag{4.8.20}$$

The decomposition of the stress into the sum of an equilibrium part and a nonequilibrium part is a characteristic of the two examples discussed in Section 4.1. Equation (4.1.3) and (4.1.27) correspond to (4.8.18). In the case of the thermoelastic solid, the stress $\mathbf{T}^e$ is identically zero.

Exercise 4.8.1

Show that the Gibbs relation for the material defined by (4.6.34) is

$$\dot{\psi} = -\eta\dot{\theta} + \frac{1}{\rho}\operatorname{tr}\mathbf{T}^0\mathbf{L} \tag{4.8.21}$$

If equations (4.8.9) and (4.8.10) are substituted into (4.8.8) it follows from the resulting equation along with (4.8.5) that

$$\begin{aligned}
&\operatorname{tr}\left(\mathbf{F}^{-1}\left(\frac{\partial \mathbf{G}^e_{\mathbf{K}_X}(\theta,\mathbf{0},\mathbf{F},\mathbf{0})}{\partial \dot{\mathbf{F}}}[\mathbf{A}]\right)\mathbf{A}\right) + \operatorname{tr}\left(\mathbf{F}^{-1}\left(\frac{\partial \mathbf{G}^e_{\mathbf{K}_X}(\theta,\mathbf{0},\mathbf{F},\mathbf{0})}{\partial \mathbf{g}}[\mathbf{a}]\right)\mathbf{A}\right)\\
&\qquad -\frac{1}{\theta}\mathbf{a}\cdot\left(\frac{\partial \mathbf{l}_{\mathbf{K}_X}(\theta,\mathbf{0},\mathbf{F},\mathbf{0})}{\partial \dot{\mathbf{F}}}[\mathbf{A}]\right) - \frac{1}{\theta}\mathbf{a}\cdot\left(\frac{\partial \mathbf{l}_{\mathbf{K}_X}(\theta,\mathbf{0},\mathbf{F},\mathbf{0})}{\partial \mathbf{g}}\mathbf{a}\right) \geq 0
\end{aligned} \tag{4.8.22}$$

for all $(\mathbf{a}, \mathbf{A})$ in $\mathscr{V} \times \mathscr{L}(\mathscr{V};\mathscr{V})$. Equations (4.8.14), (4.8.15) and (4.8.22) are results which necessarily hold in equilibrium. It is important to stress that they are generally not equivalent to the residual inequality (4.7.27). In other words, (4.7.27) contains information in addition to the results (4.8.14), (4.8.15) and (4.8.22).

Exercise 4.8.2

Use (4.8.22) and derive the inequalities (4.1.7), (4.1.8) and (4.1.9) for the material defined by (4.1.1) through (4.1.6).

Exercise 4.8.3

Show that for the material defined by (4.1.1) through (4.1.6), equation (4.8.22) is equivalent to the residual inequality (4.7.27).

Exercise 4.8.4

Show that for the material defined by (4.6.34) the thermodynamic energy equation takes the form

$$\rho\theta\dot{\eta} = -\operatorname{div}\mathbf{q} + \operatorname{tr}\mathbf{T}^e\mathbf{L} + \rho r \tag{4.8.23}$$

Exercise 4.8.5

Show that $\mathbf{T}_\mathbf{R}^0$, the equilibrium first Piola-Kirchhoff stress tensor, is given by

$$\mathbf{T}_\mathbf{R}^0 = \rho_\mathbf{R}\frac{\partial u_{\mathbf{K}_X}(\theta,\mathbf{F})}{\partial \mathbf{F}} \tag{4.8.24}$$

Exercise 4.8.6

The results in Sections 4.7 and 4.8 can be written in an alternate form by changing variables from $(\theta,\mathbf{g},\mathbf{F},\dot{\mathbf{F}})$ to $(\eta,\mathbf{g},\mathbf{F},\dot{\mathbf{F}})$. Assume equation (4.7.15) can be inverted to obtain θ as a function $\hat{t}_{\mathbf{K}_X}$ of η and $\mathbf{F}$ and show that

$$\theta = \hat{t}_{\mathbf{K}_X}(\eta,\mathbf{F}) = \frac{\partial \hat{e}_{\mathbf{K}_X}(\eta,\mathbf{F})}{\partial \eta} \tag{4.8.25}$$

and

$$\mathbf{T}^0 = \rho\mathbf{F}\frac{\partial \hat{e}_{\mathbf{K}_X}(\eta,\mathbf{F})}{\partial \mathbf{F}} \tag{4.8.26}$$

where

$$\varepsilon = \hat{e}_{\mathbf{K}_X}(\eta,\mathbf{F}) \tag{4.8.27}$$

4.9. Thermoelastic Material with Heat Conduction and Viscous Dissipation – Material Frame Indifference

In this section the axiom of material frame indifference is stated, and it is used to further restrict the form of the constitutive equations for the material defined by (4.6.34). Recall from Section 4.3 that a *change of frame* is a one-to-one mapping $\mathscr{E}\times\mathscr{R}\to\mathscr{E}\times\mathscr{R}$, defined by

$$\mathbf{x}^* = \mathbf{c}(t) + \mathbf{Q}(t)(\mathbf{x}-\mathbf{0}) \tag{4.9.1}$$

and

$$t^* = t - a \tag{4.9.2}$$

where $\mathbf{c}(t)$ is a time-dependent element of $\mathscr{E}$, $\mathbf{Q}(t)$ is a time-dependent element of $\mathcal{O}(\mathscr{V})$, and a is in $\mathscr{R}$. It is sufficient to assume here that $\mathbf{c}$ and $\mathbf{Q}$ are of class $\mathbf{C}^2$.

The axiom of material frame indifference is the following statement [Ref. 3, Sect. 19].

Axiom. The constitutive equations (4.6.8) must be invariant under changes of frame. If $(\psi,\eta,\mathbf{T},\mathbf{q})$ at (X,t) is given by a function $f_{\mathbf{K}_X}$ of the pair $(\Theta,\boldsymbol{\chi})$, the value $(\psi^*,\eta^*,\mathbf{T}^*,\mathbf{q}^*)$ at (X,t^*) is given by the same function $f_{\mathbf{K}_X}$ of the pair $(\Theta^*,\boldsymbol{\chi}^*)$, where

$$(\psi^*,\eta^*,\mathbf{T}^*,\mathbf{q}^*)=(\psi,\eta,\mathbf{Q}(t)\mathbf{T}\mathbf{Q}(t)^T,\mathbf{Q}(t)\mathbf{q}) \tag{4.9.3}$$

$$\Theta^*(X,\tau^*)=\Theta(X,\tau) \tag{4.9.4}$$

$$\boldsymbol{\chi}^*(X,\tau^*)=\mathbf{c}(\tau)+\mathbf{Q}(\tau)(\boldsymbol{\chi}(X,\tau)-\mathbf{0}) \tag{4.9.5}$$

and

$$\tau^*=\tau-a \tag{4.9.6}$$

for all a in $\mathscr{R}$, $\mathbf{Q}(\tau)$ in $\mathscr{O}(\mathscr{V})$ and $\mathbf{c}(\tau)$ in $\mathscr{E}$ such that $(\Theta^*,\boldsymbol{\chi}^*)$ is in the domain of $f_{\mathbf{K}_X}$.

As stated in Section 4.3, it is important to realize that the reference configuration plays no role in the axiom. If a reference configuration is used, it is not altered by the transformations (4.9.4), (4.9.5), and (4.9.6). The physical idea behind this axiom is the feeling that translations of the time scale and rigid motions of the body should not influence the response functions for the material. Equation (4.9.3) implies that

$$\psi^*=\psi \tag{4.9.7}$$

$$\eta^*=\eta \tag{4.9.8}$$

$$\mathbf{T}^*=\mathbf{Q}(t)\mathbf{T}\mathbf{Q}(t)^T \tag{4.9.9}$$

and

$$\mathbf{q}^*=\mathbf{Q}(t)\mathbf{q} \tag{4.9.10}$$

Equation (4.9.7) and (4.9.9) are simple restatements of (4.3.11) and (4.3.12), respectively. Equation (4.9.8) is an assumption based upon the physical assumption that the entropy density is not altered by the transformations (4.9.4), (4.9.5) and (4.9.6). Equation (4.9.10) follows by an argument similar to the one used to derive (4.9.9). This argument is outlined in the following exercise.

Exercise 4.9.1

Given (3.4.8) and $q^* = \mathbf{q}^* \cdot \mathbf{n}^*$, assume that the quantities q^* and $\mathbf{n}^*$ transform by the rules

$$q^* = q \tag{4.9.11}$$

and

$$\mathbf{n}^* = \mathbf{Q}(t)\mathbf{n} \tag{4.9.12}$$

Derive the transformation (4.9.10).

For the special material defined by (4.6.34), the constitutive functions depend only on quantities evaluated at the present time. It is possible to show that there is no loss of generality in taking $a = 0$ in (4.9.6). We shall adopt this choice in the following discussion. It follows from (4.9.4) and (4.9.5) that

$$\mathbf{g}^* = \mathbf{Q}(t)\mathbf{g} \tag{4.9.13}$$

$$\mathbf{F}^* = \mathbf{Q}(t)\mathbf{F} \tag{4.9.14}$$

and

$$\dot{\mathbf{F}}^* = \overline{\mathbf{Q}(t)\mathbf{F}}^{\cdot} = \dot{\mathbf{Q}}(t)\mathbf{F} + \mathbf{Q}(t)\dot{\mathbf{F}} \tag{4.9.15}$$

Exercise 4.9.2

Show that (4.9.14) implies that

$$\mathbf{R}^* = \mathbf{Q}(t)\mathbf{R} \tag{4.9.16}$$

$$\mathbf{U}^* = \mathbf{U} \tag{4.9.17}$$

and

$$\mathbf{V}^* = \mathbf{Q}(t)\mathbf{V}\mathbf{Q}(t)^T \tag{4.9.18}$$

Therefore, the axiom of material frame indifference states that if

$$(\psi, \eta, \mathbf{T}, \mathbf{q}) = k_{\mathbf{K}_X}\left(\theta, \mathbf{g}, \mathbf{F}, \dot{\mathbf{F}}\right) \tag{4.9.19}$$

then

$$\left(\psi,\eta,\mathbf{Q}(t)\mathbf{T}\mathbf{Q}(t)^T,\mathbf{Q}(t)\mathbf{q}\right)=k_{\mathbf{K}_X}\left(\theta,\mathbf{Q}(t)\mathbf{g},\mathbf{Q}(t)\mathbf{F},\dot{\mathbf{Q}}(t)\mathbf{F}+\mathbf{Q}(t)\dot{\mathbf{F}}\right) \tag{4.9.20}$$

for all $\mathbf{Q}(t)$ in $\mathscr{O}(\mathscr{V})$ and $\dot{\mathbf{Q}}(t)$ in $\mathscr{L}(\mathscr{V};\mathscr{V})$ with $\dot{\mathbf{Q}}(t)\mathbf{Q}(t)^T=-\left(\dot{\mathbf{Q}}(t)\mathbf{Q}(t)^T\right)^T$ such that $\left(\theta,\mathbf{Q}(t)\mathbf{g},\mathbf{Q}(t)\mathbf{F},\dot{\mathbf{Q}}(t)\mathbf{F}+\mathbf{Q}(t)\dot{\mathbf{F}}\right)$ lies in the domain of $k_{\mathbf{K}_X}$. Since the domain of $k_{\mathbf{K}_X}$ is a subset of $\mathscr{R}\times\mathscr{V}\times\mathscr{GL}(\mathscr{V})^+\times\mathscr{L}(\mathscr{V};\mathscr{V})$, it is clearly necessary for $\mathbf{Q}(t)$ in (4.9.20) to lie in $\mathscr{O}(\mathscr{V})^+$, the proper orthogonal group, introduced in Section 4.4. A formal definition of $\mathscr{O}(\mathscr{V})^+$ is

$$\mathscr{O}(\mathscr{V})^+=\left\{\mathbf{Q}\middle|\mathbf{Q}\text{ in }\mathscr{GL}(\mathscr{V})^+\text{ and }\mathbf{Q}\mathbf{Q}^T=\mathbf{I}\right\} \tag{4.9.21}$$

The condition on $\dot{\mathbf{Q}}(t)$ stated above arises because $\mathbf{Q}(t)$ is orthogonal for each t. To see this condition, simply note that

$$\overline{\mathbf{Q}(t)\mathbf{Q}(t)^T}^{\boldsymbol{\cdot}}=\dot{\mathbf{Q}}(t)\mathbf{Q}(t)^T+\left(\dot{\mathbf{Q}}(t)\mathbf{Q}(t)^T\right)^T=\mathbf{0} \tag{4.9.22}$$

We have used results like (4.9.22) in Sections 2.4 and 3.4.

Next, we shall deduce the solutions to equations (4.9.19) and (4.9.20) and, thus, derive the restrictions implied by the axiom of material frame indifference. The procedure we shall use is one originated by Noll [Ref. 4; Ref. 3, Sect. 29]. As in Section 4.3, this procedure involves first deducing *necessary* conditions from (4.9.19) and (4.9.20) The resulting necessary conditions will then be shown to be *sufficient* in that (4.9.19) and (4.9.20) will be satisfied *for all* proper orthogonal necessary condition. At any instant t we take

$$\mathbf{Q}(t)=\mathbf{R}^T \tag{4.9.23}$$

and

$$\dot{\mathbf{Q}}(t)=-\mathbf{R}^T\mathbf{W} \tag{4.9.24}$$

where $\mathbf{W}$ is the spin tensor defined by (2.4.30). Clearly $\mathbf{Q}(t)$ is in $\mathscr{O}(\mathscr{V})^+$ and $\dot{\mathbf{Q}}(t)\mathbf{Q}(t)^T=-\mathbf{R}^T\mathbf{W}\mathbf{R}$ is skew-symmetric. If (4.9.23) and (4.9.24) are substituted into (4.9.20), we can use (2.4.1a) and (2.4.29) to write the result in the form

$$\left(\psi,\eta,\mathbf{R}^T\mathbf{T}\mathbf{R},\mathbf{R}^T\mathbf{q}\right)=k_{\mathbf{K}_X}\left(\theta,\mathbf{R}^T\mathbf{g},\mathbf{U},\mathbf{R}^T\mathbf{D}\mathbf{R}\mathbf{U}\right) \tag{4.9.25}$$

If we now use (4.6.35), it follows from (4.9.19) and (4.9.25) that

$$u_{\mathbf{K}_X}\left(\theta,\mathbf{g},\mathbf{F},\dot{\mathbf{F}}\right)=u_{\mathbf{K}_X}\left(\theta,\mathbf{R}^T\mathbf{g},\mathbf{U},\mathbf{R}^T\mathbf{DRU}\right) \tag{4.9.26}$$

$$h_{\mathbf{K}_X}\left(\theta,\mathbf{g},\mathbf{F},\dot{\mathbf{F}}\right)=h_{\mathbf{K}_X}\left(\theta,\mathbf{R}^T\mathbf{g},\mathbf{U},\mathbf{R}^T\mathbf{DRU}\right) \tag{4.9.27}$$

$$\mathbf{G}_{\mathbf{K}_X}\left(\theta,\mathbf{g},\mathbf{F},\dot{\mathbf{F}}\right)=\mathbf{RG}_{\mathbf{K}_X}\left(\theta,\mathbf{R}^T\mathbf{g},\mathbf{U},\mathbf{R}^T\mathbf{DRU}\right)\mathbf{R}^T \tag{4.9.28}$$

and

$$\mathbf{l}_{\mathbf{K}_X}\left(\theta,\mathbf{g},\mathbf{F},\dot{\mathbf{F}}\right)=\mathbf{Rl}_{\mathbf{K}_X}\left(\theta,\mathbf{R}^T\mathbf{g},\mathbf{U},\mathbf{R}^T\mathbf{DRU}\right) \tag{4.9.29}$$

where the thermodynamic restrictions deduced in Sections 4.7 and 4.8 have not, as yet, been utilized. Next, we wish to show that if (4.9.26) through (4.9.29) are assumed, then (4.9.19) and (4.9.20) are satisfied for all $\mathbf{Q}(t)$ is in $\mathcal{O}(\mathscr{V})^+$ and all $\dot{\mathbf{Q}}(t)$ such that $\dot{\mathbf{Q}}(t)\mathbf{Q}(t)^T$ is skew-symmetric. When this argument is complete, we will have established that (4.9.26) through (4.9.29) are both necessary and sufficient for (4.9.19) and (4.9.20) to be satisfied. For the sake of brevity, we shall only provide the sufficient proof for (4.9.28). The reader can work out the corresponding proofs for (4.9.26), (4.9.27) and (4.9.29). From (4.9.19), (4.9.20) and (4.6.35) it follows that we must establish that

$$\mathbf{Q}(t)\mathbf{G}_{\mathbf{K}_X}\left(\theta,\mathbf{g},\mathbf{F},\dot{\mathbf{F}}\right)\mathbf{Q}(t)^T=\mathbf{G}_{\mathbf{K}_X}\left(\theta,\mathbf{Q}(t)\mathbf{g},\mathbf{Q}(t)\mathbf{F},\dot{\mathbf{Q}}(t)\mathbf{F}+\mathbf{Q}(t)\dot{\mathbf{F}}\right) \tag{4.9.30}$$

when (4.9.28) is true. Given (4.9.28), then

$$\begin{aligned}&\mathbf{G}_{\mathbf{K}_X}\left(\theta,\mathbf{Q}(t)\mathbf{g},\mathbf{Q}(t)\mathbf{F},\dot{\mathbf{Q}}(t)\mathbf{F}+\mathbf{Q}(t)\dot{\mathbf{F}}\right)\\&\quad=\mathbf{Q}(t)\mathbf{RG}_{\mathbf{K}_X}\left(\theta,\left(\mathbf{Q}(t)\mathbf{R}\right)^T\mathbf{Q}(t)\mathbf{g},\mathbf{U},\left(\mathbf{Q}(t)\mathbf{R}\right)^T\left(\mathbf{Q}(t)\mathbf{DQ}(t)^T\right)\left(\mathbf{Q}(t)\mathbf{R}\right)\mathbf{U}\right)\left(\mathbf{Q}(t)\mathbf{R}\right)^T\\&\quad=\mathbf{Q}(t)\mathbf{RG}_{\mathbf{K}_X}\left(\theta,\mathbf{R}^T\mathbf{g},\mathbf{U},\mathbf{R}^T\mathbf{DRU}\right)\mathbf{R}^T\mathbf{Q}(t)^T\end{aligned} \tag{4.9.31}$$

If we substitute (4.9.28) into the right side of (4.9.31), it is easily seen that (4.9.30) is identically satisfied.

Given (4.9.26) through (4.9.29), it is convenient in certain applications to use (2.4.1a) and (2.4.8) to define functions $\hat{u}_{\mathbf{K}_X}$, $\hat{h}_{\mathbf{K}_X}$, $\hat{\mathbf{G}}_{\mathbf{K}_X}$, and $\hat{\mathbf{l}}_{\mathbf{K}_X}$ by

$$\hat{u}_{\mathbf{K}_X}\left(\theta,\mathbf{F}^T\mathbf{g},\mathbf{C},\mathbf{F}^T\mathbf{DF}\right)=u_{\mathbf{K}_X}\left(\theta,\mathbf{C}^{-1/2}\mathbf{F}^T\mathbf{g},\mathbf{C}^{1/2},\mathbf{C}^{-1/2}\mathbf{F}^T\mathbf{DF}\right) \tag{4.9.32}$$

$$\hat{h}_{\mathbf{K}_X}\left(\theta,\mathbf{F}^T\mathbf{g},\mathbf{C},\mathbf{F}^T\mathbf{DF}\right)=h_{\mathbf{K}_X}\left(\theta,\mathbf{C}^{-1/2}\mathbf{F}^T\mathbf{g},\mathbf{C}^{1/2},\mathbf{C}^{-1/2}\mathbf{F}^T\mathbf{DF}\right) \tag{4.9.33}$$

$$\hat{\mathbf{G}}_{\mathbf{K}_X}\left(\theta,\mathbf{F}^T\mathbf{g},\mathbf{C},\mathbf{F}^T\mathbf{DF}\right)=\mathbf{C}^{-1/2}\mathbf{G}_{\mathbf{K}_X}\left(\theta,\mathbf{C}^{-1/2}\mathbf{F}^T\mathbf{g},\mathbf{C}^{1/2},\mathbf{C}^{-1/2}\mathbf{F}^T\mathbf{DF}\right)\mathbf{C}^{-1/2} \tag{4.9.34}$$

and

$$\hat{\mathbf{I}}_{\mathbf{K}_X}\left(\theta,\mathbf{F}^T\mathbf{g},\mathbf{C},\mathbf{F}^T\mathbf{DF}\right)=\mathbf{C}^{-1/2}\mathbf{I}_{\mathbf{K}_X}\left(\theta,\mathbf{C}^{-1/2}\mathbf{F}^T\mathbf{g},\mathbf{C}^{1/2},\mathbf{C}^{-1/2}\mathbf{F}^T\mathbf{DF}\right) \tag{4.9.35}$$

With these assumptions, the constitutive assumption (4.6.34) takes the form

$$\psi=\hat{u}_{\mathbf{K}_X}\left(\theta,\mathbf{F}^T\mathbf{g},\mathbf{C},\mathbf{F}^T\mathbf{DF}\right) \tag{4.9.36}$$

$$\eta=\hat{h}_{\mathbf{K}_X}\left(\theta,\mathbf{F}^T\mathbf{g},\mathbf{C},\mathbf{F}^T\mathbf{DF}\right) \tag{4.9.37}$$

$$\mathbf{T}=\mathbf{F}\hat{\mathbf{G}}_{\mathbf{K}_X}\left(\theta,\mathbf{F}^T\mathbf{g},\mathbf{C},\mathbf{F}^T\mathbf{DF}\right)\mathbf{F}^T \tag{4.9.38}$$

and

$$\mathbf{q}=\mathbf{F}\hat{\mathbf{I}}_{\mathbf{K}_X}\left(\theta,\mathbf{F}^T\mathbf{g},\mathbf{C},\mathbf{F}^T\mathbf{DF}\right) \tag{4.9.39}$$

Note that the dependence on $\dot{\mathbf{F}}$ is through the special combination $\mathbf{F}^T\mathbf{DF}$. This dependency is consistent with the first example of Section 4.1.

If we now utilize the restrictions (4.7.14) and (4.7.15), it follows that (4.9.36) and (4.9.37) are simplified to

$$\psi=\hat{u}_{\mathbf{K}_X}\left(\theta,\mathbf{C}\right) \tag{4.9.40}$$

$$\eta=\hat{h}_{\mathbf{K}_X}\left(\theta,\mathbf{C}\right)=-\frac{\partial\hat{u}_{\mathbf{K}_X}\left(\theta,\mathbf{C}\right)}{\partial\theta} \tag{4.9.41}$$

Next, we wish to show that the restriction (4.8.17) is *automatically* satisfied when ψ is given by (4.9.40). The argument given in Section 4.3 to derive (4.3.35) is essentially the same as the one to be given here. From (4.7.14) and (4.9.40), we have

$$u_{\mathbf{K}_X}\left(\theta,\mathbf{F}\right)=\hat{u}_{\mathbf{K}_X}\left(\theta,\mathbf{C}\right) \tag{4.9.42}$$

The derivative $\partial\hat{u}_{\mathbf{K}_X}/\partial\mathbf{F}$ is a linear transformation defined by [see (4.2.7)]

$$\operatorname{tr}\left(\frac{\partial u_{\kappa}\left(\theta,\mathbf{F}\right)^T}{\partial\mathbf{F}}\mathbf{A}\right)=\left.\frac{du_{\mathbf{K}_X}\left(\theta,\mathbf{F}+\lambda\mathbf{A}\right)}{d\lambda}\right|_{\lambda=0} \tag{4.9.43}$$

for all $\mathbf{A}$ in $\mathscr{L}(\mathscr{V};\mathscr{V})$. It follows from (2.4.3) that

$$u_{\mathbf{K}_X}(\theta,\mathbf{F}+\lambda\mathbf{A})=\hat{u}_{\mathbf{K}_X}\left(\theta,\mathbf{C}+\lambda\left(\mathbf{F}^T\mathbf{A}+\mathbf{A}^T\mathbf{F}\right)+\lambda^2\mathbf{A}^T\mathbf{A}\right) \tag{4.9.44}$$

The derivative $\partial\hat{u}_{\mathbf{K}_X}/\partial\mathbf{C}$ is a *symmetric* linear transformation defined by

$$\operatorname{tr}\left(\frac{\partial\hat{u}_{\mathbf{K}_X}(\theta,\mathbf{C})}{\partial\mathbf{C}}\mathbf{J}\right)=\left.\frac{d\hat{u}_{\mathbf{K}_X}(\theta,\mathbf{C}+\lambda\mathbf{J})}{d\lambda}\right|_{\lambda=0} \tag{4.9.45}$$

for all *symmetric* $\mathbf{J}$ in $\mathscr{L}(\mathscr{V};\mathscr{V})$. Given (4.9.43) and (4.9.45), it follows from (4.9.44) that

$$\operatorname{tr}\left(\frac{\partial u_{\mathbf{K}_X}(\theta,\mathbf{F})^T}{\partial\mathbf{F}}\mathbf{A}\right)=\operatorname{tr}\left(\frac{\partial\hat{u}_{\mathbf{K}_X}(\theta,\mathbf{C})}{\partial\mathbf{C}}\left(\mathbf{A}^T\mathbf{F}+\mathbf{F}^T\mathbf{A}\right)\right) \tag{4.9.46}$$

Because (4.9.46) holds for all $\mathbf{A}$ in $\mathscr{L}(\mathscr{V};\mathscr{V})$, it follows that

$$\frac{\partial u_{\mathbf{K}_X}(\theta,\mathbf{F})^T}{\partial\mathbf{F}}=2\frac{\partial\hat{u}_{\mathbf{K}_X}(\theta,\mathbf{C})}{\partial\mathbf{C}}\mathbf{F}^T \tag{4.9.47}$$

Therefore, as in Section 4.3, material frame indifference insures that (4.8.17) is satisfied. It also follows from (4.9.47) and (4.8.19) that

$$\mathbf{T}^0=2\rho\mathbf{F}\frac{\partial\hat{u}_{\mathbf{K}_X}(\theta,\mathbf{C})}{\partial\mathbf{C}}\mathbf{F}^T \tag{4.9.48}$$

The result developed in Exercises 4.3.6 and 4.3.7 can be established in the context of the material defined by (4.6.34). This result is that, given (4.8.15), the following equations are equivalent:

$$\mathbf{G}_{\mathbf{K}_X}(\theta,\mathbf{0},\mathbf{F},\mathbf{0})=\mathbf{G}_{\mathbf{K}_X}(\theta,\mathbf{0},\mathbf{F},\mathbf{0})^T \tag{4.9.49}$$

$$\mathbf{G}_{\mathbf{K}_X}(\theta,\mathbf{0},\mathbf{QF},\mathbf{0})=\mathbf{Q}\mathbf{G}_{\mathbf{K}_X}(\theta,\mathbf{0},\mathbf{F},\mathbf{0})\mathbf{Q}^T \tag{4.9.50}$$

for all $\mathbf{Q}$ in $\mathscr{O}(\mathscr{V})^+$, and

$$u_{\mathbf{K}_X}(\theta,\mathbf{QF})=u_{\mathbf{K}_X}(\theta,\mathbf{F}) \tag{4.9.51}$$

for all $\mathbf{Q}$ in $\mathscr{O}(\mathscr{V})^+$ [Ref. 4, Sect.4; Ref. 25].

Exercise 4.9.3

Show that the equilibrium value of the second Piola-Kirchhoff stress tensor is given by

$$\tilde{\mathbf{T}}^{0} = 2\frac{\partial \Sigma_{\mathbf{K}_X}(\theta, \mathbf{C})}{\partial \mathbf{C}} \tag{4.9.52}$$

where

$$\Sigma_{\mathbf{K}_X}(\theta, \mathbf{C}) = \rho_R u_{\mathbf{K}_X}(\theta, \mathbf{C}) \tag{4.9.53}$$

The quantity $\Sigma_{\mathbf{K}_X}(\theta, \mathbf{C})$ is the *strain energy* per unit of *undeformed* volume.

Exercise 4.9.4

Show that the material defined by (4.1.1) through (4.1.4) satisfies material frame indifference.

Exercise 4.9.5

Show that the material defined by (4.1.25) through (4.1.28) does *not* satisfy material frame indifference.

Exercise 4.9.6

Show that the material defined by (4.1.25) through (4.1.28) does satisfy a restricted version of material frame indifference where $\mathbf{Q}$ is small in some sense. Explain the physical meaning of this result.

Exercise 4.9.7

Consider a material for which

$$\mathbf{q} = \mathbf{l}_{\mathbf{K}_X}(\theta, \mathbf{g}, \dot{\mathbf{g}}, \mathbf{F}) \tag{4.9.54}$$

Show that the axiom of material frame indifference applied to $\mathbf{l}_{\mathbf{K}_X}$ yields

$$\mathbf{q} = \mathbf{R}\mathbf{l}_{\mathbf{K}_X}(\theta, \mathbf{R}^T\mathbf{g}, \mathbf{R}^T\mathbf{d}, \mathbf{U}) \tag{4.9.55}$$

where,

$$\mathbf{d} = \left(\left(\frac{\mathbf{g}}{\|\mathbf{g}\|}\right)\cdot\dot{\mathbf{g}}\right)\frac{\mathbf{g}}{\|\mathbf{g}\|} \tag{4.9.56}$$

[Refs. 26, 27].

Exercise 4.9.8

Given (4.9.9), show that

$$\overset{\circ}{\mathbf{T}}{}^{*} = \mathbf{Q}(t)\overset{\circ}{\mathbf{T}}\mathbf{Q}(t)^{T} \tag{4.9.57}$$

where $\overset{\circ}{\mathbf{T}}$ is a stress rate given by either one of the following formulas:

$$\overset{\circ}{\mathbf{T}} = \dot{\mathbf{T}} - \mathbf{WT} + \mathbf{TW} \tag{4.9.58}$$

or

$$\overset{\circ}{\mathbf{T}} = \dot{\mathbf{T}} + \mathbf{L}^{T}\mathbf{T} + \mathbf{TL} \tag{4.9.59}$$

The quantity $\overset{\circ}{\mathbf{T}}{}^{*}$ is given by formulas identical to (4.9.58) or (4.9.59) with $\mathbf{T}$ replaced by $\mathbf{T}^{*}$ and $\mathbf{L}$ replaced by $\mathbf{L}^{*}$. The stress rate (4.9.58) is called the *corotational stress rate* and (4.9.59) is called the *convected stress rate*. These stress rates arise when one studies materials of the rate type such as the Maxwellian material [Ref. 3, Sect. 36]. Maxwellian materials will be discussed in Chapter V.

4.10. Thermoelastic Material with Heat Conduction and Viscous Dissipation – Material Symmetry

In Section 4.4 the concept of material symmetry was investigated for the isothermal elastic material. These ideas are expanded in this section and applied to the material defined by (4.6.34). The material in this section is discussed in greater detail in Refs. 3, 5, 20, and 21.

It is useful to attempt to assign a mathematical meaning to the intuitive idea that two particles are identical. The only *physical properties* of the particle X in $\mathscr{B}$ are its density ρ and its response function $k_{\mathbf{K}_X}$. Therefore, in asserting that two particles are identical, it is reasonable to express this condition in terms of their densities and their response functions. In the following discussion, it is convenient *not* to make use of the restrictions obtained in Sections 4.7, 4.8 and 4.9.

Definition. Consider two particles X and Y in the body $\mathscr{B}$. The particles X and Y are *identical* if local reference configuration $\mathbf{K}_X$ for X and $\mathbf{J}_Y$ for Y exist such that

$$\rho_R(\mathbf{X}) = \rho_R(\mathbf{Y}) \tag{4.10.1}$$

where $\rho_R(\mathbf{X})$ is the density of X at $\mathbf{X}=\boldsymbol{\kappa}(X)$ in any $\boldsymbol{\kappa}$ in $\mathbf{K}_X$ and $\hat{\rho}_R(\mathbf{Y})$ is the density of Y at $\mathbf{Y}=\hat{\boldsymbol{\kappa}}(\mathbf{Y})$ in any $\hat{\boldsymbol{\kappa}}$ in $\mathbf{J}_Y$, and

$$k_{\mathbf{K}_X}\left(\theta,\mathbf{g},\mathbf{F},\dot{\mathbf{F}}\right)=k_{\mathbf{J}_Y}\left(\theta,\mathbf{g},\mathbf{F},\dot{\mathbf{F}}\right) \tag{4.10.2}$$

for all $\left(\theta,\mathbf{g},\mathbf{F},\dot{\mathbf{F}}\right)$ in the domain of $k_{\mathbf{K}_X}$ and $k_{\mathbf{J}_Y}$.

Physically, the above definition attempts to convey the statement that X and Y are identical if we can find a local reference configuration for X and a local reference configuration for Y such that X and Y have the same density, and, for the same values of θ, $\mathbf{g}$, $\mathbf{F}$, and $\dot{\mathbf{F}}$, we obtain the same values of ψ, η, $\mathbf{T}$, and $\mathbf{q}$. The geometric arrangement that applies here is shown in the Figure 4.10.1.

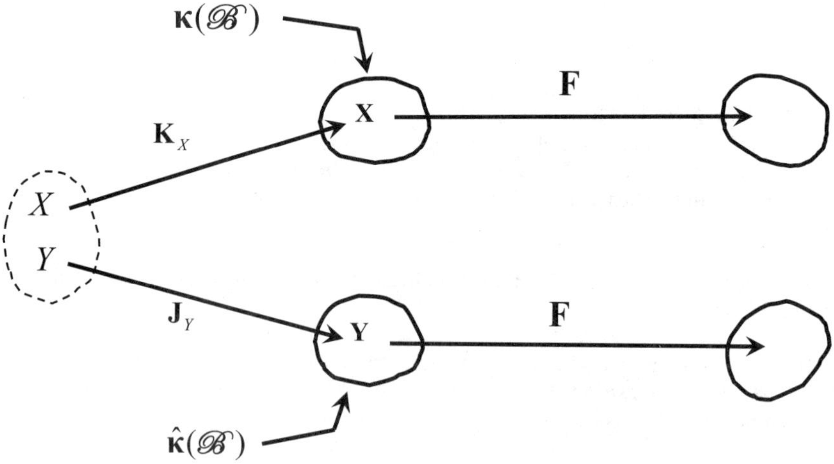

Figure 1.10.1

When two particles are identical in the sense of the above definition, they are said to be *materially isomorphic*. The local reference configurations $\mathbf{K}_X$ and $\mathbf{J}_X$ are called the *materially isomorphic local reference configurations* for X and Y, respectively.

Next we shall investigate the case where X in $\mathscr{B}$ is materially isomorphic to *itself*. If $\boldsymbol{\kappa}$ is in $\mathbf{K}_X$ and $\hat{\boldsymbol{\kappa}}$ is in $\mathbf{J}_X$, then

$$\hat{\mathbf{X}}=\hat{\boldsymbol{\kappa}}\circ\boldsymbol{\kappa}^{-1}(\mathbf{X})=\boldsymbol{\lambda}(\mathbf{X}) \tag{4.10.3}$$

relates the position of X in $\boldsymbol{\kappa}$ to its position in $\hat{\boldsymbol{\kappa}}$ (see (4.4.3) and (4.6.25)). For X to be materially isomorphic to itself, it follows from (4.10.2) that

$$k_{\mathbf{K}_X}\left(\theta,\mathbf{g},\mathbf{F},\dot{\mathbf{F}}\right)=k_{\mathbf{J}_X}\left(\theta,\mathbf{g},\mathbf{F},\dot{\mathbf{F}}\right) \tag{4.10.4}$$

Next we need a formula which relates the response function $k_{\mathbf{J}_X}$ to $k_{\mathbf{K}_X}$. The necessary formula has been developed in Section 4.6. From (4.6.30) and (4.6.33)

$$k_{\mathbf{J}_X}\left(\theta,\mathbf{g},\mathbf{F}\mathbf{P}^{-1},\dot{\mathbf{F}}\mathbf{P}^{-1}\right)=k_{\mathbf{K}_X}\left(\theta,\mathbf{g},\mathbf{F},\dot{\mathbf{F}}\right) \tag{4.10.5}$$

where $\mathbf{P}$ is defined by (4.6.28). As explained in Section 4.6, equation (4.10.5) reflects the fact that the values of the response function are independent of the special choice of local reference configuration. Figure 4.10.2 is useful when trying to justify (4.10.5).

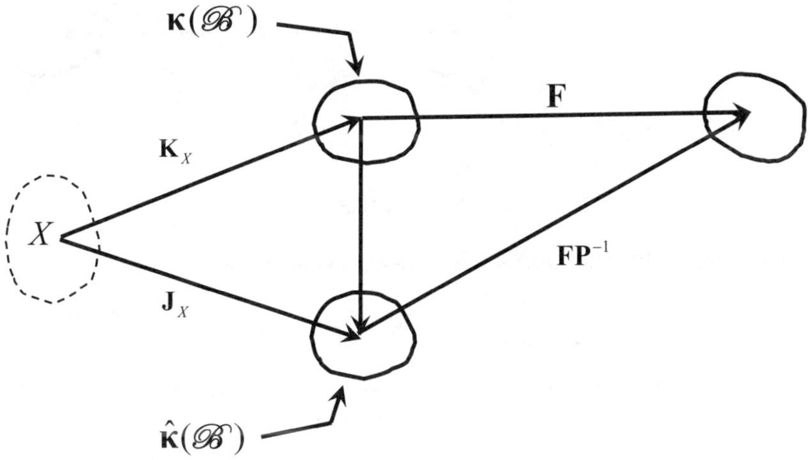

Figure 4.10.2

Since (4.10.5) holds for all $\left(\theta,\mathbf{g},\mathbf{F},\dot{\mathbf{F}}\right)$ in the domain of $k_{\mathbf{K}_X}$, we can formally replace $\mathbf{F}$ by $\mathbf{FH}$, where $\mathbf{H}=\mathbf{P}$, to obtain

$$k_{\mathbf{J}_X}\left(\theta,\mathbf{g},\mathbf{F},\dot{\mathbf{F}}\right)=k_{\mathbf{K}_X}\left(\theta,\mathbf{g},\mathbf{FH},\dot{\mathbf{F}}\mathbf{H}\right) \tag{4.10.6}$$

Equations (4.10.4) and (4.10.6) combine to yield the important result

$$k_{\mathbf{K}_X}\left(\theta,\mathbf{g},\mathbf{F},\dot{\mathbf{F}}\right)=k_{\mathbf{K}_X}\left(\theta,\mathbf{g},\mathbf{FH},\dot{\mathbf{F}}\mathbf{H}\right) \tag{4.10.7}$$

for all $(\theta, \mathbf{g}, \mathbf{F}, \dot{\mathbf{F}})$ such that $(\theta, \mathbf{g}, \mathbf{FH}, \dot{\mathbf{F}}\mathbf{H})$ is in the domain of $k_{\mathbf{K}_X}$. This equation and the condition (4.10.1) characterize the circumstance where X is materially isomorphic to itself. Figure 4.10.3 indicates the geometric interpretation of (4.10.7).

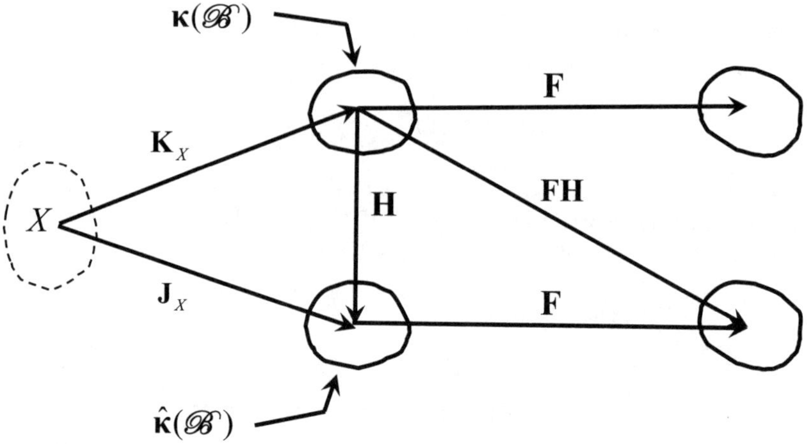

Figure 4.10.3

In order that (4.10.1) be satisfied it is necessary and sufficient that

$$|\det \mathbf{H}| = 1 \tag{4.10.8}$$

To see this result simply use (3.1.12) for the deformation (4.10.3) and force the density at $\hat{\mathbf{X}}$ to equal the density at $\mathbf{X}$. Since the domain of $k_{\mathbf{K}_X}$ is a subset of $\mathscr{R}^+ \times \mathscr{V} \times \mathscr{GL}(\mathscr{V})^+ \times \mathscr{L}(\mathscr{V};\mathscr{V})$, we must replace (4.10.8) by the more restrictive condition

$$\det \mathbf{H} = 1 \tag{4.10.9}$$

or otherwise $\mathbf{FH}$ in (4.10.7) would be in $\mathscr{GL}(\mathscr{V})^-$.

As explained in Section 4.4, the subset of $\mathscr{GL}(\mathscr{V})$ such that (4.10.8) is satisfied is called the *unimodular group*. It is a subgroup of $\mathscr{GL}(\mathscr{V})$. We shall continue to denote it by $\mathscr{U}(\mathscr{V})$. Again, as in Section 4.4, the *proper unimodular group* or the *special linear group* is the subgroup of $\mathscr{U}(\mathscr{V})$ defined by

$$\mathscr{U}(\mathscr{V})^+ = \{\mathbf{H} \mid \mathbf{H} \text{ in } \mathscr{GL}(\mathscr{V}),\ \det \mathbf{H} = 1\} \tag{4.10.10}$$

Definition. The *symmetry group* of X in $\mathscr{B}$ relative to the local reference configuration $\mathbf{K}_X$ is the group of material isomorphisms of X with itself.

We shall denote the symmetry group of X relative to $\mathbf{K}_X$ by $\mathscr{G}_{\mathbf{K}_X}$. It follows from (4.10.7) and (4.10.9) that

$$\mathscr{G}_{\mathbf{K}_X} = \left\{ \mathbf{H} \middle| \mathbf{H} \text{ in } \mathscr{U}(\mathscr{V})^{+} \text{ and } k_{\mathbf{K}_X}\left(\theta, \mathbf{g}, \mathbf{F}, \dot{\mathbf{F}}\right) = k_{\mathbf{K}_X}\left(\theta, \mathbf{g}, \mathbf{FH}, \dot{\mathbf{F}}\mathbf{H}\right) \right\} \tag{4.10.11}$$

Exercise 4.10.1

Show that $\mathscr{G}_{\mathbf{K}_X}$ is indeed a *subgroup* of $\mathscr{U}(\mathscr{V})^{+}$.

As the definition (4.10.11) indicates, the symmetry group $\mathscr{G}_{\mathbf{K}_X}$ depends upon the local reference configuration $\mathbf{K}_X$. In order to characterize this dependence let $\boldsymbol{\kappa}$ be in $\mathbf{K}_X$ and $\hat{\boldsymbol{\kappa}}$ be in $\hat{\mathbf{K}}_X$. It follows from (4.10.5) that

$$k_{\mathbf{K}_X}\left(\theta, \mathbf{g}, \mathbf{F}, \dot{\mathbf{F}}\right) = k_{\hat{\mathbf{K}}_X}\left(\theta, \mathbf{g}, \mathbf{F}\mathbf{P}^{-1}, \dot{\mathbf{F}}\mathbf{P}^{-1}\right) \tag{4.10.12}$$

where $\mathbf{P}$ is defined by (4.6.28). Given (4.10.12), we can formally replace $\mathbf{F}$ by $\mathbf{FH}$ to obtain

$$k_{\mathbf{K}_X}\left(\theta, \mathbf{g}, \mathbf{FH}, \dot{\mathbf{F}}\mathbf{H}\right) = k_{\hat{\mathbf{K}}_X}\left(\theta, \mathbf{g}, \mathbf{FHP}^{-1}, \dot{\mathbf{F}}\mathbf{HP}^{-1}\right) \tag{4.10.13}$$

For $\mathbf{H}$ in $\mathscr{G}_{\mathbf{K}_X}$, it follows from the definition (4.10.11), and equations (4.10.12) and (4.10.13) that

$$k_{\hat{\mathbf{K}}_X}\left(\theta, \mathbf{g}, \mathbf{F}\mathbf{P}^{-1}, \dot{\mathbf{F}}\mathbf{P}^{-1}\right) = k_{\hat{\mathbf{K}}_X}\left(\theta, \mathbf{g}, \mathbf{FHP}^{-1}, \dot{\mathbf{F}}\mathbf{HP}^{-1}\right) \tag{4.10.14}$$

If we formally replace $\mathbf{F}$ by $\mathbf{FP}$ in (4.10.14), we see that

$$k_{\hat{\mathbf{K}}_X}\left(\theta, \mathbf{g}, \mathbf{F}, \dot{\mathbf{F}}\right) = k_{\hat{\mathbf{K}}_X}\left(\theta, \mathbf{g}, \mathbf{FPHP}^{-1}, \dot{\mathbf{F}}\mathbf{PHP}^{-1}\right) \tag{4.10.15}$$

It follows from the definition (4.10.11), that $\mathscr{G}_{\hat{\mathbf{K}}_X}$ is defined by

$$\mathscr{G}_{\hat{\mathbf{K}}_X} = \left\{ \hat{\mathbf{H}} \middle| \hat{\mathbf{H}} \text{ in } \mathscr{U}(\mathscr{V})^{+} \text{ and } k_{\hat{\mathbf{K}}_X}\left(\theta, \mathbf{g}, \mathbf{F}, \dot{\mathbf{F}}\right) = k_{\hat{\mathbf{K}}_X}\left(\theta, \mathbf{g}, \mathbf{F}\hat{\mathbf{H}}, \dot{\mathbf{F}}\hat{\mathbf{H}}\right) \right\} \tag{4.10.16}$$

This result and (4.10.15) show that when $\mathbf{H}$ is in $\mathscr{G}_{\mathbf{K}_X}$ the corresponding element $\hat{\mathbf{H}}$ in $\mathscr{G}_{\hat{\mathbf{K}}_X}$ is given by

$$\hat{\mathbf{H}} = \mathbf{P}\mathbf{H}\mathbf{P}^{-1} \tag{4.10.17}$$

It is convenient to express (4.10.17) by the relationship

$$\mathscr{G}_{\hat{\mathbf{K}}_X} = \mathbf{P}\mathscr{G}_{\mathbf{K}_X}\mathbf{P}^{-1} \tag{4.10.18}$$

The result (4.10.18) was first proven by Noll [Ref. 4]. In group theory, the group $\mathscr{G}_{\hat{\mathbf{K}}_X}$ would be called the *conjugate group* to $\mathscr{G}_{\mathbf{K}_X}$. We established a special case of (4.10.18) in Section 4.4. However, in that section we did not stress the dependence on the *local* reference configuration.

Exercise 4.10.2

Show that if $\mathbf{H}$ is in $\mathscr{O}(\mathscr{V})^+$, then $\hat{\mathbf{H}}$ is not generally orthogonal.

The symmetry group of a particle X is generally not related in any special way to the symmetry group of a particle Y. However, if X and Y are materially isomorphic, then we shall show that the two symmetry groups are equal providing the appropriate local reference configurations for X and Y are used. Recall that when X and Y are materially isomorphic, there exist local reference configurations $\mathbf{K}_X$ for X and $\mathbf{J}_Y$ for Y, such that (4.10.1) and (4.10.2) hold. What we shall show is that

$$\mathscr{G}_{\mathbf{K}_X} = \mathscr{G}_{\mathbf{J}_Y} \tag{4.10.19}$$

The proof is elementary. From (4.10.11) we can write

$$\mathscr{G}_{\mathbf{J}_Y} = \left\{\mathbf{H}\,\middle|\,\mathbf{H} \text{ in } \mathscr{U}(\mathscr{V})^+ \text{ and } k_{\mathbf{J}_Y}\left(\theta,\mathbf{g},\mathbf{F},\dot{\mathbf{F}}\right) = k_{\mathbf{J}_Y}\left(\theta,\mathbf{g},\mathbf{FH},\dot{\mathbf{F}}\mathbf{H}\right)\right\} \tag{4.10.20}$$

From (4.10.2), this definition can be written

$$\mathscr{G}_{\mathbf{J}_Y} = \left\{\mathbf{H}\,\middle|\,\mathbf{H} \text{ in } \mathscr{U}(\mathscr{V})^+ \text{ and } k_{\mathbf{K}_X}\left(\theta,\mathbf{g},\mathbf{F},\dot{\mathbf{F}}\right) = k_{\mathbf{K}_X}\left(\theta,\mathbf{g},\mathbf{FH},\dot{\mathbf{F}}\mathbf{H}\right)\right\} \tag{4.10.21}$$

A comparison of (4.10.21) and (4.10.11) shows that (4.10.19) holds.

Definition. A local reference configuration $\mathbf{K}_X$ is *undistorted* if $\mathscr{G}_{\mathbf{K}_X}$ contains $\mathscr{O}(\mathscr{V})^+$ or $\mathscr{O}(\mathscr{V})^+$ contains $\mathscr{G}_{\mathbf{K}_X}$.

Another way to state this definition is to say that $\mathscr{G}_{\mathbf{K}_X}$ is *comparable* to $\mathscr{O}(\mathscr{V})^+$ with respect to *inclusion*.

Definition. A particle X is *hemitropic* if there exists an undistorted local reference configuration $\mathbf{K}_X$ such that $\mathcal{O}(\mathscr{V})^+$ is in $\mathscr{G}_{\mathbf{K}_X}$.

Definition. A particle is *isotropic* if it is hemitropic and if

$$\mathbf{G}_{\mathbf{K}_X}\left(\theta,\mathbf{g},\mathbf{F},\dot{\mathbf{F}}\right)=\mathbf{G}_{\mathbf{K}_X}\left(\theta,-\mathbf{g},\mathbf{F},\dot{\mathbf{F}}\right) \tag{4.10.22}$$

and

$$\mathbf{l}_{\mathbf{K}_X}\left(\theta,\mathbf{g},\mathbf{F},\dot{\mathbf{F}}\right)=-\mathbf{l}_{\mathbf{K}_X}\left(\theta,-\mathbf{g},\mathbf{F},\dot{\mathbf{F}}\right) \tag{4.10.23}$$

Thus, X is isotropic if there exists a local reference configuration such that any proper rotation is in the symmetry group and if $\mathbf{G}_{\mathbf{K}_X}$ is an even function of $\mathbf{g}$ and $\mathbf{l}_{\mathbf{K}_X}$ is an odd function of $\mathbf{g}$.

Definition. The particle X is a *solid* particle if there exists an undistorted local reference configuration $\mathbf{K}_X$ such that $\mathscr{G}_{\mathbf{K}_X}$ is contained in $\mathcal{O}(\mathscr{V})^+$.

It is important to stress that in order to assert that X is a solid particle, it is necessary to put it in a special local reference configuration. From the last exercise, it follows that if $\mathscr{G}_{\mathbf{K}_X}$ is contained in $\mathcal{O}(\mathscr{V})^+$, the conjugate group $\mathscr{G}_{\hat{\mathbf{K}}_X}=\mathbf{P}\mathscr{G}_{\mathbf{K}_X}\mathbf{P}^{-1}$ is not generally in $\mathcal{O}(\mathscr{V})^+$. It follows from the above definitions that for an *isotropic solid* particle $\mathscr{G}_{\mathbf{K}_X}=\mathcal{O}(\mathscr{V})^+$ and (4.10.22) and (4.10.23) hold.

Definition. A particle X is a *fluid* if it is isotropic and if $\mathscr{G}_{\mathbf{K}_X}(\mathscr{V})=\mathscr{U}(\mathscr{V})^+$.

Exercise 4.10.3

Show that if X is a fluid particle with respect to $\mathbf{K}_X$, it is a fluid particle with respect to *every* local reference configuration. Thus a fluid particle does not have a preferred reference configuration.

It is a fact in group theory that if $\mathcal{O}(\mathscr{V})^+$ is contained in $\mathscr{G}_{\mathbf{K}_X}$ and $\mathscr{G}_{\mathbf{K}_X}$ is contained in $\mathscr{U}(\mathscr{V})^+$, then either $\mathscr{G}_{\mathbf{K}_X}=\mathcal{O}(\mathscr{V})^+$ or $\mathscr{G}_{\mathbf{K}_X}=\mathscr{U}(\mathscr{V})^+$ [Refs. 6, 7]. Therefore, among the materials defined by (4.6.34), the only isotropic ones are either fluids or solids.

Next we wish to derive the explicit restrictions imposed by material symmetry for the cases where X is an isotropic solid particle and where X is a fluid particle. If we incorporate the definition (4.10.11) with the restrictions deduced in Sections 4.7, 4.8, and 4.9, it follows from (4.9.39), (4.9.40), (4.9.38) and (4.10.11) that $\mathscr{G}_{\mathbf{K}_X}$ is a subset of $\mathscr{U}(\mathscr{V})^+$ such that

$$\hat{u}_{\mathbf{K}_X}(\theta,\mathbf{C}) = \hat{u}_{\mathbf{K}_X}(\theta,\mathbf{H}^T\mathbf{C}\mathbf{H}) \tag{4.10.24}$$

$$\hat{\mathbf{G}}_{\mathbf{K}_X}(\theta,\mathbf{F}^T\mathbf{g},\mathbf{C},\mathbf{F}^T\mathbf{D}\mathbf{F}) = \mathbf{H}\hat{\mathbf{G}}_{\mathbf{K}_X}(\theta,\mathbf{H}^T\mathbf{F}^T\mathbf{g},\mathbf{H}^T\mathbf{C}\mathbf{H},\mathbf{H}^T\mathbf{F}^T\mathbf{D}\mathbf{F}\mathbf{H})\mathbf{H}^T \tag{4.10.25}$$

and

$$\hat{\mathbf{l}}_{\mathbf{K}_X}(\theta,\mathbf{F}^T\mathbf{g},\mathbf{C},\mathbf{F}^T\mathbf{D}\mathbf{F}) = \mathbf{H}\hat{\mathbf{l}}_{\mathbf{K}_X}(\theta,\mathbf{H}^T\mathbf{F}^T\mathbf{g},\mathbf{H}^T\mathbf{C}\mathbf{H},\mathbf{H}^T\mathbf{F}^T\mathbf{D}\mathbf{F}\mathbf{H}) \tag{4.10.26}$$

If we assume that X is an isotropic solid and $\mathbf{K}_X$ is its undistorted reference configuration, then (4.10.24), (4.10.25), and (4.10.26) hold for all $\mathbf{H}$ is $\mathcal{O}(\mathscr{V})^+$ and, in addition, $\hat{\mathbf{G}}_{\mathbf{K}_X}$ is an *even* function of $\mathbf{g}$ (see (4.10.22)) and $\hat{\mathbf{l}}_{\mathbf{K}_X}$ is an *odd* function of $\mathbf{g}$ (see (4.10.23)). These facts will be used to establish the following results:

For an isotropic solid particle in its undistorted reference configuration,

$$\psi = \hat{u}_{\mathbf{K}_X}(\theta,\mathbf{B}) \tag{4.10.27}$$

$$\mathbf{T} = \bar{\mathbf{G}}_{\mathbf{K}_X}(\theta,\mathbf{g},\mathbf{B},\mathbf{D}) \tag{4.10.28}$$

and

$$\mathbf{q} = \bar{\mathbf{l}}_{\mathbf{K}_X}(\theta,\mathbf{g},\mathbf{B},\mathbf{D}) \tag{4.10.29}$$

where $\mathbf{B}$ is the left Cauchy-Green tensor defined by (2.4.2) and, for all $\mathbf{Q}$ in $\mathcal{O}(\mathscr{V})^+$,

$$\hat{u}_{\mathbf{K}_X}(\theta,\mathbf{B}) = \hat{u}_{\mathbf{K}_X}(\theta,\mathbf{Q}\mathbf{B}\mathbf{Q}^T) \tag{4.10.30}$$

$$\mathbf{Q}\bar{\mathbf{G}}_{\mathbf{K}_X}(\theta,\mathbf{g},\mathbf{B},\mathbf{D})\mathbf{Q}^T = \bar{\mathbf{G}}_{\mathbf{K}_X}(\theta,\mathbf{Q}\mathbf{g},\mathbf{Q}\mathbf{B}\mathbf{Q}^T,\mathbf{Q}\mathbf{D}\mathbf{Q}^T) \tag{4.10.31}$$

and

$$\mathbf{Q}\bar{\mathbf{l}}_{\mathbf{K}_X}(\theta,\mathbf{g},\mathbf{B},\mathbf{D}) = \bar{\mathbf{l}}_{\mathbf{K}_X}(\theta,\mathbf{Q}\mathbf{g},\mathbf{Q}\mathbf{B}\mathbf{Q}^T,\mathbf{Q}\mathbf{D}\mathbf{Q}^T) \tag{4.10.32}$$

In addition $\bar{\mathbf{G}}_{\mathbf{K}_X}$ is an even function of $\mathbf{g}$ and $\bar{\mathbf{l}}_{\mathbf{K}_X}$ is an odd function of $\mathbf{g}$.

In order to establish (4.10.27) through (4.10.32) we follow an argument similar to the one used to establish. (4.4.35) We take $\mathbf{H} = \mathbf{R}^T$ in (4.10.24) and use (2.4.6) to obtain the necessary condition

$$\hat{u}_{\mathbf{K}_X}(\theta,\mathbf{C}) = \hat{u}_{\mathbf{K}_X}(\theta,\mathbf{B}) \tag{4.10.33}$$

Thus, in the case under discussion, $\hat{u}_{\mathbf{K}_X}$ has the same value on $(\theta,\mathbf{B})$ as on $(\theta,\mathbf{C})$. Given (4.10.33), it follows from (4.10.24) and (2.4.6) that, *for all* $\mathbf{Q}$ in $\mathscr{O}(\mathscr{V})^+$,

$$\hat{u}_{\mathbf{K}_X}(\theta,\mathbf{QBQ}^T) = \hat{u}_{\mathbf{K}_X}(\theta,\mathbf{QRCR}^T\mathbf{Q}^T) = \hat{u}_{\mathbf{K}_X}(\theta,\mathbf{C}) = \hat{u}_{\mathbf{K}_X}(\theta,\mathbf{B}) \tag{4.10.34}$$

because $\mathbf{QR}$ is also in $\mathscr{O}(\mathscr{V})^+ = \mathscr{G}_{\mathbf{K}_X}$. Conversely, given (4.10.30) and (4.10.33) it follows that (4.10.24) holds. Consequently, (4.10.30) and (4.10.33) are equivalent to (4.10.24). By an entirely similar argument, it follows from (4.10.25) and (4.10.26) that a necessary condition is

$$\mathbf{F}\hat{\mathbf{G}}_{\mathbf{K}_X}(\theta,\mathbf{F}^T\mathbf{g},\mathbf{C},\mathbf{F}^T\mathbf{DF})\mathbf{F}^T = \mathbf{B}^{1/2}\hat{\mathbf{G}}_{\mathbf{K}_X}(\theta,\mathbf{B}^{1/2}\mathbf{g},\mathbf{B},\mathbf{B}^{1/2}\mathbf{DB}^{1/2})\mathbf{B}^{1/2} \tag{4.10.35}$$

and

$$\mathbf{F}\hat{\mathbf{l}}_{\mathbf{K}_X}(\theta,\mathbf{F}^T\mathbf{g},\mathbf{C},\mathbf{F}^T\mathbf{DF}) = \mathbf{B}^{1/2}\hat{\mathbf{l}}_{\mathbf{K}_X}(\theta,\mathbf{B}^{1/2}\mathbf{g},\mathbf{B},\mathbf{B}^{1/2}\mathbf{DB}^{1/2}) \tag{4.10.36}$$

Next we *define* the functions $\bar{\mathbf{G}}_{\mathbf{K}_X}$ and $\bar{\mathbf{l}}_{\mathbf{K}_X}$ by

$$\bar{\mathbf{G}}_{\mathbf{K}_X}(\theta,\mathbf{g},\mathbf{B},\mathbf{D}) = \mathbf{B}^{1/2}\hat{\mathbf{G}}_{\mathbf{K}_X}(\theta,\mathbf{B}^{1/2}\mathbf{g},\mathbf{B},\mathbf{B}^{1/2}\mathbf{DB}^{1/2})\mathbf{B}^{1/2} \tag{4.10.37}$$

and

$$\bar{\mathbf{l}}_{\mathbf{K}_X}(\theta,\mathbf{g},\mathbf{B},\mathbf{D}) = \mathbf{B}^{1/2}\hat{\mathbf{l}}_{\mathbf{K}_X}(\theta,\mathbf{B}^{1/2}\mathbf{g},\mathbf{B},\mathbf{B}^{1/2}\mathbf{DB}^{1/2}) \tag{4.10.38}$$

Equation (4.10.31) follows easily because by (4.10.37) and (4.10.35),

$$\begin{aligned}
&\bar{\mathbf{G}}_{\mathbf{K}_X}(\theta,\mathbf{Qg},\mathbf{QBQ}^T,\mathbf{QDQ}^T)\\
&\quad= \mathbf{QB}^{1/2}\mathbf{Q}^T\hat{\mathbf{G}}_{\mathbf{K}_X}(\theta,\mathbf{QB}^{1/2}\mathbf{Q}^T\mathbf{Qg},\mathbf{QBQ}^T,\mathbf{QB}^{1/2}\mathbf{Q}^T\mathbf{QDQ}^T\mathbf{QB}^{1/2}\mathbf{Q}^T)\mathbf{QB}^{1/2}\mathbf{Q}^T\\
&\quad= \mathbf{QB}^{1/2}\mathbf{Q}^T\hat{\mathbf{G}}_{\mathbf{K}_X}(\theta,\mathbf{QB}^{1/2}\mathbf{g},\mathbf{QBQ}^T,\mathbf{QB}^{1/2}\mathbf{DB}^{1/2}\mathbf{Q}^T)\mathbf{QB}^{1/2}\mathbf{Q}^T
\end{aligned} \tag{4.10.39}$$

Because (4.10.25) holds for all $\mathbf{Q}$ in $\mathscr{O}(\mathscr{V})^+$, (4.10.39) can be rewritten as

$$\begin{aligned}
&\bar{\mathbf{G}}_{\mathbf{K}_X}(\theta,\mathbf{Qg},\mathbf{QBQ}^T,\mathbf{QDQ}^T)\\
&\quad= \mathbf{QB}^{1/2}\hat{\mathbf{G}}_{\mathbf{K}_X}(\theta,\mathbf{B}^{1/2}\mathbf{g},\mathbf{B},\mathbf{B}^{1/2}\mathbf{DB}^{1/2})\mathbf{B}^{1/2}\mathbf{Q}^T
\end{aligned} \tag{4.10.40}$$

with the definition (4.10.37), (4.10.40) becomes the desired result (4.10.31). Conversely, given (4.10.31), (4.10.35), and the definition (4.10.37), equation (4.10.25) can easily be shown to hold. Therefore, with the definition (4.10.37), (4.10.31), and (4.10.35) are equivalent to (4.10.25). Equation (4.10.32) follows by an identical argument. Equation (4.10.27) follows from (4.10.40) and (4.10.33). Equation (4.10.28) follows from (4.10.37), (4.10.35), and (4.10.38). Finally, (4.10.29) follows from (4.10.38), (4.10.36), and (4.9.39). The function $\bar{\mathbf{G}}_{\mathbf{K}_X}$ is an even function of $\mathbf{g}$ because $\hat{\mathbf{G}}_{\mathbf{K}_X}$ is even. Likewise $\bar{\mathbf{I}}_{\mathbf{K}_X}$ is an odd function of $\mathbf{g}$ because $\hat{\mathbf{I}}_{\mathbf{K}_X}$ is an odd function. Thus, the assertion is established. Notice that the even and odd dependence on $\mathbf{g}$ follows from (4.10.31) and (4.10.32) if we extend the allowable orthogonal linear transformations to include all of $\mathcal{O}(\mathscr{V})$. To see this assertion, simply take $\mathbf{Q} = -\mathbf{I}$ in (4.10.31) and (4.10.32). If (4.10.30) through (4.10.32) are viewed as holding *for all* $\mathbf{Q}$ in $\mathcal{O}(\mathscr{V})$, the functions $\hat{u}_{\mathbf{K}_X}$, $\bar{\mathbf{G}}_{\mathbf{K}_X}$ and $\bar{\mathbf{I}}_{\mathbf{K}_X}$ are said to be *isotropic*.

Exercise 4.10.4

Show that for an isotropic solid particle in its undistorted local reference configuration

$$\eta = -\frac{\partial \hat{u}_{\mathbf{K}_X}(\theta, \mathbf{B})}{\partial \theta} \tag{4.10.41}$$

and

$$\mathbf{T}^0 = 2\rho \mathbf{B}\frac{\partial \hat{u}_{\mathbf{K}_X}(\theta, \mathbf{B})}{\partial \mathbf{B}} = 2\rho \frac{\partial \hat{u}_{\mathbf{K}_X}(\theta, \mathbf{B})}{\partial \mathbf{B}}\mathbf{B} \tag{4.10.42}$$

As we used in Section 4.4, equation (4.10.30) implies the existence of a function $\overset{*}{u}_{\mathbf{K}_X}$ of the fundamental invariants of $\mathbf{B}$ in addition to the temperature θ. Therefore,

$$\psi = \hat{u}_{\mathbf{K}_X}(\theta, \mathbf{B}) = \overset{*}{u}_{\mathbf{K}_X}(\theta, I_{\mathbf{B}}, II_{\mathbf{B}}, III_{\mathbf{B}}) \tag{4.10.43}$$

Exercise 4.10.5

Use (4.10.43) and (4.10.42) and show that

$$\mathbf{T}^0 = \alpha_0(\theta, I_{\mathbf{B}}, II_{\mathbf{B}}, III_{\mathbf{B}})\mathbf{I} + \alpha_1(\theta, I_{\mathbf{B}}, II_{\mathbf{B}}, III_{\mathbf{B}})\mathbf{B} + \alpha_2(\theta, I_{\mathbf{B}}, II_{\mathbf{B}}, III_{\mathbf{B}})\mathbf{B}^2 \tag{4.10.44}$$

where

$$\alpha_0(\theta, I_{\mathbf{B}}, II_{\mathbf{B}}, III_{\mathbf{B}}) = 2\rho III_{\mathbf{B}}\frac{\partial \overset{*}{u}_{\mathbf{K}_X}(\theta, I_{\mathbf{B}}, II_{\mathbf{B}}, III_{\mathbf{B}})}{\partial III_{\mathbf{B}}} \tag{4.10.45}$$

$$\alpha_1\left(\theta, I_{\mathbf{B}}, II_{\mathbf{B}}, III_{\mathbf{B}}\right)=2\rho\frac{\partial u^*_{\mathbf{K}_X}\left(\theta, I_{\mathbf{B}}, II_{\mathbf{B}}, III_{\mathbf{B}}\right)}{\partial I_{\mathbf{B}}}+2\rho I_{\mathbf{B}}\frac{\partial u^*_{\mathbf{K}_X}\left(\theta, I_{\mathbf{B}}, II_{\mathbf{B}}, III_{\mathbf{B}}\right)}{\partial II_{\mathbf{B}}} \tag{4.10.46}$$

and

$$\alpha_2\left(\theta, I_{\mathbf{B}}, II_{\mathbf{B}}, III_{\mathbf{B}}\right)=-2\rho\frac{\partial u^*_{\mathbf{K}_X}\left(\theta, I_{\mathbf{B}}, II_{\mathbf{B}}, III_{\mathbf{B}}\right)}{\partial II_{\mathbf{B}}} \tag{4.10.47}$$

Note that $\mathbf{T}^0$ is necessarily hydrostatic in the undistorted local reference configuration.

Exercise 4.10.6

Show that (4.10.32) is satisfied in the state $(\theta, \mathbf{g}, \mathbf{B}, \mathbf{0})$ by the representation

$$\bar{\mathbf{l}}_{\mathbf{K}_X}\left(\theta, \mathbf{g}, \mathbf{B}, \mathbf{0}\right)=\left(\sigma_0\mathbf{I}+\sigma_1\mathbf{B}+\sigma_2\mathbf{B}^2\right)\mathbf{g} \tag{4.10.48}$$

where

$$\sigma_\tau=\sigma_\tau\left(\theta, I_{\mathbf{B}}, II_{\mathbf{B}}, III_{\mathbf{B}}, \mathbf{g}\cdot\mathbf{g}, \mathbf{g}\cdot(\mathbf{B}\mathbf{g}), \mathbf{g}\cdot\left(\mathbf{B}^2\mathbf{g}\right)\right), \tag{4.10.49}$$

for $\tau=0,1,2$ [Ref. 29]. Note that a similar representation could be stated for $\bar{\mathbf{l}}_{\mathbf{K}_X}$ in the state $(\theta, \mathbf{g}, \mathbf{I}, \mathbf{D})$ by formally replacing $\mathbf{B}$ by $\mathbf{D}$. Equation (4.10.48) shows that when $\mathbf{D}=\mathbf{0}$ in the undistorted local reference configuration, $\mathbf{q}$ and $\mathbf{g}$ have a common line of action.

If we assume X is a *fluid particle*, then (4.10.24), (4.10.25), and (4.10.26) hold for all $\mathbf{H}$ in $\mathscr{U}(\mathscr{V})^+$. In addition, $\hat{\mathbf{G}}_{\mathbf{K}_X}$ is an even function of $\mathbf{g}$ and $\hat{l}_{\mathbf{K}_X}$ is an odd function of $\mathbf{g}$. These facts will next be used to establish the following results:

For a fluid particle

$$\psi=\tilde{u}_{\mathbf{K}_X}\left(\theta, \rho\right) \tag{4.10.50}$$

$$\mathbf{T}=\tilde{\mathbf{G}}_{\mathbf{K}_X}\left(\theta, \mathbf{g}, \rho, \mathbf{D}\right) \tag{4.10.51}$$

and

$$\mathbf{q}=\tilde{\mathbf{l}}_{\mathbf{K}_X}\left(\theta, \mathbf{g}, \rho, \mathbf{D}\right) \tag{4.10.52}$$

where, for all $\mathbf{Q}$ in $\mathscr{O}(\mathscr{V})^+$,

$$\mathbf{Q}\tilde{\mathbf{G}}_{\mathbf{K}_X}\left(\theta,\mathbf{g},\rho,\mathbf{D}\right)\mathbf{Q}^T = \tilde{\mathbf{G}}_{\mathbf{K}_X}\left(\theta,\mathbf{Qg},\rho,\mathbf{QDQ}^T\right) \qquad (4.10.53)$$

and

$$\mathbf{Q}\tilde{\mathbf{l}}_{\mathbf{K}_X}\left(\theta,\mathbf{g},\rho,\mathbf{D}\right) = \tilde{\mathbf{l}}_{\mathbf{K}_X}\left(\theta,\mathbf{Qg},\rho,\mathbf{QDQ}^T\right) \qquad (4.10.54)$$

In addition, $\tilde{\mathbf{G}}_{\mathbf{K}_X}$ is an even function of $\mathbf{g}$ and $\tilde{\mathbf{l}}_{\mathbf{K}_X}$ is an odd function of $\mathbf{g}$. These conditions, along with (4.10.53) and (4.10.54), make $\tilde{\mathbf{G}}_{\mathbf{K}_X}$ and $\tilde{\mathbf{l}}_{\mathbf{K}_X}$ *isotropic* functions.

It is convenient in this case not to work with (4.10.24), (4.10.25), and (4.10.26) as in the proof of (4.10.27) through (4.10.32). We shall work with the definition (4.10.11). Given the thermodynamic restrictions derived in Section 4.7, it follows that $\mathscr{G}_{\mathbf{K}_X}$ is the set of linear transformations $\mathbf{H}$ such that $\det \mathbf{H} = 1$ and

$$u_{\mathbf{K}_X}\left(\theta,\mathbf{F}\right) = u_{\mathbf{K}_X}\left(\theta,\mathbf{FH}\right) \qquad (4.10.55)$$

$$\mathbf{G}_{\mathbf{K}_X}\left(\theta,\mathbf{g},\mathbf{F},\dot{\mathbf{F}}\right) = \mathbf{G}_{\mathbf{K}_X}\left(\theta,\mathbf{g},\mathbf{FH},\dot{\mathbf{F}}\mathbf{H}\right) \qquad (4.10.56)$$

and

$$\mathbf{l}_{\mathbf{K}_X}\left(\theta,\mathbf{g},\mathbf{F},\dot{\mathbf{F}}\right) = \mathbf{l}_{\mathbf{K}_X}\left(\theta,\mathbf{g},\mathbf{FH},\dot{\mathbf{F}}\mathbf{H}\right) \qquad (4.10.57)$$

for all $\left(\theta,\mathbf{g},\mathbf{F},\dot{\mathbf{F}}\right)$ in the domain of $k_{\mathbf{K}_X}$. Following the argument used in Section 4.4, we deduce a necessary condition by taking

$$\mathbf{H} = \mathbf{F}^{-1}\left(\det \mathbf{F}\right)^{1/3} \qquad (4.10.58)$$

If (4.10.58) is substituted in (4.10.55), (4.10.56) and (4.10.57), the results are

$$u_{\mathbf{K}_X}\left(\theta,\mathbf{F}\right) = u_{\mathbf{K}_X}\left(\theta,\left(\det \mathbf{F}\right)^{1/3}\mathbf{I}\right) \qquad (4.10.59)$$

$$\mathbf{G}_{\mathbf{K}_X}\left(\theta,\mathbf{g},\mathbf{F},\dot{\mathbf{F}}\right) = \mathbf{G}_{\mathbf{K}_X}\left(\theta,\mathbf{g},\left(\det \mathbf{F}\right)^{1/3}\mathbf{I},\left(\det \mathbf{F}\right)^{1/3}\mathbf{L}\right) \qquad (4.10.60)$$

and

$$\mathbf{l}_{\mathbf{K}_X}\left(\theta,\mathbf{g},\mathbf{F},\dot{\mathbf{F}}\right) = \mathbf{l}_{\mathbf{K}_X}\left(\theta,\mathbf{g},\left(\det \mathbf{F}\right)^{1/3}\mathbf{I},\left(\det \mathbf{F}\right)^{1/3}\mathbf{L}\right) \qquad (4.10.61)$$

where (2.2.24) has been used. Conversely, if (4.10.59) through (4.10.61) are assumed, then (4.10.55) through (4.10.57) are satisfied for all $\mathbf{H}$ in $\mathscr{G}_{\mathbf{K}_X} = \mathscr{U}(\mathscr{V})^+$. Next, we must insure that material frame indifference is satisfied. For the function $u_{\mathbf{K}_X}$, equations (4.9.19), (4.9.20) and (4.6.35) show that we must have

$$u_{\mathbf{K}_X}(\theta,\mathbf{F}) = u_{\mathbf{K}_X}(\theta,\mathbf{QF}) \tag{4.10.62}$$

Since $\det\mathbf{QF} = \det\mathbf{Q}\det\mathbf{F} = \det\mathbf{F}$, (4.10.59) shows that (4.10.62) is satisfied. If we use (3.1.12) to define a function $\tilde{u}_{\mathbf{K}_X}$ of (θ,ρ) by

$$\tilde{u}_{\mathbf{K}_X}(\theta,\rho) = u_{\mathbf{K}_X}\left(\theta,\rho_R^{1/3}\rho^{-1/3}\mathbf{I}\right) \tag{4.10.63}$$

then (4.10.50) is established. Material frame indifference for $\mathbf{G}_{\mathbf{K}_X}$ is the restriction (4.9.30). From (4.10.60), this restriction takes the form

$$\begin{aligned}
&\mathbf{Q}(t)\mathbf{G}_{\mathbf{K}_X}\left(\theta,\mathbf{g},(\det\mathbf{F})^{1/3}\mathbf{I},(\det\mathbf{F})^{1/3}\mathbf{L}\right)\mathbf{Q}(t)^T \\
&\quad= \mathbf{G}_{\mathbf{K}_X}\left(\theta,\mathbf{Q}(t)\mathbf{g},(\det\mathbf{F})^{1/3}\mathbf{I},(\det\mathbf{F})^{1/3}\mathbf{Q}(t)\mathbf{L}\mathbf{Q}(t)^T + (\det\mathbf{F})^{1/3}\mathbf{Q}(t)\dot{\mathbf{Q}}(t)^T\right)
\end{aligned} \tag{4.10.64}$$

By an argument similar to the one used in Section 4.9, we take $\mathbf{Q}(t)=\mathbf{I}$ and $\dot{\mathbf{Q}}(t) = -\mathbf{W}$ in (4.10.64). With (2.4.29) and (2.4.30), the result can be written

$$\mathbf{G}_{\mathbf{K}_X}\left(\theta,\mathbf{g},(\det\mathbf{F})^{1/3}\mathbf{I},(\det\mathbf{F})^{1/3}\mathbf{L}\right) = \mathbf{G}_{\mathbf{K}_X}\left(\theta,\mathbf{g},(\det\mathbf{F})^{1/3}\mathbf{I},(\det\mathbf{F})^{1/3}\mathbf{D}\right) \tag{4.10.65}$$

If this result is substituted back into (4.10.64), it follows that

$$\begin{aligned}
&\mathbf{Q}\mathbf{G}_{\mathbf{K}_X}\left(\theta,\mathbf{g},(\det\mathbf{F})^{1/3}\mathbf{I},(\det\mathbf{F})^{1/3}\mathbf{D}\right)\mathbf{Q}^T \\
&\quad= \mathbf{G}_{\mathbf{K}_X}\left(\theta,\mathbf{Q}\mathbf{g},(\det\mathbf{F})^{1/3}\mathbf{I},(\det\mathbf{F})^{1/3}\mathbf{Q}(t)\mathbf{D}\mathbf{Q}(t)^T\right)
\end{aligned} \tag{4.10.66}$$

If we define $\tilde{\mathbf{G}}_{\mathbf{K}_X}$ by

$$\tilde{\mathbf{G}}_{\mathbf{K}_X}(\theta,\mathbf{g},\rho,\mathbf{D}) = \mathbf{G}_{\mathbf{K}_X}\left(\theta,\mathbf{g},\rho_R^{1/3}\rho^{-1/3}\mathbf{I},\rho_R^{1/3}\rho^{-1/3}\mathbf{D}\right) \tag{4.10.67}$$

then (4.10.51) follows from (4.10.60), (4.10.65), and (4.10.67). Equation (4.10.53) follows from (4.10.66) and (4.10.67). The derivation of (4.10.52) and (4.10.54) follows by an identical procedure.

The function $\tilde{\mathbf{G}}_{\mathbf{K}_X}$ is an even function of $\mathbf{g}$ because of (4.10.22). Likewise, $\tilde{\mathbf{l}}_{\mathbf{K}_X}$ is an odd function of $\mathbf{g}$ because of (4.10.23). As with the case of an isotropic solid particle, for the fluid case the even and odd dependence on $\mathbf{g}$ follows from (4.10.53) and (4.10.54) if we use *all* $\mathbf{Q}$ in $\mathcal{O}(\mathcal{V})$. In order to verify this assertion simply take $\mathbf{Q} = -\mathbf{I}$ in (4.10.53) and (4.10.54).

Given (4.10.50), it follows from (4.7.15) that

$$\eta = \tilde{h}_{\mathbf{K}_X}(\theta, \rho) = -\frac{\partial \tilde{u}_{\mathbf{K}_X}(\theta, \rho)}{\partial \theta} \tag{4.10.68}$$

for a fluid particle X. Equation (4.10.68) is equivalent to (4.1.5). From (4.8.19) and (4.10.50), it follows that

$$\mathbf{T}^0 = \rho \mathbf{F} \frac{\partial u_{\mathbf{K}_X}(\theta, \mathbf{F})^T}{\partial \mathbf{F}} = \rho \frac{\partial \tilde{u}_{\mathbf{K}_X}(\theta, \rho)}{\partial \rho} \mathbf{F} \left(\frac{\partial \rho}{\partial \mathbf{F}} \right) \tag{4.10.69}$$

From (3.1.12) and (2.2.29),

$$\left(\frac{\partial \rho}{\partial \mathbf{F}} \right)^T = \rho_R \left(\frac{\partial (\det \mathbf{F})^{-1}}{\partial \mathbf{F}} \right)^T = \rho_R (\det \mathbf{F})^{-2} \left(\frac{\partial (\det \mathbf{F})}{\partial \mathbf{F}} \right)^T$$

$$= \rho_R (\det \mathbf{F})^{-1} \mathbf{F}^{-1} = -\rho \mathbf{F}^{-1} \tag{4.10.70}$$

With (4.10.70), (4.10.69) can be written

$$\mathbf{T}^0 = -\pi \mathbf{I} \tag{4.10.71}$$

where

$$\pi = \rho^2 \frac{\partial \tilde{u}_{\mathbf{K}_X}(\theta, \rho)}{\partial \rho} \tag{4.10.72}$$

is the hydrostatic pressure. Equation (4.10.72) is equivalent to (4.1.6). As in Section 4.4, equation (4.10.71) shows that the definition of a fluid particle given here is consistent with the intuitive idea that a fluid will not support a shear stress in equilibrium.

By a representation theorem first proven by Noll, (4.10.54) and the condition that $\tilde{\mathbf{l}}_{\mathbf{K}_X}$ be an odd function of $\mathbf{g}$ can be shown to yield

$$\mathbf{q} = \tilde{\mathbf{l}}_{\mathbf{K}_X}(\theta, \mathbf{g}, \rho, \mathbf{D}) = \left(\sigma_0 \mathbf{I} + \sigma_1 \mathbf{D} + \sigma_2 \mathbf{D}^2 \right) \mathbf{g} \tag{4.10.73}$$

where

$$\sigma_\tau = \sigma_\tau\left(\theta, I_\mathbf{D}, II_\mathbf{D}, III_\mathbf{D}, \mathbf{g} \cdot \mathbf{g}, \mathbf{g} \cdot (\mathbf{Dg}), \mathbf{g} \cdot (\mathbf{D}^2\mathbf{g})\right) \qquad (4.10.74)$$

for $\tau = 0, 1, 2$ [Ref. 29]. The last six arguments in (4.10.74) are the six *joint invariants* of $(\mathbf{g}, \mathbf{D})$.

Exercise 4.10.7

Show that (4.10.73), along with (4.10.74), forces $\tilde{\mathbf{l}}_{\mathbf{K}_X}$ to be an isotropic function.

In closing this section we shall indicate how one extends a symmetric property of a *particle* to that of a *body*. We have previously shown that if X and Y are materially isomorphic there exist local reference configurations $\mathbf{K}_X$ and $\mathbf{J}_Y$ such that

$$\mathscr{G}_{\mathbf{K}_X} = \mathscr{G}_{\mathbf{J}_Y} \qquad (4.10.75)$$

Definition. A body $\mathscr{B}$ is *materially uniform* if each pair of particles is materially isomorphic.

We shall assume that $\mathscr{B}$ is materially uniform in the remainder of these lectures. Therefore, for each particle X there corresponds a local reference configuration $\mathbf{K}_X$ used to establish the material isomorphism with any other particle. Following Noll, we define a function $\mathbf{K}$ on $\mathscr{B}$ by

$$\mathbf{K}(X) = \mathbf{K}_X \qquad (4.10.76)$$

for all X in $\mathscr{B}$, were $\mathbf{K}_X$ is the materially isomorphic local reference configuration for X [Ref. 23]. The function $\mathbf{K}$ is called a *uniform* local reference configuration. Actually, not all materially uniform bodies have a uniform local reference configuration as introduced here. However, in this work we shall not complicate the discussion by considering such bodies [Ref. 24].

Given that $\mathscr{B}$ is materially uniform, an equation such as (4.10.2) holds for all pairs of particles. Consequently, we can define a function $k_\mathbf{K}$ by

$$k_\mathbf{K}\left(\theta, \mathbf{g}, \mathbf{F}, \dot{\mathbf{F}}\right) = k_{\mathbf{K}_X}\left(\theta, \mathbf{g}, \mathbf{F}, \dot{\mathbf{F}}\right) \qquad (4.10.77)$$

for all $\left(\theta, \mathbf{g}, \mathbf{F}, \dot{\mathbf{F}}\right)$ in the domain of $k_{\mathbf{K}_X}$. Because of (4.10.2), the function $k_\mathbf{K}$ is *independent* of the particle. Thus, for a materially uniform body there is a *single* function $k_\mathbf{K}$ which determines the response of *every* particle in the body. With an obvious change of notation, the reader can easily use (4.10.77) to simplify the results in Sections 4.7 through 4.10 to the case of a materially uniform body. An immediate simplification, which follows from (4.10.75), is that for a materially uniform body in its uniform local reference configuration there is a *single symmetry group* for the body. We shall denote this group by $\mathscr{G}_{\mathbf{K}}$. It is called the *symmetry group* for $\mathscr{B}$ *relative* to $\mathbf{K}$. It should be clear to the reader that the definitions given earlier which involve the symmetry group of a particle X relative to a local reference configuration can be restated as definitions for a materially uniform $\mathscr{B}$ by replacing "local reference configuration $\mathbf{K}_X$" by "uniform local reference configuration $\mathbf{K}$", "particle X" by "body $\mathscr{B}$" and "particle" by "body".

Given a reference configuration $\hat{\boldsymbol{\kappa}}$ for $\mathscr{B}$, we can select a particle X and regard $\hat{\boldsymbol{\kappa}}$ as in $\hat{\mathbf{K}}_X$. For a different particle, $\hat{\boldsymbol{\kappa}}$ would be in a different local reference configuration. Given the particle X and the corresponding materially uniform local reference configuration $\mathbf{K}(X)$, we can find a $\boldsymbol{\kappa}$ in $\mathbf{K}(X)$ and define a linear transformation $\mathbf{P}_X$ in $\mathscr{L}(\mathscr{V};\mathscr{V})$ by [see (4.6.25) and (4.6.28)]

$$\mathbf{P}_X = \operatorname{GRAD}\left(\hat{\boldsymbol{\kappa}} \circ \boldsymbol{\kappa}^{-1}(\mathbf{X})\right) \tag{4.10.78}$$

We can carry out this construction for every X in $\mathscr{B}$. While a single $\hat{\boldsymbol{\kappa}}$ can be used for the body, we could be forced to find a different $\boldsymbol{\kappa}$ for each particle of the body. Without further assumptions, there need not be a *single* function $\boldsymbol{\kappa}$ such that $\boldsymbol{\kappa}$ is in $\mathbf{K}(X)$ *for all* X in $\mathscr{B}$. In any case, by carrying out the above construction for every X in $\mathscr{B}$ we can define a function $\mathbf{P}:\mathscr{B} \to \mathscr{L}(\mathscr{V};\mathscr{V})$ by

$$\mathbf{P}(X) = \mathbf{P}_X \tag{4.10.79}$$

for every X in $\mathscr{B}$. Since there does not necessarily exist a single function $\boldsymbol{\kappa}$ such that $\boldsymbol{\kappa}$ is in $\mathbf{K}(X)$ for all X in $\mathscr{B}$, we *cannot* combine (4.10.78) and (4.10.79) to obtain a *field equation*

$$\mathbf{P}\left(\boldsymbol{\kappa}^{-1}(\cdot)\right) = \operatorname{GRAD}\left(\hat{\boldsymbol{\kappa}} \circ \boldsymbol{\kappa}^{-1}(\cdot)\right) \tag{4.10.80}$$

Given the reference configuration $\hat{\boldsymbol{\kappa}}$, then from (4.6.27)

$$\hat{\mathbf{F}}(\cdot,t) = \operatorname{GRAD}\boldsymbol{\chi}_{\hat{\boldsymbol{\kappa}}}(\cdot,t) \tag{4.10.81}$$

This equation is a field equation holding for every $\hat{\mathbf{X}}$ in $\hat{\boldsymbol{\kappa}}(\mathscr{B})$. Given an X in $\mathscr{B}$ and a $\boldsymbol{\kappa}$ in $\mathbf{K}(X)$, we can use (4.6.26), (4.10.79) and (4.10.81) to write

$$\mathbf{F}(X,t) = \left(\operatorname{GRAD}\boldsymbol{\chi}_{\hat{\boldsymbol{\kappa}}}\left(\hat{\mathbf{X}},t\right)\right)\mathbf{P}(X) \tag{4.10.82}$$

By constructing (4.10.82) at every X in $\mathscr{B}$, we obtain a field equation

$$\mathbf{F}(\cdot,t)=\left[\mathrm{GRAD}\boldsymbol{\chi}_{\hat{\boldsymbol{\kappa}}}\left(\hat{\boldsymbol{\kappa}}(\cdot),t\right)\right]\mathbf{P}(\cdot) \tag{4.10.83}$$

Because (4.10.80) *does not* hold, we *cannot* write a field equation

$$\mathbf{F}(\cdot,t)=\mathrm{GRAD}\boldsymbol{\chi}_{\boldsymbol{\kappa}}\left(\boldsymbol{\kappa}(\cdot),t\right) \tag{4.10.84}$$

Thus, we see from (4.10.83) the important fact that for a materially uniform body, the quantity $\mathbf{F}$ in the argument of the response function $k_{\mathbf{K}}$ is generally not a gradient of a single function $\boldsymbol{\chi}_{\boldsymbol{\kappa}}$ defined on some subset of $\mathscr{E}$. This fact gives rise to theories of *continuous distributions of dislocations* [Refs. 22, 23].

Exercise 4.10.8

Consider an isotropic solid body in its undistorted uniform local reference configuration. By the argument above, the strain is measured by the tensor $\mathbf{B}$. Show that for each $\hat{\mathbf{X}}$ in $\hat{\boldsymbol{\kappa}}(\mathscr{B})$, the strain is measured by

$$\mathbf{B}(\hat{\mathbf{X}},t)=\mathrm{GRAD}\boldsymbol{\chi}_{\hat{\boldsymbol{\kappa}}}(\hat{\mathbf{X}},t)\mathbf{G}(\hat{\mathbf{X}})\mathrm{GRAD}\boldsymbol{\chi}_{\hat{\boldsymbol{\kappa}}}(\hat{\mathbf{X}},t)^{T} \tag{4.10.85}$$

where $\mathbf{G}(\hat{\mathbf{X}})$ is a positive definite symmetric linear transformation. How is $\mathbf{G}(\hat{\mathbf{X}})$ related to $\mathbf{P}(\cdot)$ in (4.10.83)?

Clearly, some formal simplification would arise if we could adopt (4.10.80) and thus (4.10.84). Whenever there exists a single reference configuration $\boldsymbol{\kappa}$ such that $\boldsymbol{\kappa}$ is in $\mathbf{K}(X)$ for all X in $\mathscr{B}$ the materially uniform body $\mathscr{B}$ is said to be *homogeneous*. Thus, it is for *homogeneous materially uniform* bodies that (4.10.84) holds. It follows from (4.10.77) and (4.10.84), that the constitutive equations for a materially uniform homogeneous body are

$$\begin{aligned}&\left(\psi(\cdot,t),\eta(\cdot,t),\mathbf{T}(\cdot,t),\mathbf{q}(\cdot,t)\right)\\&=k_{K}\left(\Theta_{\boldsymbol{\kappa}}(\cdot,t),\mathbf{g}(\cdot,t),\mathrm{GRAD}\boldsymbol{\chi}_{\boldsymbol{\kappa}}(\cdot,t),\overline{\mathrm{GRAD}\boldsymbol{\chi}_{\boldsymbol{\kappa}}(\cdot,t)}^{\boldsymbol{\cdot}}\right)\end{aligned} \tag{4.10.86}$$

Equation (4.10.86) holds for every $\mathbf{X}$ in $\boldsymbol{\kappa}(\mathscr{B})$. The *single function* k_{K} characterizes the response of each particle of the materially uniform, homogeneous, body.

This textbook is concerned with the introduction of a wide variety of concepts which arise in continuum mechanics. Not all of these concepts will be exploited here. In particular, in the remainder of this work we shall only be concerned with materially uniform homogeneous bodies.

4.11. Constitutive Equations for a Compressible, Conducting, Viscous Fluid

From the results of Section 4.10, for a fluid body we have

$$\psi = \tilde{u}_{\mathbf{K}}(\theta, \rho) \tag{4.11.1}$$

$$\eta = \tilde{h}_{\mathbf{K}}(\theta, \rho) = -\frac{\partial \tilde{u}_{\mathbf{K}}(\theta, \rho)}{\partial \theta} \tag{4.11.2}$$

$$\mathbf{T} = -\pi \mathbf{I} + \tilde{\mathbf{G}}^e_{\mathbf{K}}(\theta, \mathbf{g}, \rho, \mathbf{D}) \tag{4.11.3}$$

and

$$\mathbf{q} = \tilde{\mathbf{l}}_{\mathbf{K}}(\theta, \mathbf{g}, \rho, \mathbf{D}) \tag{4.11.4}$$

where $\tilde{\mathbf{G}}^e_{\mathbf{K}}$ and $\tilde{l}_{\mathbf{K}}$ are isotropic functions and π is given by (4.10.72). The defining properties for isotropic functions are

$$\mathbf{Q}\tilde{\mathbf{G}}^e_{\mathbf{K}}(\theta, \mathbf{g}, \rho, \mathbf{D})\mathbf{Q}^T = \tilde{\mathbf{G}}^e_{\mathbf{K}}(\theta, \mathbf{Q}\mathbf{g}, \rho, \mathbf{Q}\mathbf{D}\mathbf{Q}^T) \tag{4.11.5}$$

and

$$\mathbf{Q}\tilde{\mathbf{l}}_{\mathbf{K}}(\theta, \mathbf{g}, \rho, \mathbf{D}) = \tilde{\mathbf{l}}_{\mathbf{K}}(\theta, \mathbf{Q}\mathbf{g}, \rho, \mathbf{Q}\mathbf{D}\mathbf{Q}^T) \tag{4.11.6}$$

for all $\mathbf{Q}$ in $\mathscr{O}(\mathscr{V})^+$, and

$$\tilde{\mathbf{G}}^e_{\mathbf{K}}(\theta, \mathbf{g}, \rho, \mathbf{D}) = \tilde{\mathbf{G}}^e_{\mathbf{K}}(\theta, -\mathbf{g}, \rho, \mathbf{D}) \tag{4.11.7}$$

and

$$\tilde{\mathbf{l}}_{\mathbf{K}}(\theta, \mathbf{g}, \rho, \mathbf{D}) = -\tilde{\mathbf{l}}_{\mathbf{K}}(\theta, -\mathbf{g}, \rho, \mathbf{D}) \tag{4.11.8}$$

Because of (4.7.27), we must have

$$\operatorname{tr}\left(\tilde{\mathbf{G}}^e_{\mathbf{K}}(\theta, \mathbf{g}, \rho, \mathbf{D})\mathbf{D}\right) - \mathbf{g} \cdot \tilde{\mathbf{l}}_{\mathbf{K}}(\theta, \mathbf{g}, \rho, \mathbf{D})/\theta \geq 0 \tag{4.11.9}$$

From the results of Section 4.8 [see (4.8.14) and (4.8.15)], it follows from (4.11.9) that

$$\tilde{\mathbf{G}}^e_{\mathbf{K}}(\theta, \mathbf{0}, \rho, \mathbf{0}) = \mathbf{0} \tag{4.11.10}$$

and

$$\tilde{\mathbf{I}}_{\mathbf{K}}\left(\theta,\mathbf{0},\rho,\mathbf{0}\right)=\mathbf{0} \tag{4.11.11}$$

Of course, (4.11.11) is automatically satisfied by the stronger result

$$\tilde{\mathbf{I}}_{\mathbf{K}}\left(\theta,\mathbf{0},\rho,\mathbf{D}\right)=\mathbf{0} \tag{4.11.12}$$

which follows from (4.11.8) evaluated at $\mathbf{g}=\mathbf{0}$.

Given the results (4.11.1) through (4.11.12), in this section we wish to specialize the constitutive equations to the case where the departure from equilibrium is *small*. This argument will produce (4.1.3), (4.1.4), (4.1.7), (4.1.8), and (4.1.9). For one-dimensional materials, Section 1.9 contains a calculation similar to the one given here. If we define a nonnegative number α by

$$\alpha^2=\mathbf{g}\cdot\mathbf{g}+\operatorname{tr}\mathbf{D}^2 \tag{4.11.13}$$

it follows that $\alpha=0$ if and only if $\mathbf{D}=\mathbf{0}$ and $\mathbf{g}=\mathbf{0}$. Consequently, α measures the departure from the equilibrium state. It follows that

$$\begin{aligned}\tilde{\mathbf{G}}^{e}_{\mathbf{K}}\left(\theta,\mathbf{g},\rho,\mathbf{D}\right)&=\tilde{\mathbf{G}}^{e}_{\mathbf{K}}\left(\theta,\mathbf{0},\rho,\mathbf{0}\right)+\frac{\partial\tilde{\mathbf{G}}_{\mathbf{K}}\left(\theta,\mathbf{0},\rho,\mathbf{0}\right)}{\partial\mathbf{g}}[\mathbf{g}]\\&+\frac{\partial\tilde{\mathbf{G}}_{\mathbf{K}}\left(\theta,\mathbf{0},\rho,\mathbf{0}\right)}{\partial\mathbf{D}}[\mathbf{D}]+O\left(\alpha^2\right)\end{aligned} \tag{4.11.14}$$

and

$$\tilde{\mathbf{I}}_{\mathbf{K}}\left(\theta,\mathbf{g},\rho,\mathbf{D}\right)=\tilde{\mathbf{I}}_{\mathbf{K}}\left(\theta,\mathbf{0},\rho,\mathbf{0}\right)+\frac{\partial\tilde{\mathbf{I}}_{\mathbf{K}}\left(\theta,\mathbf{0},\rho,\mathbf{0}\right)}{\partial\mathbf{g}}\mathbf{g}+\frac{\partial\tilde{\mathbf{I}}_{\mathbf{K}}\left(\theta,\mathbf{0},\rho,\mathbf{0}\right)}{\partial\mathbf{D}}[\mathbf{D}]+O\left(\alpha^2\right) \tag{4.11.15}$$

From (4.11.10) and (4.11.11), the leading term in each expansion is zero. In addition, it follows from (4.11.7) that

$$\frac{\partial\tilde{\mathbf{G}}_{\mathbf{K}}\left(\theta,\mathbf{g},\rho,\mathbf{D}\right)}{\partial\mathbf{g}}=-\frac{\partial\tilde{\mathbf{G}}_{\mathbf{K}}\left(\theta,-\mathbf{g},\rho,\mathbf{D}\right)}{\partial\mathbf{g}} \tag{4.11.16}$$

and, from (4.11.12), that

$$\frac{\partial\tilde{\mathbf{I}}_{\mathbf{K}}\left(\theta,\mathbf{0},\rho,\mathbf{D}\right)}{\partial\mathbf{D}}=\mathbf{0} \tag{4.11.17}$$

These two equations imply that

$$\frac{\partial \tilde{\mathbf{G}}_{\mathbf{K}}(\theta,\mathbf{0},\rho,\mathbf{0})}{\partial \mathbf{g}}=0 \tag{4.11.18}$$

and

$$\frac{\partial \tilde{\mathbf{I}}_{\mathbf{K}}(\theta,\mathbf{0},\rho,\mathbf{0})}{\partial \mathbf{D}}=\mathbf{0} \tag{4.11.19}$$

Therefore, the expansions (4.11.14) and (4.11.15) simplify to

$$\tilde{\mathbf{G}}_{\mathbf{K}}^{e}(\theta,\mathbf{g},\rho,\mathbf{D})=\mathbf{\Lambda}(\theta,\rho)[\mathbf{D}]+O(\alpha^{2}) \tag{4.11.20}$$

and

$$\tilde{\mathbf{I}}_{\mathbf{K}}(\theta,\mathbf{g},\rho,\mathbf{D})=-\mathbf{K}(\theta,\rho)\mathbf{g}+O(\alpha^{2}) \tag{4.11.21}$$

where

$$\mathbf{\Lambda}(\theta,\rho)=\frac{\partial \tilde{\mathbf{G}}_{\mathbf{K}}(\theta,\mathbf{0},\rho,\mathbf{0})}{\partial \mathbf{D}} \tag{4.11.22}$$

and

$$\mathbf{K}(\theta,\rho)=-\frac{\partial \tilde{\mathbf{I}}_{\mathbf{K}}(\theta,\mathbf{0},\rho,\mathbf{0})}{\partial \mathbf{g}} \tag{4.11.23}$$

The coefficients $\mathbf{\Lambda}(\theta,\rho)$ and $\mathbf{K}(\theta,\rho)$ must also satisfy the restrictions implied by (4.11.5) and (4.11.6). It follows from (4.11.20) and (4.11.5) that $\mathbf{\Lambda}(\theta,\rho)$ must obey

$$\mathbf{Q}\left(\mathbf{\Lambda}(\theta,\rho)[\mathbf{D}]\right)\mathbf{Q}^{T}=\mathbf{\Lambda}(\theta,\rho)[\mathbf{Q}\mathbf{D}\mathbf{Q}^{T}] \tag{4.11.24}$$

for all $\mathbf{Q}$ in $\mathcal{O}(\mathscr{V})^{+}$ and all symmetric linear transformations $\mathbf{D}$. It follows from (4.11.6) and (4.11.21) that $\mathbf{K}(\theta,\rho)$ must obey

$$\mathbf{Q}\mathbf{K}(\theta,\rho)=\mathbf{K}(\theta,\rho)\mathbf{Q} \tag{4.11.25}$$

for all $\mathbf{Q}$ in $\mathscr{O}(\mathscr{V})^{+}$. Equation (4.11.24) shows that $\mathbf{\Lambda}(\theta,\rho)$ is a *fourth-order isotropic tensor.* Equation (4.11.25) shows that $\mathbf{K}(\theta,\rho)$ is a *second-order isotropic tensor.* As shown in Appendix B, $\mathbf{\Lambda}(\theta,\rho)[\mathbf{D}]$ takes the form

$$\mathbf{\Lambda}(\theta,\rho)[\mathbf{D}] = \lambda(\theta,\rho)(\operatorname{tr}\mathbf{D})\mathbf{I} + 2\mu(\theta,\rho)\mathbf{D} \tag{4.11.26}$$

and $\mathbf{K}(\theta,\rho)$ takes the form

$$\mathbf{K}(\theta,\rho) = \kappa(\theta,\rho)\mathbf{I} \tag{4.11.27}$$

Equations (4.11.26), (4.11.20), and (4.11.3) combine to yield

$$\mathbf{T} = -\pi\mathbf{I} + \lambda(\theta,\rho)(\operatorname{tr}\mathbf{D})\mathbf{I} + 2\mu(\theta,\rho)\mathbf{D} + O(\alpha^2) \tag{4.11.28}$$

Equations (4.11.27), (4.11.21) and (4.11.4) combine to yield

$$\mathbf{q} = -\kappa(\theta,\rho)\mathbf{g} + O(\alpha^2) \tag{4.11.29}$$

Clearly (4.1.3) follows from (4.11.28) by neglecting the terms $O(\alpha^2)$. Likewise, (4.1.4) follows from (4.11.29) by neglecting the terms $O(\alpha^2)$. If (4.11.28) and (4.11.29) are substituted into (4.11.9), the result can be written

$$\begin{aligned}\left(\lambda(\theta,\rho)+\tfrac{2}{3}\mu(\theta,\rho)\right)(\operatorname{tr}\mathbf{D})^2 + 2\mu(\theta,\rho)\operatorname{tr}\left(\mathbf{D}-\tfrac{1}{3}(\operatorname{tr}\mathbf{D})\mathbf{I}\right)^2 \\ + \frac{\kappa(\theta,\rho)}{\theta}\mathbf{g}\cdot\mathbf{g} + O(\alpha^3) \geq 0\end{aligned} \tag{4.11.30}$$

In the same way that (4.8.2) implies (4.8.22), (4.11.30) yields (4.1.7), (4.1.8), and (4.1.9). Because of the terms $O(\alpha^3)$ in (4.11.30), (4.1.7), (4.1.8), and (4.1.9) are *not* equivalent to (4.11.30). However, if the linearized theory happens to be *exact*, the terms $O(\alpha^2)$ in (4.11.28) and (4.11.29) do not appear. In this case, we obtain

$$\begin{aligned}\left(\lambda(\theta,\rho)+\tfrac{2}{3}\mu(\theta,\rho)\right)(\operatorname{tr}\mathbf{D})^2 + 2\mu(\theta,\rho)\operatorname{tr}\left(\mathbf{D}-\tfrac{1}{3}(\operatorname{tr}\mathbf{D})\mathbf{I}\right)^2 \\ + \frac{\kappa(\theta,\rho)}{\theta}\mathbf{g}\cdot\mathbf{g} \geq 0\end{aligned} \tag{4.11.31}$$

instead of (4.11.30). Equation (4.11.31) is equivalent to (4.1.7), (4.1.8), and (4.1.9) because each term in (4.11.31) is algebraically independent of the others.

Exercise 4.11.1

Neglect $O(\alpha^2)$ terms in (4.11.28) and (4.11.29) and derive (4.1.18) and (4.1.20).

Exercise 4.11.2

Given (4.11.1), (4.11.2), and (4.10.72), change variables from (θ,ρ) to (η,ρ) and show that

$$\varepsilon = \varepsilon(\eta,\rho) \tag{4.11.32}$$

$$\theta = \frac{\partial \varepsilon(\eta,\rho)}{\partial \eta} \tag{4.11.33}$$

and

$$\pi = \rho^2 \frac{\partial \varepsilon(\eta,\rho)}{\partial \rho} \tag{4.11.34}$$

where, for notational simplicity, the function whose value is ε has also been denoted by ε.

Exercise 4.11.3

The enthalpy density χ and the Gibbs function ζ are defined by

$$\chi = \varepsilon + \frac{\pi}{\rho} \tag{4.11.35}$$

and

$$\zeta = \psi + \frac{\pi}{\rho} \tag{4.11.36}$$

respectively. Show that

$$\chi = \chi(\eta,\pi) \tag{4.11.37}$$

$$\theta = \frac{\partial \chi(\eta,\pi)}{\partial \eta} \tag{4.11.38}$$

$$\frac{1}{\rho} = \frac{\partial \chi(\eta,\pi)}{\partial \pi} \tag{4.11.39}$$

$$\zeta = \zeta(\theta, \pi) \tag{4.11.40}$$

$$\eta = -\frac{\partial \zeta(\theta, \pi)}{\partial \theta} \tag{4.11.41}$$

and

$$\frac{1}{\rho} = \frac{\partial \zeta(\theta, \pi)}{\partial \pi} \tag{4.11.42}$$

Exercise 4.11.4

Show that

$$\frac{\partial \pi(\eta, \rho)}{\partial \eta} = \rho^2 \frac{\partial \theta(\eta, \rho)}{\partial \rho} \tag{4.11.43}$$

$$\frac{\partial \theta(\eta, \pi)}{\partial \pi} = \frac{\partial \frac{1}{\rho}(\eta, \pi)}{\partial \eta} \tag{4.11.44}$$

$$\frac{\partial \pi(\theta, \rho)}{\partial \theta} = -\rho^2 \frac{\partial \eta(\theta, \rho)}{\partial \rho} \tag{4.11.45}$$

and

$$\frac{\partial \eta(\theta, \pi)}{\partial \pi} = -\frac{\partial \frac{1}{\rho}(\theta, \pi)}{\partial \theta} \tag{4.11.46}$$

Equations (4.11.43) through (4.11.46) are examples of *Maxwell* relations in thermodynamics.

Exercise 4.11.5

By suitable changes of variables one can obtain

$$\varepsilon = \varepsilon(\theta, \rho) \tag{4.11.47}$$

and

$$\chi = \chi(\theta, \pi) \tag{4.11.48}$$

Show that

$$\rho^2 \frac{\partial \varepsilon(\theta,\rho)}{\partial \rho} = \pi - \theta \frac{\partial \pi(\theta,\rho)}{\partial \theta} \tag{4.11.49}$$

and

$$\frac{\partial \chi(\theta,\pi)}{\partial \pi} = \frac{1}{\rho} - \theta \frac{\partial \frac{1}{\rho}(\theta,\pi)}{\partial \theta} \tag{4.11.50}$$

Exercise 4.11.6

The specific heat at constant volume, c_v, and the specific heat at constant pressure, c_p, are defined by

$$c_v = \frac{\partial \varepsilon(\theta,\rho)}{\partial \theta} \tag{4.11.51}$$

and

$$c_p = \frac{\partial \chi(\theta,\pi)}{\partial \theta} \tag{4.11.52}$$

respectively. Show that

$$c_v = \theta \frac{\partial \eta(\theta,\rho)}{\partial \theta} = -\theta \frac{\partial^2 \psi(\theta,\rho)}{\partial \theta^2} \tag{4.11.53}$$

$$c_p = \theta \frac{\partial \eta(\theta,\pi)}{\partial \theta} = -\theta \frac{\partial^2 \zeta(\theta,\pi)}{\partial \theta^2} \tag{4.11.54}$$

$$\frac{\partial c_v(\theta,\rho)}{\partial \rho} = -\frac{\theta}{\rho^2} \frac{\partial^2 \pi(\theta,\rho)}{\partial \theta^2} \tag{4.11.55}$$

$$\frac{\partial c_p(\theta,\pi)}{\partial \pi} = -\theta \frac{\partial^2 \frac{1}{\rho}(\theta,\pi)}{\partial \theta^2} \tag{4.11.56}$$

$$c_p - c_v = \theta \frac{\partial \frac{1}{\rho}(\theta,\pi)}{\partial \theta} \frac{\partial \pi(\theta,\rho)}{\partial \theta} \tag{4.11.57}$$

and

$$\frac{\partial\pi(\eta,\rho)}{\partial\rho}=\frac{c_p}{c_v}\frac{\partial\pi(\theta,\rho)}{\partial\rho} \tag{4.11.58}$$

Equations (4.11.49), (4.11.50) and (4.11.55) through (4.11.58) are useful in thermodynamics when one is given a *thermal equation of state* $\pi(\theta,\rho)$.

In the next collection of exercises the fluid is assumed to be nonviscous and a nonconductor. Therefore, it is to be assumed that

$$\lambda=\mu=\kappa=0 \tag{4.11.59}$$

These assumptions reduce (4.1.18) and (4.1.20) to

$$\rho\ddot{\mathbf{x}}=-\operatorname{grad}\pi \tag{4.11.60}$$

and

$$\rho\theta\dot{\eta}=0 \tag{4.11.61}$$

where, in addition, we have taken $\mathbf{b}$ and r to be zero. Equation (4.11.61) shows that the motion is *isentropic*, i.e., the entropy density does not change for a particle along its path. However, $\dot{\eta}=0$ implies $\eta=\eta(\mathbf{X})$ and entropy changes from particle to particle are allowed. If, in fact, the entropy density is constant, then the motion is said to be *homentropic*.

Exercise 4.11.7

Show that (4.11.60) can be written

$$\ddot{\mathbf{x}}=-\operatorname{grad}\chi+\theta\operatorname{grad}\eta \tag{4.11.62}$$

Exercise 4.11.8

Use (4.11.60), (4.11.61) and any necessary thermodynamic identities to show that

$$\overline{\left(\chi+\tfrac{1}{2}\dot{\mathbf{x}}^2\right)}^{\cdot}=\frac{1}{\rho}\frac{\partial\pi}{\partial t} \tag{4.11.63}$$

The combination $\chi+\frac{1}{2}\dot{\mathbf{x}}^2$ is called the *stagnation enthalpy*. Equation (4.11.63) shows that it is a constant along particle paths in steady flow problems.

Exercise 4.11.9

Show that an alternate form of (4.11.62) is

$$\frac{\partial \dot{\mathbf{x}}}{\partial t} + \operatorname{grad}\left(\chi + \tfrac{1}{2}\dot{\mathbf{x}}^2\right) + 2\mathbf{W}\dot{\mathbf{x}} = \theta \operatorname{grad} \eta \tag{4.11.64}$$

For steady motion, (4.11.64) is known as the *Crocco-Vazsonyi* theorem. In gasdynamics, this theorem is used to establish that an irrotational flow $(\mathbf{W} = \mathbf{0})$ does not remain irrotational when it passes through a curved shock wave.

Exercise 4.11.10

In Appendix A, the *curl* of a vector field was not defined. If $\mathbf{g}(\mathbf{x})$ is a vector field, its *curl* is a vector field defined by

$$\left(\operatorname{curl} \mathbf{g}(\mathbf{x})\right) \times \mathbf{u} = \left(\operatorname{grad} \mathbf{g}(\mathbf{x})\right)\mathbf{u} - \left(\operatorname{grad} \mathbf{g}(\mathbf{x})\right)^T \mathbf{u} \tag{4.11.65}$$

for all $\mathbf{u}$ in $\mathscr{V}$. Equation (4.11.65) yields the following component formula:

$$\operatorname{curl} \mathbf{g}(\mathbf{x}) = -\varepsilon_{jkq} \frac{\partial g_k(\mathbf{x})}{\partial x_q} \mathbf{i}_j \tag{4.11.66}$$

For example, the definition (4.11.65) allows us to write the term $2\mathbf{W}\dot{\mathbf{x}}$ in (4.11.64) in the form

$$\zeta \times \dot{\mathbf{x}} = 2\mathbf{W}\dot{\mathbf{x}} \tag{4.11.67}$$

where ζ is the *vorticity vector* defined by

$$\zeta = \operatorname{curl} \dot{\mathbf{x}}(\mathbf{x}, t) \tag{4.11.68}$$

Utilize the definition of the curl and show that

$$\operatorname{curl} \ddot{\mathbf{x}} = \operatorname{grad} \theta \times \operatorname{grad} \eta = -\operatorname{grad} \frac{1}{\rho} \times \operatorname{grad} \pi \tag{4.11.69}$$

Exercise 4.11.11

Use (2.2.12) and (3.1.14) and show that

$$\operatorname{curl} \ddot{\mathbf{x}} = \rho \overline{\left(\frac{\zeta}{\rho}\right)}^{\boldsymbol{\cdot}} - \mathbf{L}\zeta \tag{4.11.70}$$

and, as a result,

$$\rho\overline{\left(\mathbf{F}^{-1}\frac{\boldsymbol{\zeta}}{\rho}\right)}^{\cdot}=\mathbf{F}^{-1}\operatorname{curl}\ddot{\mathbf{x}} \tag{4.11.71}$$

Equation (4.11.70) is known as *Beltrami's* equation. When combined with (4.11.69) it yields the vorticity equation of *Vazsonyi*. For a homentropic flow, (4.11.69) and (4.11.71) yield

$$\mathbf{F}^{-1}\frac{\boldsymbol{\zeta}}{\rho}=\frac{\boldsymbol{\zeta}_{\mathbf{R}}}{\rho_{\mathbf{R}}} \tag{4.11.72}$$

where $\boldsymbol{\zeta}_{\mathrm{R}}$ is the vorticity in the reference configuration. If $\boldsymbol{\zeta}_{\mathbf{R}}=\mathbf{0}$, then the resulting flow is irrotational.

Exercise 4.11.12

If the velocity field $\dot{\mathbf{x}}$ is given by

$$\dot{\mathbf{x}}(\mathbf{x},t)=\operatorname{grad}\phi(\mathbf{x},t) \tag{4.11.73}$$

where ϕ is the *velocity potential*, then an elementary calculation shows that $\boldsymbol{\zeta}(\mathbf{x},t)=\mathbf{0}$. Given (4.11.73), assume the flow is homentropic and derive the following compressible *Bernoulli* equation:

$$\frac{\partial\phi}{\partial t}+\chi+\tfrac{1}{2}\dot{\mathbf{x}}^2=C \tag{4.11.74}$$

where C is a constant.

Exercise 4.11.13

Show that the velocity potential ϕ is a solution of

$$a^2\Delta\phi=\frac{\partial^2\phi}{\partial t^2}+2\operatorname{grad}\phi\cdot\operatorname{grad}\frac{\partial\phi}{\partial t}+\operatorname{grad}\phi\cdot\left\{\left(\operatorname{grad}\left(\operatorname{grad}\phi\right)\right)\operatorname{grad}\phi\right\} \tag{4.11.75}$$

where a^2 is the squared speed of sound defined by

$$a^2=\frac{\partial\pi(\eta,\rho)}{\partial\rho} \tag{4.11.76}$$

4.12. Constitutive Equations for an Isotropic Linear Thermoelastic Solid with Heat Conduction

If we omit the dependence on $\dot{\mathbf{F}}$ in the constitutive assumption (4.6.34), then from the results of Exercise 4.7.1 and from those of Section 4.10, it follows that for an isotropic solid body in its undistorted uniform local reference configuration we have

$$\psi = \hat{u}_{\mathbf{K}}(\theta,\mathbf{B}) \tag{4.12.1}$$

$$\eta = \bar{h}_{\mathbf{K}}(\theta,\mathbf{B}) = -\frac{\partial \hat{u}_{\mathbf{K}}(\theta,\mathbf{B})}{\partial \theta} \tag{4.12.2}$$

$$\mathbf{T} = \bar{\mathbf{G}}_{\mathbf{K}}(\theta,\mathbf{B}) = 2\rho\mathbf{B}\frac{\partial \hat{u}_{\mathbf{K}}(\theta,\mathbf{B})}{\partial \mathbf{B}} \tag{4.12.3}$$

and

$$\mathbf{q} = \bar{\mathbf{l}}_{\mathbf{K}}(\theta,\mathbf{g},\mathbf{B}) \tag{4.12.4}$$

where

$$\hat{u}_{\mathbf{K}}(\theta,\mathbf{B}) = \hat{u}_{\mathbf{K}}(\theta,\mathbf{QBQ}^T) \tag{4.12.5}$$

and

$$\mathbf{Q}\bar{\mathbf{l}}_{\mathbf{K}}(\theta,\mathbf{g},\mathbf{B}) = \bar{\mathbf{l}}_{\mathbf{K}}(\theta,\mathbf{Qg},\mathbf{QBQ}^T) \tag{4.12.6}$$

for all $\mathbf{Q}$ in $\mathcal{O}(\mathcal{V})^{+}$. In addition, $\bar{\mathbf{l}}_{\mathbf{K}}$ is an odd function of $\mathbf{g}$. For the case under discussion, the condition which corresponds to (4.10.31) is automatically satisfied because of (4.12.5) and (4.12.3). The proof of this assertion is essentially the same as the proof that (4.3.24) implies (4.3.25). From (4.7.21), the function $\bar{\mathbf{l}}_{\mathbf{K}}$ in (4.12.4) is restricted by

$$-\mathbf{g}\cdot\bar{\mathbf{l}}_{\mathbf{K}}(\theta,\mathbf{g},\mathbf{B})/\theta \geq 0 \tag{4.12.7}$$

In the case under discussion (4.8.14) is equivalent to

$$\bar{\mathbf{l}}_{\mathbf{K}}(\theta,\mathbf{0},\mathbf{B}) = \mathbf{0} \tag{4.12.8}$$

Equation (4.12.8) is also implied by the fact that $\bar{\mathbf{l}}_{\mathbf{K}}$ is an odd function of $\mathbf{g}$.

Given the results (4.12.1) through (4.12.7), in this section we wish to specialize the constitutive equations to the case where the displacement gradients, temperature gradients, and temperature variations are small. This argument will produce (4.1.25) through (4.1.29). In carrying out this derivation, it is algebraically more convenient to utilize the function $\hat{u}_{\mathbf{K}}$ in (4.12.1) rather than the function $u^*_{\mathbf{K}}$ in (4.10.43). We shall carry out the specialization in two distinct steps. First we define a positive number α by

$$\alpha^2 = \tfrac{1}{4}\operatorname{tr}(\mathbf{B}-\mathbf{I})^2 + \mathbf{g}\cdot\mathbf{g} + (\theta-\theta_0)^2 \tag{4.12.9}$$

where θ_0 is a constant. From its definition, $\alpha = 0$ if and only if $\mathbf{B} = \mathbf{I}$, $\mathbf{g} = \mathbf{0}$ and $\theta = \theta_0$. As a first step, we shall derive approximate formulas for ψ, $\mathbf{T}$, and $\mathbf{q}$ valid for small α. The next step involves regarding $\mathbf{B}$ to be a function of $\mathbf{H}$ [see (2.4.14)] and then deducing an approximate relation for $\mathbf{B}$ in terms of $\mathbf{H}$. The resulting approximate equations will be (4.1.25) through (4.1.29).

By expansion of (4.12.1) about the state $\mathbf{B} = \mathbf{I}$ and $\theta = \theta_0$, it follows that

$$\begin{aligned}\psi &= \hat{u}_{\mathbf{K}}(\theta_0,\mathbf{I}) + \frac{\partial \hat{u}_{\mathbf{K}}(\theta_0,\mathbf{I})}{\partial\theta}(\theta-\theta_0) + \operatorname{tr}\left(\frac{\partial \hat{u}_{\mathbf{K}}(\theta_0,\mathbf{I})}{\partial\mathbf{B}}(\mathbf{B}-\mathbf{I})\right) + \frac{1}{2}\frac{\partial^2 \hat{u}_{\mathbf{K}}(\theta_0,\mathbf{I})}{\partial\theta^2}(\theta-\theta_0)^2 \\ &+ \operatorname{tr}\left(\frac{\partial^2 \hat{u}_{\mathbf{K}}(\theta_0,\mathbf{I})}{\partial\theta\partial\mathbf{B}}(\mathbf{B}-\mathbf{I})\right)(\theta-\theta_0) + \frac{1}{2}\operatorname{tr}\left(\left(\frac{\partial^2 \hat{u}_{\mathbf{K}}(\theta_0,\mathbf{I})}{\partial\mathbf{B}^2}[\mathbf{B}-\mathbf{I}]\right)(\mathbf{B}-\mathbf{I})\right) + O(\alpha^3)\end{aligned} \tag{4.12.10}$$

From (4.12.5), it follows that the tensor coefficients in (4.12.10) must be isotropic. It then follows from the same arguments which produced (4.11.26) and (4.11.27) that the tensor coefficients have the representations

$$2\rho_R \frac{\partial \hat{u}_{\mathbf{K}}(\theta_0,\mathbf{I})}{\partial\mathbf{B}} = -p\mathbf{I} \tag{4.12.11}$$

$$2\rho_R \frac{\partial \hat{u}_{\mathbf{K}}(\theta_0,\mathbf{I})}{\partial\theta\,\partial\mathbf{B}} = -\beta\mathbf{I} \tag{4.12.12}$$

and

$$4\rho_R \frac{\partial^2 \hat{u}_{\mathbf{K}}(\theta_0,\mathbf{I})}{\partial\mathbf{B}^2}[\mathbf{B}-\mathbf{I}] = \lambda\left(\operatorname{tr}(\mathbf{B}-\mathbf{I})\right)\mathbf{I} + 2\mu(\mathbf{B}-\mathbf{I}) \tag{4.12.13}$$

where p, β, λ, and μ are constants. In order to simplify the notation somewhat, we also define constants η_0 and c_v by

$$\eta_0 = -\frac{\partial \hat{u}_{\mathbf{K}}(\theta_0, \mathbf{I})}{\partial \theta} \tag{4.12.14}$$

and

$$c_v = -\theta_0 \frac{\partial^2 \hat{u}_{\mathbf{K}}(\theta_0, \mathbf{I})}{\partial \theta^2} \tag{4.12.15}$$

Given (4.12.11) through (4.12.15), we can rewrite (4.12.10) as

$$\begin{aligned} \rho_R \psi = {} & \rho_R \hat{u}_{\mathbf{K}}(\theta_0, \mathbf{I}) - \rho_R \eta_0 (\theta - \theta_0) - \tfrac{1}{2} p \operatorname{tr}(\mathbf{B} - \mathbf{I}) - \frac{\rho_R c_v}{2\theta_0}(\theta - \theta_0)^2 \\ & - \tfrac{1}{2}\beta \operatorname{tr}(\mathbf{B} - \mathbf{I})(\theta - \theta_0) + \frac{\lambda}{2}\left(\operatorname{tr}\tfrac{1}{2}(\mathbf{B} - \mathbf{I})\right)^2 + \tfrac{1}{4}\mu \operatorname{tr}\left((\mathbf{B} - \mathbf{I})^2\right) + O(\alpha^3) \end{aligned} \tag{4.12.16}$$

From equations (4.12.16) and (4.12.2),

$$\rho_R \eta = \rho_R \eta_0 + \rho_R c_v \left(\frac{\theta - \theta_0}{\theta_0} \right) + \tfrac{1}{2}\beta \operatorname{tr}(\mathbf{B} - \mathbf{I}) + O(\alpha^2) \tag{4.12.17}$$

It also follows from (4.12.16) that

$$\rho_R \frac{\partial \hat{u}_{\mathbf{K}}(\theta, \mathbf{B})}{\partial \mathbf{B}} = -\tfrac{1}{2} p \mathbf{I} - \tfrac{1}{2}\beta(\theta - \theta_0)\mathbf{I} + \tfrac{1}{4}\lambda\left(\operatorname{tr}(\mathbf{B} - \mathbf{I})\right)\mathbf{I} + \tfrac{1}{2}\mu(\mathbf{B} - \mathbf{I}) + O(\alpha^2) \tag{4.12.18}$$

By use of (3.1.12) and (2.4.2), equation (4.12.3) can be written

$$\mathbf{T} = 2\rho_R (\det \mathbf{B})^{1/2} \mathbf{B} \frac{\partial \hat{u}_{\mathbf{K}}(\theta, \mathbf{B})}{\partial \mathbf{B}} \tag{4.12.19}$$

Exercise 4.12.1

Show that

$$(\det \mathbf{B})^{1/2} = 1 - \tfrac{1}{2} \operatorname{tr}(\mathbf{B} - \mathbf{I}) + O(\alpha^2) \tag{4.12.20}$$

If (4.12.18) and (4.12.20) are substituted into (4.12.19), the result is

$$\mathbf{T} = -p\mathbf{I} + \tfrac{1}{2}(\lambda + p)\left(\operatorname{tr}(\mathbf{B} - \mathbf{I})\right)\mathbf{I} + (\mu - p)(\mathbf{B} - \mathbf{I}) - \beta(\theta - \theta_0)\mathbf{I} + O(\alpha^2) \tag{4.12.21}$$

The isothermal version of (4.12.21) is the earlier result (4.4.52). Equation (4.12.17) shows that η_0 is the entropy density in the state $\mathbf{B} = \mathbf{I}$ and $\theta = \theta_0$. Equation (4.12.21) shows that p is the pressure in the state $\mathbf{B} = \mathbf{I}$ and $\theta = \theta_0$.

From (4.12.4) and (4.12.8), it follows that

$$\mathbf{q} = \frac{\partial \bar{\mathbf{l}}_{\mathbf{K}} (\theta_0, \mathbf{0}, \mathbf{I})}{\partial \mathbf{g}} \mathbf{g} + O(\alpha^2) \tag{4.12.22}$$

Because of (4.12.6), the tensor $\partial \bar{\mathbf{l}}_{\mathbf{K}} (\theta_0, \mathbf{0}, \mathbf{I}) / \partial \mathbf{g}$ is a second-order isotropic tensor. It then follows that

$$\frac{\partial \bar{\mathbf{l}}_{\mathbf{K}} (\theta_0, \mathbf{0}, \mathbf{I})}{\partial \mathbf{g}} = -\kappa \mathbf{I} \tag{4.12.23}$$

Therefore, (4.12.22) can be rewritten

$$\mathbf{q} = -\kappa \mathbf{g} + O(\alpha^2) \tag{4.12.24}$$

If (4.12.24) is substituted into (4.12.7), the result can be written

$$\frac{\kappa}{\theta_0} \mathbf{g} \cdot \mathbf{g} + O(\alpha^3) \geq 0 \tag{4.12.25}$$

This result implies

$$\kappa \geq 0 \tag{4.12.26}$$

However, (4.12.26) does not imply (4.12.25) unless the term $O(\alpha^2)$ in (4.12.24) is omitted.

If the terms $O(\alpha^2)$ in (4.12.17), (4.12.21), and (4.12.24) are omitted, the resulting equations correspond to a theory of thermoelasticity which is linear in $\theta - \theta_0$ and $\mathbf{B} - \mathbf{I}$. From (2.4.14), $\mathbf{B} - \mathbf{I}$ is a *nonlinear* function of $\mathbf{H}$. In order to bring equations (4.12.16), (4.12.17), (4.12.21), and (4.12.24) closer to (4.1.25) through (4.1.28), we shall now define a positive number σ by

$$\sigma^2 = \operatorname{tr} \mathbf{HH}^T + \mathbf{g} \cdot \mathbf{g} + (\theta - \theta_0)^2 \tag{4.12.27}$$

If $\sigma = 0$, it is necessary and sufficient that $\mathbf{g} = \mathbf{0}$, $\theta = \theta_0$, and $\mathbf{H} = \mathbf{0}$. Since $\mathbf{H}$ has the representation (2.5.17), $\mathbf{H} = \mathbf{0}$ implies $\tilde{\mathbf{E}} = \tilde{\mathbf{R}} = \mathbf{0}$. Thus, a small σ corresponds to small infinitesimal strains, small infinitesimal rotations, small temperature gradients, and small temperature variations. It follows from (2.5.11) that

$$\mathbf{B} = \mathbf{I} + 2\tilde{\mathbf{E}} + O(\sigma^2) \tag{4.12.28}$$

It follows from (4.12.28), (4.12.27), and (4.12.9) that

$$\alpha^2 = O(\sigma^2) \tag{4.12.29}$$

and, as a result,

$$O(\alpha^2) = O(\sigma^2) \tag{4.12.30}$$

If we use (4.12.28) and (4.12.30), equations (4.12.16), (4.12.17), (4.12.21), (4.12.24), and (4.12.25) become

$$\begin{aligned}\rho_R\psi = {}& \rho_R\hat{u}_{\mathbf{K}}(\theta_0,\mathbf{I}) - \rho_R\eta_0(\theta-\theta_0) - p\operatorname{tr}\tilde{\mathbf{E}} - \frac{\rho_R c_v}{2\theta_0}(\theta-\theta_0)^2 \\ & -\beta(\operatorname{tr}\tilde{\mathbf{E}})(\theta-\theta_0) + \frac{\lambda}{2}(\operatorname{tr}\tilde{\mathbf{E}})^2 + \mu\operatorname{tr}(\tilde{\mathbf{E}})^2 + O(\sigma^3)\end{aligned} \tag{4.12.31}$$

$$\rho_R\eta = \rho_R\eta_0 + \rho_R c_v\left(\frac{(\theta-\theta_0)}{\theta_0}\right) + \beta\operatorname{tr}\tilde{\mathbf{E}} + O(\sigma^2) \tag{4.12.32}$$

$$\mathbf{T} = -p\mathbf{I} + (\lambda+p)(\operatorname{tr}\tilde{\mathbf{E}})\mathbf{I} + 2(\mu-p)\tilde{\mathbf{E}} - \beta(\theta-\theta_0)\mathbf{I} + O(\sigma^2) \tag{4.12.33}$$

$$\mathbf{q} = -\kappa\mathbf{g} + O(\sigma^2) \tag{4.12.34}$$

and

$$\frac{\kappa}{\theta_0}\mathbf{g}\cdot\mathbf{g} + O(\sigma^3) \geq 0 \tag{4.12.35}$$

Given (2.2.21), (2.5.18), (3.2.27), (3.4.18) and (4.6.14) it follows that

$$\mathbf{T}_R = \mathbf{T} + p\left[\mathbf{H}^T - (\operatorname{tr}\tilde{\mathbf{E}})\mathbf{I}\right] + O(\sigma^2) \tag{4.12.36}$$

$$\mathbf{q}_R = \mathbf{q} + O(\sigma^2) \tag{4.12.37}$$

and

$$\text{GRAD}\,\theta = \mathbf{g} + O\left(\sigma^2\right) \tag{4.12.38}$$

Therefore,

$$\mathbf{T}_R = -p\left(\mathbf{I} - \mathbf{H}^T\right) + \lambda\left(\text{tr}\,\tilde{\mathbf{E}}\right)\mathbf{I} + 2(\mu - p)\tilde{\mathbf{E}} - \beta(\theta - \theta_0)\mathbf{I} + O\left(\sigma^2\right) \tag{4.12.39}$$

and

$$\mathbf{q}_R = -\kappa\,\text{GRAD}\,\theta + O\left(\sigma^2\right) \tag{4.12.40}$$

Equations (4.1.25) through (4.1.28) follow from (4.12.31), (4.12.32), (4.12.39), and (4.12.40) by omitting the terms $O\left(\sigma^2\right)$, by selecting the datum for the free energy density to be such that $\hat{u}_{\mathbf{K}_X}\left(\theta_0, \mathbf{I}\right) = 0$, by selecting the datum for the entropy density to be such that $\eta_0 = 0$ and, finally, by assuming the undistorted state is a *natural state*, i. e. $p = 0$.

Exercise 4.12.2

Use (4.10.44) and show that

$$-p = \alpha_0\left(\theta_0, 3, 3, 1\right) + \alpha_1\left(\theta_0, 3, 3, 1\right) + \alpha_2\left(\theta_0, 3, 3, 1\right) \tag{4.12.41}$$

$$\lambda + p = 2\left(\frac{\partial}{\partial I_{\mathbf{B}}} + \frac{\partial}{2\partial II_{\mathbf{B}}} + \frac{\partial}{\partial III_{\mathbf{B}}}\right)\left(\alpha_0 + \alpha_1 + \alpha_2\right)\Bigg|_{\substack{\theta=\theta_0, I_{\mathbf{B}}=3\\ II_{\mathbf{B}}=3, III_{\mathbf{B}}=1}} \tag{4.12.42}$$

$$\mu - p = \alpha_1\left(\theta_0, 3, 3, 1\right) + 2\alpha_2\left(\theta_0, 3, 3, 1\right) \tag{4.12.43}$$

and

$$-\beta = \frac{\partial}{\partial \theta}\left(\alpha_0 + \alpha_1 + \alpha_2\right)\Bigg|_{\substack{\theta=\theta_0, I_{\mathbf{B}}=3\\ II_{\mathbf{B}}=3, III_{\mathbf{B}}=1}} \tag{4.12.44}$$

Exercise 4.12.3

Show that the linearized equations of motion and energy take the forms

$$\rho_R \ddot{\mathbf{w}} = (\lambda + \mu)\text{GRAD}\left(\text{Div}\,\mathbf{w}\right) + (\mu - p)\Delta\mathbf{w} - \beta\,\text{GRAD}\,\theta \tag{4.12.45}$$

and

$$\rho_R c_v \dot{\theta} + \beta\theta_0 \operatorname{Div}\dot{\mathbf{w}} = \kappa\Delta\theta \tag{4.12.46}$$

Exercise 4.12.4

Show that

$$\begin{aligned}\rho_R\left(\psi+\eta\left(\theta+\theta_0\right)\right) &= \rho_R\hat{u}_{\mathbf{K}}\left(\theta_0,\mathbf{I}\right) - p\operatorname{tr}\tilde{\mathbf{E}} + \frac{\rho_R c_v}{2\theta_0}\left(\theta-\theta_0\right)^2 + \frac{\lambda}{2}\left(\operatorname{tr}\tilde{\mathbf{E}}\right)^2 \\ &\quad + \mu\operatorname{tr}\left(\tilde{\mathbf{E}}\right)^2 + O\left(\sigma^3\right)\end{aligned} \tag{4.12.47}$$

Exercise 4.12.5

Use the result (4.12.47) and show that (3.5.17) reduces to

$$\frac{\partial}{\partial t}\int_{\kappa(\mathscr{P})}\left(\Sigma+\tfrac{1}{2}\dot{\mathbf{w}}^2\right)dV \le \oint_{\partial\kappa(\mathscr{P})}\dot{\mathbf{w}}\cdot\left(\left(\mathbf{T}_R+p\mathbf{I}\right)d\mathbf{S}\right) - \oint_{\partial\kappa(\mathscr{P})}\frac{\theta-\theta_0}{\theta_0}\mathbf{q}_R\cdot d\mathbf{S} \tag{4.12.48}$$

where

$$\Sigma = \frac{\rho_R c_v}{2\theta_0}\left(\theta-\theta_0\right)^2 + \frac{\lambda}{2}\left(\operatorname{tr}\tilde{\mathbf{E}}\right)^2 + \mu\operatorname{tr}\left(\tilde{\mathbf{E}}\right)^2 \tag{4.12.49}$$

and terms $O\left(\sigma^3\right)$ are dropped. The inequality (4.12.48) is the three dimensional version of the inequality which follows from (1.11.23) by noting that $\kappa \ge 0$ in (1.11.23)

Exercise 4.12.6

Use the result of Exercise 4.12.5 and prove that for an isolated body the total energy decreases in time. This assertion is a rough statement of a precise theorem in thermoelasticity. It is rough because one must provide definitions of an isolated body and of total energy.

Exercise 4.12.7

In Exercise 1.11.10 it was pointed out how an energy argument implies uniqueness of certain types of initial-boundary value problems in one-dimensional thermoelasticity. The three-dimensional version is due to *Wiener* and was published in 1957. As in Section 1.11, one must make certain assumptions about the material constants. These assumptions are restrictions which are not provided by the second law of thermodynamics. For the three-dimensional theory, these restrictions follow from the requirement that Σ in (4.12.49) is a positive definite function of θ and $\tilde{\mathbf{E}}$. Show that this assumption implies that

$$c_v > 0 \tag{4.12.50}$$

$$\lambda + \tfrac{2}{3}\mu > 0 \tag{4.12.51}$$

and

$$\mu > 0 \tag{4.12.52}$$

A full discussion of uniqueness in thermoelasticity can be found in Ref. 30. Stability arguments in thermoelasticity are discussed in Ref. 31.

Exercise 4.12.8

Decouple (4.12.45) and (4.12.46) (with $p = 0$) and show that θ and $\mathbf{w}$ obey

$$\frac{\partial}{\partial t}\left(\frac{\partial^2\theta}{\partial t^2} - a^{*2}\Delta\theta\right) - \frac{\kappa}{\rho_R c_v}\Delta\left(\frac{\partial^2\theta}{\partial t^2} - a^2\Delta\theta\right) = 0 \tag{4.12.53}$$

and

$$\left(\frac{\partial}{\partial t}\left(\frac{\partial^2}{\partial t^2} - a^{*2}\Delta\right) - \frac{\kappa}{\rho_R c_v}\Delta\left(\frac{\partial^2}{\partial t^2} - a^2\Delta\right)\right)\left(\frac{\partial^2}{\partial t^2} - \frac{\mu}{\rho}\Delta\right)\mathbf{w} = \mathbf{0} \tag{4.12.54}$$

where a is the *isothermal longitudinal wave speed* defined by

$$a^2 = \frac{\lambda + 2\mu}{\rho_R} \tag{4.12.55}$$

a^* is the *isentropic longitudinal wave speed* defined by

$$a^{*2} = \frac{1}{\rho_R}\left(\lambda + \frac{\theta_0}{\rho_R c_v}\beta^2 + 2\mu\right) \tag{4.12.56}$$

The ratio μ/ρ_R is the squared *transverse wave speed*. Equations (4.12.53) and (4.12.54) generalize to three dimensions the result (1.11.14). They generalize to thermoelasticity the result (4.4.58). Equations (4.12.53) and (4.12.54) were first obtained by *Cristea* in 1952 [Ref. 30].

References

1. TRUESDELL, C., and R.A. TOUPIN, *The Classical Field Theories*, Handbuch der Physik, Vol. III/1 (ed. S. Flugge), Springer-Verlag, Berlin (1960).
2. COLEMAN, B.D., and W. NOLL, The Thermodynamics of Elastic Materials with Heat Conduction and Viscosity, *Arch. Ration. Mech. Anal.*, **13**, 245-261 (1963).

3. TRUESDELL, C., and W. NOLL, *The Non-Linear Field Theories of Mechanics*, Handbuch der Physik, Vol. III/3 (ed. S. Flugge), Springer-Verlag (1965).
4. NOLL, W., On the Continuity of the Solid and Fluid States, *J. Rational Mech. Anal.*, **4**, 13-81 (1955).
5. NOLL, W., A Mathematical Theory of the Mechanical Behavior of Continuous Media, *Arch. Ration. Mech. Anal.*, **2**, 197-226, (1958).
6. NOLL, W., Proof of the Maximality of the Orthogonal Group in the Unimodular Group, *Arch. Ration. Mech. Anal.*, **18**, 100-102, (1965).
7. OGDEN, R. W., *Non-Linear Elastic Deformations*, Dover Publications, Inc., Mineola (1997).
8. BRAUER, R., On the Relation Between the Orthogonal Group and the Unimodular Group, *Arch. Ration. Mech. Anal.*, **18**, 97-99, (1965).
9. GURTIN, M.E., *The Linear Theory of Elasticity*, Handbuch der Physik, Vol. VIa/2, Springer-Verlag, Berlin (1972).
10. TRUESDELL, C., A Theorem on the Isotropy Groups of a Hyperelastic Material, *Proc. Nat. Acad. Sci U.S.A.*, **52**, 1081-1083 (1964).
11. SYNGE, J.L., *Classical Dynamics*, Handbuch der Physik, Vol. III/1 (ed. S. Flugge), Springer-Verlag, Berlin (1960).
12. GELFAND, I.M., and S.V. FOMIN, *Calculus of Variations* (trans. by R.A. Silverman), Prentice Hall, Englewood Cliffs (1963).
13. COHEN, H., and C.-C. WANG, On the Response and Symmetry of Elastic Materials with Internal Constraints, *Arch. Ration. Mech. Anal.*, **99**, 1-36(1987).
14. COHEN, H., and C.-C. WANG, On the Stored Energy Functions of Hyperelastic Materials with Internal Constraints, *Arch. Ration. Mech. Anal.*, **102**, 119-140 (1988).
15. COLEMAN, B.D., and V.J. MIZEL, Existence of Caloric Equations of State in Thermodynamics, *J. Chem. Phys.*, **40**, 1116-1125 (1964).
16. COLEMAN, B.D., and W. NOLL, Foundations of Linear Viscoelasticity, *Rev. Mod. Phys.*, **33**, 239-249 (1961).
17. COLEMAN, B.D., Thermodynamics of Materials with Memory, *Arch. Ration. Mech. Anal.*, **17**, 1-46 (1964).
18. LEITMAN, M.J., and G.M. FISHER, *The Linear Theory of Viscoelasticity*, Handbuch der Physik, Vol. VI a/3 (ed. C. Truesdell), Springer-Verlag, Berlin (1973).
19. COLEMAN, B.D., and V.J. MIZEL, Thermodynamics and Departures from Fourier's Law of Heat Conduction, *Arch. Ration. Mech. Anal.*, **13**, 245-261 (1963).
20. GURTIN, M.E., Thermodynamics and the Possibility of Long-Range Interaction in Rigid Heat Conductors, *Arch. Ration. Mech. Anal.*, **18**, 335-342 (1965).
21. GURTIN, M.E., Thermodynamics and the Possibility of Spatial Interaction in Elastic Materials, *Arch. Ration. Mech. Anal.*, **19**, 339-352 (1965).
22. WANG, C.-C., and R.M. BOWEN, On the Thermodynamics of Non-Linear Materials with Quasi-Elastic Response, *Arch. Ration. Mech. Anal.*, **22**, 79-99 (1966).
23. NOLL, W., Materially Uniform Simple Bodies with Inhomogeneities *Arch. Ration. Mech. Anal.*, **27**, 1-32 (1967).
24. WANG, C.-C., On the Geometric Structure of Simple Bodies, A Mathematical Foundation for the Theory of Continuous Distributions of Dislocations, *Arch. Ration. Mech. Anal.*, **27**, 33-94 (1967).
25. BOGY, D.B., and P.M. NAGHDI, On Heat Conduction and Wave Propagation in Rigid Solids, *J. Math. Phys.*, **11**, 919-923 (1970).

26. WANG, C.-C., and C. TRUESDELL, *Introduction to Rational Elasticity*, Noordhoff International Publishing, Leyden (1973).
27. ADKINS, J. E., Nonlinear Diffusion, II Constitutive Equations for Mixtures of Isotropic Fluids, *Phil. Trans. Roy. Soc. London A*, **255**, 635-648 (1963).
28. BEDFORD, A., On the Relative Velocity in a Mixture Theory for Composite Materials, *J. Appl. Mech.*, **42**, 502-503 (1975).
29. NOLL, W., Representations of Certain Isotropic Tensor Functions, Department Math. Carnegie Institute of Technology Report, June 1967.
30. CARLSON, D.E., *Linear Thermoelasticity*, Handbuch der Physik, Vol. VI a/2 (ed. C. Truesdell), Springer-Verlag, Berlin (1972).

Bibliography

31. KNOPS, R.J., and E.W. WILKES, *Theory of Elastic Stability*, Handbuch der Physik, Vol. VI a/3 (ed. C. Truesdell), Springer-Verlag, Berlin (1973).
32. TRUESDELL, C., Rational Mechanics of Deformation and Flow, Proc. 4th Intl. Congr. Rheo., 3-31 (1965).
33. GREEN, A.E., P.M. NAGHDI, and J.A. TRAPP, Thermodynamics of a Continuum with Internal Constraints, *Intl. J. Engr. Sci.*, **8**, 891-908 (1970).
34. GURTIN, M.E., and P.P. GUIDUGLI, The Thermodynamics of Constrained Materials, *Arch. Ration. Mech. Anal.*, **51**, 192-208 (1973).
35. COLEMAN, B.D., Thermodynamics of Materials with Memory, *Arch. Ration. Mech. Anal.*, **17**, 1-46 (1964).
36. COLEMAN, B.D., On Thermodynamics, Strain Impulses and Viscoelasticity, *Arch. Ration. Mech. Anal.*, **17**, 230-254 (1964).
37. NOWINSKI, J.L., *Theory of Thermoelasticity with Applications*, Sijthoff & Noordhoff International Publishers. Alphen aan den Rijn (1978).
38. NOWACKI, W., *Thermoelasticity*, Addison-Wesley, Reading (1962).
39. ACHENBACH, J.D., *Wave Propagation in Elastic Solids*, North-Holland, Amsterdam (1975).
40. FUNG, Y.C., *Foundations of Solid Mechanics*, Prentice-Hall, Englewood Cliffs (1965).
41. SERRIN, J., *Mathematical Principles of Classical Fluid Mechanics*, Handbuch der Physik, Vol. VIII/1, (ed. C. Truesdell), Springer-Verlag, Berlin (1959).
42. TRUESDELL, C., *Rational Thermodynamics*, Second Edition, Springer-Verlag, New York (1984).
43. COLEMAN, B.D., and M.E. GURTIN, Equipresence and Constitutive Equations for Rigid Head Conductors, *Z. Angew. Math. Phys.*, **18**, 199-208 (1967).
44. COLEMAN, B.D., and V.J. MIZEL, A General Theory of Dissipation in Materials with Memory, *Arch. Ration. Mech. Anal.*, **27**, 255-274 (1968).
45. COLEMAN, B.D., and E.H. DILL, On the Thermodynamics of Electromagnetic Fields in Materials with Memory, *Arch. Ration. Mech. Anal.*, **41**, 132-162 (1971).
46. COLEMAN, B.D., and D.R. OWEN, A Mathematical Foundation for Thermodynamics, *Arch. Ration. Mech. Anal.*, **54**, 1-104 (1974).
47. COLEMAN, B.D., and D.R. OWEN, On Thermodynamics and Elastic-Plastic Materials, *Arch. Ration. Mech. Anal.*, **59**, 25-51 (1976).

48. COLEMAN, B.D., and D.R. OWEN, On the Thermodynamics of Elastic-Plastic Materials with Temperature-Dependent Moduli and Yield Stresses, *Arch. Ration. Mech. Anal.*, **70**, 339-354 (1979).
49. COLEMAN, B.D., and V.J. MIZEL, Existence of Entropy as a Consequence of Asymptotic Stability, *Arch. Ration. Mech. Anal.*, **25**, 243-270 (1967).
50. COLEMAN, B.D., and J.M. GREENBERG, Thermodynamics and the Stability of Fluid Motion, *Arch. Ration. Mech. Anal.*, **25**, 321-341 (1967).
51. COLEMAN, B.D., and V.J. MIZEL, On Thermodynamic Conditions for the Stability of Evolving Systems, *Arch. Ration. Mech. Anal.*, **29**, 105-113 (1968).
52. DUNN, J.E., and R.L. FOSDICK, Thermodynamics, Stability, and Boundness of Fluids of Complexity 2 and Fluids of Second Grade, *Arch. Ration. Mech. Anal.*, **56**, 191-252 (1974).
53. DAFERMOS, C. M., Stability of Motions of Thermoelastic Fluids, *J. of Thermal Stresses*, **2**, 127-134 (1979).
54. GURTIN, M.E., On the Thermodynamics of Materials with Memory, *Arch. Ration. Mech. Anal.*, **28**, 40-50 (1968).
55. CHEN, P.J., and M.E. GURTIN, On a Theory of Heat Conduction Involving Two Temperatures, *Z. Angew. Math. Phys.*, **19**, 614-627 (1968).
56. CHEN, P.J., GURTIN, M.E., and W.O. WILLIAMS, A Note on Non-Simple Heat Conduction, *Z. Angew. Math. Phys.*, **19**, 969-970 (1968).
57. CHEN, P.J., GURTIN, M.E., and W.O. WILLIAMS, On the Thermodynamics of Non-simple Elastic Materials with Two Temperatures, *Z. Angew. Math. Phys.*, **20**, 107-112 (1969).
58. GURTIN, M.E., Thermodynamics and the Griffith Criterion for Brittle Fracture, *Int. J. Solids Struct.*, **15**, 553-560 (1979).
59. GURTIN, M.E., Thermodynamics and the Cohesive Zone in Fracture, *Z. Angew. Math. Phys.*, **30**, 991-1003 (1979).
60. DAY, W.A., and M.E. GURTIN, On the Symmetry of the Conductivity Tensor and other Restrictions in the Nonlinear Theory of Heat Conduction, *Arch. Ration. Mech. Anal.*, **33**, 26-32 (1969).
61. GURTIN, M.E., Time-Reversal and Symmetry in the Thermodynamics of Materials with Memory, *Arch. Ration. Mech. Anal.*, **44**, 387-399 (1972).
62. DAY, W.A., Time-Reversal and Symmetry of the Relation Function of a Linear Viscoelastic Material, *Arch. Ration. Mech. Anal.*, **40**, 155-159 (1971).
63. DAY, W.A., A Theory of Thermodynamics for Materials with Memory, *Arch. Ration. Mech. Anal.*, **34**, 85-96 (1969).
64. DAY, W.A., Thermodynamics Based on a Work Axiom, *Arch. Ration. Mech. Anal.*, **31**, 1-34 (1968).
65. TRUESDELL, C., *A First Course in Rational Continuum Mechanics*, Vol. 1, Academic Press, New York (1977).
66. GURTIN, M. E., *An Introduction to Continuum Mechanics*, Academic Press, New York (1981).
67. CHADWICK, P., *Continuum Mechanics*, Concise Theory and Problems, George Allen & Unwin Ltd., London (1976).
68. SPENCER, A. J. M., *Continuum Mechanics*, Longman, London (1980).
69. MALVERN, L. E., *Introduction to the Mechanics of a Continuous Medium*, Prentice-Hall, Englewood Cliffs (1969).
70. LEIGH, D. C., *Nonlinear Continuum Mechanics*, McGraw-Hill, New York (1968).

71. JAUNZEMIS, W., *Continuum Mechanics*, Macmillan, New York (1967).

5

MATERIALS WITH INTERNAL STATE VARIABLES

So as to illustrate a continuum model which possesses a relaxation phenomena, this chapter is concerned with a class of materials known as *materials with internal state variables*. Such models have been found to be useful in a large number of cases. They turn out to contain the Maxwellian material and the Maxwell-Cattaneo heat conductor mentioned in Section 1.7. The mathematical model of a relaxing gas is another example [Ref. 1]. Roughly speaking, one introduces additional "state variables" and requires that they obey a constitutive relation of the form of an ordinary differential equation. Depending upon the phenomena to be analyzed, the state variables have many different physical interpretations. Admittedly, in some cases, the physical interpretation of the state variables can be vague. One example where this is not the case, is when one wishes to model a chemically reacting mixture without diffusion. It turns out that the equations of balance given in Chapter III remain valid in this case. In addition, one has additional equations of balance which gives the exchange of mass between the constituents. In this model, the state variables are called "extents of reaction" and measure the progress of the independent chemical reactions taking place in the mixture. Background on this particular model of a material with internal state variables can be found in References 2 and 3.

The fundamental results for materials with internal state variables are presented in Section 5.1. With few exceptions, the results in Section 5.1 are taken from References 4 and 5. Section 5.2 shows how the results in 5.1 can be specialized so as to obtain the model of a Maxwell-Cattaneo heat conductor. Section 5.3 contains a brief discussion of Maxwellian materials. Finally, Section 5.4 contains a discussion of several proposals for an entropy inequality which generalizes equation (3.5.8).

5.1. Constitutive Assumptions and Thermodynamic Results

The internal state variable shall be represented by the following R-tuple:

$$\xi=\left(\xi_1, \xi_2, \xi_3, \ldots, \xi_R\right) \tag{5.1.1}$$

It is assumed to be given by the function Ξ_κ of $(\mathbf{X}, t)$ such that

$$\xi=\Xi_\kappa\left(\mathbf{X}, t\right) \tag{5.1.2}$$

The explicit tensor character of the state variable is not important at this point. Its components could be components of vectors or tensors or, as in the chemical reaction cases, they are convenient collections of scalar quantities.

A *thermodynamic process* for a material with internal state variables is the set consisting of the three functions $\theta_\kappa, \chi_\kappa$, and Ξ_κ and the seven functions of $(\mathbf{X},t)$ whose values are

$\psi,\eta,\mathbf{T},\mathbf{q},\rho,r,$ and **b** which satisfy balance of mass (3.1.12), balance of linear momentum (3.2.20), and balance of thermodynamic energy (3.5.20). The constitutive equations which define the material to be studied here are generalizations of the ones introduced in Section 4.6. It is assumed that

$$\left(\psi,\eta,\mathbf{T},\mathbf{q},\dot{\xi}\right)=f\left(\theta,\mathbf{g},\mathbf{F},\dot{\mathbf{F}},\xi\right) \tag{5.1.3}$$

The reader can see that (5.1.3) allows for the same effects presented in the model analyzed in Chapter IV. In addition $\psi,\eta,\mathbf{T}$, and $\mathbf{q}$ are allowed to depend in a general way on the internal state variable ξ. This quantity is, in turn, given by a *relaxation rate law* of the general form

$$\dot{\xi}=\boldsymbol{\omega}\left(\theta,\mathbf{g},\mathbf{F},\dot{\mathbf{F}},\xi\right) \tag{5.1.4}$$

Equation (5.1.4) is a consequence of (5.1.3) and our defining the component functions of f by

$$f=\left(u,h,\mathbf{G},\mathbf{l},\boldsymbol{\omega}\right) \tag{5.1.5}$$

For notational simplicity, the dependence of f on the reference configuration and the particle has not been shown explicitly in equations (5.1.3), (5.1.4), and (5.1.5). If necessary, the reader can make such dependencies explicit as was done in Chapter IV.

By definition, an *admissible thermodynamic process* is a thermodynamic process which is consistent with the constitutive assumption (5.1.3). As in Sections 1.8, 4.2, and 4.6, one can next prove a theorem to the effect that certain functions can be selected arbitrarily in the construction of admissible thermodynamic processes. In the case under discussion, we must make a technical assumption prior to the statement of this theorem. The assumption is that the ordinary differential equation (5.1.4) subject to the arbitrary initial condition

$$\Xi_{\kappa}\left(\mathbf{X},t\right)=\xi_{o} \tag{5.1.6}$$

has an unique solution on some interval $[t_o,t_o+\delta]$ where δ is a positive number. This assumption is assumed to hold for all $\theta_{\kappa},\chi_{\kappa},$ and ξ_o. As a consequence of this assumption, it is now true that for every choice of $\theta_{\kappa},\chi_{\kappa},$ and ξ_o there exists a unique admissible thermodynamic process. The proof of this assertion is the same as the corresponding assertion in Sections 1.8, 4.2, and 4.6. Given $\theta_{\kappa},\chi_{\kappa},$ and ξ_o then Ξ_{κ} can be calculated from (5.1.4). Equation (5.1.3) determines $\psi,\eta,\mathbf{T},\mathbf{q}$. Given this information, ρ can be calculated from (3.1.12), **b** can be calculated from (3.2.20) and r can be calculated from (3.5.20). As earlier, the reference density ρ_R is regarded as given. As the reader should have anticipated, we shall require that balance of angular momentum (3.3.10) and the entropy inequality (3.5.15) hold for every admissible thermodynamic process. The symmetry of the stress tensor is achieved by selecting **G** to have symmetric values. The restrictions implied by (3.5.15) shall be examined next.

Given (5.1.3), it follows that

$$\psi = u\left(\theta, \mathbf{g}, \mathbf{F}, \dot{\mathbf{F}}, \xi\right) \tag{5.1.7}$$

and, as a result,

$$\dot{\psi} = \frac{\partial u}{\partial \theta}\dot{\theta} + \frac{\partial u}{\partial \mathbf{g}}\cdot\dot{\mathbf{g}} + \operatorname{tr}\left(\frac{\partial u}{\partial \mathbf{F}}^T \dot{\mathbf{F}}\right) + \operatorname{tr}\left(\frac{\partial u}{\partial \dot{\mathbf{F}}}^T \ddot{\mathbf{F}}\right) - \boldsymbol{\sigma}\cdot\dot{\xi} \tag{5.1.8}$$

where $\boldsymbol{\sigma}$ is the R-tuple defined by

$$\boldsymbol{\sigma} = -\frac{\partial u\left(\theta, \mathbf{g}, \mathbf{F}, \dot{\mathbf{F}}, \xi\right)}{\partial \xi} \tag{5.1.9}$$

and, in addition,

$$\sigma\cdot\dot{\xi} = \sum_{\alpha=1}^{R} \sigma_\alpha \dot{\xi}_\alpha \tag{5.1.10}$$

In theories of chemical reactions, $\boldsymbol{\sigma}$ is known as the *chemical affinity*. Here, we shall refer to it simply as the *affinity*. We shall see later that it is an important parameter for internal state variable models. Given (5.1.8), the inequality (3.5.15) becomes

$$-\rho\left(\frac{\partial u}{\partial \theta} + h\right)\dot{\theta} - \rho\frac{\partial u}{\partial \mathbf{g}}\cdot\dot{\mathbf{g}} - \rho\operatorname{tr}\left(\frac{\partial u}{\partial \dot{\mathbf{F}}}^T \ddot{\mathbf{F}}\right) + \rho\boldsymbol{\sigma}\cdot\dot{\xi} + \operatorname{tr}\left(\mathbf{F}^{-1}\left(\mathbf{G} - \rho\mathbf{F}\frac{\partial u}{\partial \mathbf{F}}^T\right)\dot{\mathbf{F}}\right) - \mathbf{l}\cdot\mathbf{g}/\theta \geq 0 \tag{5.1.11}$$

where the argument of each function is $\left(\theta, \mathbf{g}, \mathbf{F}, \dot{\mathbf{F}}, \xi\right)$. By an argument similar to the one used in Section 4.6, (5.1.11) yields

$$\psi = u\left(\theta, \mathbf{F}, \xi\right) \tag{5.1.12}$$

$$\eta = h\left(\theta, \mathbf{F}, \xi\right) = -\frac{\partial u\left(\theta, \mathbf{F}, \xi\right)}{\partial \theta} \tag{5.1.13}$$

and

$$\rho\boldsymbol{\sigma}\cdot\dot{\xi} + \operatorname{tr}\left(\mathbf{F}^{-1}\left(\mathbf{G} - \rho\mathbf{F}\frac{\partial u}{\partial \mathbf{F}}^T\right)\dot{\mathbf{F}}\right) - \mathbf{l}\cdot\mathbf{g}/\theta \geq 0 \tag{5.1.14}$$

Exercise 5.1.1

If the response functions are independent of $\dot{\mathbf{F}}$ show that (5.1.14) implies

$$\mathbf{T}=\mathbf{G}\left(\theta,\mathbf{F},\dot{\boldsymbol{\xi}}\right)=\rho\mathbf{F}\frac{\partial u\left(\theta,\mathbf{F},\boldsymbol{\xi}\right)^T}{\partial\mathbf{F}} \tag{5.1.15}$$

and

$$\rho\boldsymbol{\sigma}\cdot\dot{\boldsymbol{\xi}}-\mathbf{l}\cdot\mathbf{g}/\theta\geq 0 \tag{5.1.16}$$

Exercise 5.1.2

If, in addition, the response functions are independent of **g** show that (5.1.16) yields

$$\mathbf{q}=\mathbf{0} \tag{5.1.17}$$

and

$$\rho\boldsymbol{\sigma}\cdot\dot{\boldsymbol{\xi}}\geq 0 \tag{5.1.18}$$

In Section 4.8 we defined a thermodynamic equilibrium state as the state where $\mathbf{g}=\mathbf{0}$ and $\dot{\mathbf{F}}=\mathbf{0}$. In this state, we were able to extract additional information from the residual entropy inequality because the function defined by (4.8.1) was a minimum at the thermodynamic equilibrium state. We now wish to extend the argument of Section 4.8 so as to apply to the material defined by (5.1.3). The residual entropy inequality is (5.1.14). Therefore, we define a function Φ of $\left(\theta,\mathbf{g},\mathbf{F},\dot{\mathbf{F}},\boldsymbol{\xi}\right)$

$$\Phi\left(\theta,g,\mathbf{F},\dot{\mathbf{F}},\boldsymbol{\xi}\right)=\rho\boldsymbol{\sigma}\cdot\dot{\boldsymbol{\xi}}+\operatorname{tr}\left(\mathbf{F}^{-1}\left(\mathbf{G}-\rho\mathbf{F}\frac{\partial u}{\partial\mathbf{F}}^T\right)\dot{\mathbf{F}}\right)-\mathbf{l}\cdot\mathbf{g}/\theta\geq 0 \tag{5.1.19}$$

This function has the property that

$$\Phi\left(\theta,\mathbf{g},\mathbf{F},\dot{\mathbf{F}},\boldsymbol{\xi}\right)\geq 0 \tag{5.1.20}$$

for all $\left(\theta,\mathbf{g},\mathbf{F},\dot{\mathbf{F}},\boldsymbol{\xi}\right)$ in the domain of the constitutive functions. Unfortunately, the function Φ does not vanish in the state $\left(\theta,\mathbf{0},\mathbf{F},\mathbf{0},\boldsymbol{\xi}\right)$. We have the slightly more complicated situation where one must characterize the thermodynamic equilibrium state by requiring the term $\boldsymbol{\sigma}\cdot\dot{\boldsymbol{\xi}}$ to vanish in addition to **g** and $\dot{\mathbf{F}}$ being zero. In theories of chemical reactions $\boldsymbol{\omega}\left(\theta,\mathbf{g},\mathbf{F},\dot{\mathbf{F}},\boldsymbol{\xi}\right)$ is the chemical reaction rate. Therefore, it is reasonable to require $\boldsymbol{\omega}$ to yield a zero value in the state called the thermodynamic equilibrium state. Following Truesdell, we shall call a state $\left(\theta^+,\mathbf{0},\mathbf{F}^+,\mathbf{0},\boldsymbol{\xi}^+\right)$ such that

$$\boldsymbol{\omega}\left(\theta^+,\mathbf{0},\mathbf{F}^+,\mathbf{0},\boldsymbol{\xi}^+\right)=\mathbf{0} \tag{5.1.21}$$

a state of *weak thermodynamic equilibrium*.[Ref. 6, Chap. 6] On the assumption that such a state exists, (5.1.19) yields

$$\Phi\left(\theta^{+},\mathbf{0},\mathbf{F}^{+},\mathbf{0},\boldsymbol{\xi}^{+}\right)=0 \tag{5.1.22}$$

Therefore, in a state of weak thermodynamic equilibrium,

$$\left.\frac{d\Phi\left(\theta^{+}+\lambda\alpha,\lambda\mathbf{a},\mathbf{F}^{+}+\lambda\mathbf{J},\lambda\mathbf{A},\boldsymbol{\xi}^{+}+\lambda\boldsymbol{\eta}\right)}{d\lambda}\right|_{\lambda=0}=0 \tag{5.1.23}$$

and

$$\left.\frac{d^{2}\Phi\left(\theta^{+}+\lambda\alpha,\lambda\mathbf{a},\mathbf{F}^{+}+\lambda\mathbf{J},\lambda\mathbf{A},\boldsymbol{\xi}^{+}+\lambda\boldsymbol{\eta}\right)}{d\lambda}\right|_{\lambda=0}\geq 0 \tag{5.1.24}$$

for all real numbers α, all vectors $\mathbf{a}$, all linear transformations $\mathbf{J}$ and $\mathbf{A}$ and all R-tuples $\boldsymbol{\eta}$.

Exercise 5.1.3

Show that (5.1.23) implies that

$$\boldsymbol{\sigma}\left(\theta^{+},\mathbf{F}^{+},\boldsymbol{\xi}^{+}\right)\cdot\frac{\partial\boldsymbol{\omega}\left(\theta^{+},\mathbf{0},\mathbf{F}^{+},\mathbf{0},\boldsymbol{\xi}^{+}\right)}{\partial\theta}=0 \tag{5.1.25}$$

$$\boldsymbol{\sigma}\left(\theta^{+},\mathbf{F}^{+},\boldsymbol{\xi}^{+}\right)\cdot\frac{\partial\boldsymbol{\omega}\left(\theta^{+},\mathbf{0},\mathbf{F}^{+},\mathbf{0},\boldsymbol{\xi}^{+}\right)}{\partial\mathbf{F}}=\mathbf{0} \tag{5.1.26}$$

$$\frac{\partial\boldsymbol{\omega}\left(\theta^{+},\mathbf{0},\mathbf{F}^{+},\mathbf{0},\boldsymbol{\xi}^{+}\right)^{T}}{\partial\boldsymbol{\xi}}\boldsymbol{\sigma}\left(\theta^{+},\mathbf{F}^{+},\boldsymbol{\xi}^{+}\right)=\mathbf{0} \tag{5.1.27}$$

$$\mathbf{q}\left(\theta^{+},\mathbf{F}^{+},\mathbf{0},\boldsymbol{\xi}^{+}\right)=\rho^{+}\boldsymbol{\sigma}\left(\theta^{+},\mathbf{0},\mathbf{F}^{+},\mathbf{0},\boldsymbol{\xi}^{+}\right)\cdot\frac{\partial\boldsymbol{\omega}\left(\theta^{+},\mathbf{0},\mathbf{F}^{+},\mathbf{0},\boldsymbol{\xi}^{+}\right)}{\partial\mathbf{g}} \tag{5.1.28}$$

and

$$\mathbf{G}\left(\theta^{+},\mathbf{0},\mathbf{F}^{+},\boldsymbol{\xi}^{+}\right)=\rho^{+}\mathbf{F}^{+}\left(\frac{\partial u\left(\theta^{+},\mathbf{0},\mathbf{F}^{+},\boldsymbol{\xi}^{+}\right)^{T}}{\partial\mathbf{F}}-\boldsymbol{\sigma}\left(\theta^{+},\mathbf{F}^{+},\mathbf{0},\boldsymbol{\xi}^{+}\right)\cdot\frac{\partial\boldsymbol{\omega}\left(\theta^{+},\mathbf{0},\mathbf{F}^{+},\mathbf{0},\boldsymbol{\xi}^{+}\right)}{\partial\dot{\mathbf{F}}}\right) \tag{5.1.29}$$

Relative to the results of Section 4.8, equations (5.1.28) and (5.1.29) might be unexpected. We *cannot* conclude that the heat flux vanishes and that the stress is determined by the free energy density whenever the material is in weak thermodynamic equilibrium. If, however, the equilibrium state is such that

$$\boldsymbol{\sigma}\left(\theta^{+},\mathbf{F}^{+},\boldsymbol{\xi}^{+}\right)=\mathbf{0} \tag{5.1.30}$$

in *addition* to (5.1.21), then (5.1.25), (5.1.26), and (5.1.27) are satisfied, and (5.1.28) and (5.1.29) reduce to results similar to those in Section 4.8. Again following Truesdell, the state $\left(\theta^{+},\mathbf{0},\mathbf{F}^{+},\mathbf{0},\boldsymbol{\xi}^{+}\right)$ such that *both* (5.1.21) and (5.1.30) hold is called a *strong* thermodynamic equilibrium state [Ref.6, Chap.6]. Equation (5.1.27) contains the important result that a weak equilibrium state is necessarily a strong one if the $R\times R$ matrix $\partial\boldsymbol{\omega}\left(\theta^{+},\mathbf{0},\mathbf{F}^{+},\mathbf{0},\boldsymbol{\xi}^{+}\right)/\partial\boldsymbol{\xi}$ is regular. In chemistry a state where the reaction rate vanishes and the chemical affinity does not is called a *false* equilibrium state [Ref. 7].

We now see that without additional assumptions a weak equilibrium state is not necessarily a strong equilibrium state. It is also interesting to question whether or not the vanishing of the affinity implies anything about the function $\boldsymbol{\omega}$. We assume that there exists a state $\left(\theta',\mathbf{0},\mathbf{F}',\mathbf{0},\boldsymbol{\xi}'\right)$ such that

$$\boldsymbol{\sigma}\left(\theta',\mathbf{F}',\boldsymbol{\xi}'\right)=\mathbf{0} \tag{5.1.31}$$

Given (5.1.31), it is necessarily true that

$$\left.\frac{d\Phi\left(\theta'+\lambda\alpha,\lambda\mathbf{a},\mathbf{F}'+\lambda\mathbf{J},\lambda\mathbf{A},\boldsymbol{\xi}'+\lambda\boldsymbol{\eta}\right)}{d\lambda}\right|_{\lambda=0}=0 \tag{5.1.32}$$

$$\left.\frac{d^{2}\Phi\left(\theta^{+}+\lambda\alpha,\lambda\mathbf{a},\mathbf{F}'+\lambda\mathbf{J},\lambda\mathbf{A},\boldsymbol{\xi}'+\lambda\boldsymbol{\eta}\right)}{d\lambda^{2}}\right|_{\lambda=0}\geq 0 \tag{5.1.33}$$

Exercise 5.1.4

Show that (5.1.32) implies that

$$\frac{\partial\boldsymbol{\sigma}\left(\theta',\mathbf{F}',\boldsymbol{\xi}'\right)}{\partial\theta}\cdot\boldsymbol{\omega}\left(\theta',\mathbf{0},\ \mathbf{F},\mathbf{0},\boldsymbol{\xi}'\right)=0 \tag{5.1.34}$$

$$\frac{\partial\boldsymbol{\sigma}\left(\theta',\mathbf{F}',\boldsymbol{\xi}'\right)}{\partial\mathbf{F}}\cdot\boldsymbol{\omega}\left(\theta',\mathbf{0},\ \mathbf{F},\mathbf{0},\boldsymbol{\xi}'\right)=\mathbf{0} \tag{5.1.35}$$

$$\frac{\partial \boldsymbol{\sigma}(\theta',\mathbf{F}',\boldsymbol{\xi}')}{\partial \boldsymbol{\xi}}\cdot\boldsymbol{\omega}(\theta',\mathbf{0},\ \mathbf{F},\mathbf{0},\boldsymbol{\xi}')=\mathbf{0} \tag{5.1.36}$$

$$\mathbf{l}(\theta',\mathbf{0},\ \mathbf{F}',\mathbf{0},\boldsymbol{\xi}')=\mathbf{0} \tag{5.1.37}$$

and

$$\mathbf{G}(\theta',\ \mathbf{0},\ \mathbf{F}',\boldsymbol{\xi}')=\rho'\mathbf{F}'\frac{\partial u(\theta',\mathbf{0},\ \mathbf{F}',\boldsymbol{\xi}')}{\partial \mathbf{F}}^{T} \tag{5.1.38}$$

The state $(\theta',\mathbf{0},\mathbf{F}',\mathbf{0},\boldsymbol{\xi}')$ such that (5.1.31) is satisfied is another candidate for the name equilibrium. This alternate definition is appealing since, in this state, the heat flux vanishes and the stress is determined by the free energy density. However, as (5.1.34), (5.1.35), and (5.1.36) indicate, the relaxation rate $\boldsymbol{\omega}(\theta',\mathbf{0},\mathbf{F}',\mathbf{0},\boldsymbol{\xi}')$ need not be zero. Equation (5.1.36) contains an important result. It asserts that the vanishing of the affinity when $\mathbf{g}$ and $\dot{\mathbf{F}}$ are zero implies $\boldsymbol{\omega}(\theta',\mathbf{0},\mathbf{F}',\mathbf{0},\boldsymbol{\xi}')$ that must vanish if the $R\times R$ matrix $\partial\boldsymbol{\sigma}(\theta',\mathbf{F}',\boldsymbol{\xi}')/\partial\boldsymbol{\xi}$ is regular. It follows from (5.1.9) that

$$\frac{\partial \boldsymbol{\sigma}(\theta,\mathbf{F},\boldsymbol{\xi})}{\partial \boldsymbol{\xi}}=-\frac{\partial^2 u(\theta,\mathbf{F},\boldsymbol{\xi})}{\partial \boldsymbol{\xi}^2} \tag{5.1.39}$$

and, thus, $\partial\boldsymbol{\sigma}/\partial\boldsymbol{\xi}$ is a symmetric matrix. In chemistry, a chemical reaction which takes place with a zero chemical affinity is called a *spinodal decomposition*. For many of the classical applications of this subject, the symmetric matrix $\partial\boldsymbol{\sigma}(\theta,\mathbf{F},\boldsymbol{\xi})/\partial\boldsymbol{\xi}$ is taken to be negative definite and, thus, regular. In summary, the formulas in Exercises 5.1.3 and 5.1.4 combine to tell us that if at a state $\boldsymbol{\omega}(\theta,\mathbf{0},\mathbf{F},\mathbf{0},\boldsymbol{\xi})$ both of the $R\times R$ matrices $\partial\boldsymbol{\omega}(\theta,\mathbf{0},\mathbf{F},\mathbf{0},\boldsymbol{\xi})/\partial\boldsymbol{\xi}$ and $\partial\boldsymbol{\sigma}(\theta,\mathbf{F},\boldsymbol{\xi})/\partial\boldsymbol{\xi}$ are regular, then the vanishing of $\boldsymbol{\sigma}(\theta,\mathbf{F},\boldsymbol{\xi})$ is equivalent to the vanishing of $\boldsymbol{\omega}(\theta,\mathbf{0},\mathbf{F},\mathbf{0},\boldsymbol{\xi})$.

In the applications, it is frequently assumed that (5.1.9) can be inverted to obtain

$$\boldsymbol{\xi}=\hat{\boldsymbol{\xi}}(\theta,\mathbf{F},\boldsymbol{\sigma}) \tag{5.1.40}$$

This equation allows for the elimination of $\boldsymbol{\xi}$ in favor of $\boldsymbol{\sigma}$ as an independent variable. For example, (5.1.4) can be replaced by

$$\dot{\boldsymbol{\xi}}=\hat{\boldsymbol{\omega}}(\theta,\mathbf{g},\mathbf{F},\dot{\mathbf{F}},\boldsymbol{\sigma}) \tag{5.1.41}$$

where

$$\hat{\boldsymbol{\omega}}(\theta,\mathbf{g},\mathbf{F},\dot{\mathbf{F}},\boldsymbol{\sigma})=\boldsymbol{\omega}(\theta,\mathbf{g},\mathbf{F},\dot{\mathbf{F}},\hat{\boldsymbol{\xi}}(\theta,\mathbf{F},\boldsymbol{\sigma})) \tag{5.1.42}$$

The inversion leading to (5.1.40) forces the $R \times R$ matrix $\partial \boldsymbol{\sigma}(\theta, \mathbf{F}, \boldsymbol{\xi})/\partial \boldsymbol{\xi}$ to be regular for all $(\theta, \mathbf{F}, \boldsymbol{\xi})$ in the domain of $\boldsymbol{\sigma}$. Thus, the results of Exercise 5.1.4 tell us that

$$\hat{\boldsymbol{\omega}}(\theta, \mathbf{0}, \mathbf{F}, \mathbf{0}, \mathbf{0}) = \mathbf{0} \tag{5.1.43}$$

Exercise 5.1.5

Define functions $\hat{u}, \hat{\mathbf{G}}$, and $\hat{\mathbf{I}}$ by

$$\hat{u}(\theta, \mathbf{F}, \boldsymbol{\sigma}) = u\left(\theta, \mathbf{F}, \hat{\boldsymbol{\xi}}(\theta, \mathbf{F}, \boldsymbol{\sigma})\right) \tag{5.1.44}$$

$$\hat{\mathbf{G}}(\theta, \mathbf{g}, \mathbf{F}, \dot{\mathbf{F}}, \boldsymbol{\sigma}) = \mathbf{G}\left(\theta, \mathbf{g}, \mathbf{F}, \dot{\mathbf{F}}, \hat{\boldsymbol{\xi}}(\theta, \mathbf{F}, \boldsymbol{\sigma})\right) \tag{5.1.45}$$

and

$$\hat{\mathbf{I}}(\theta, \mathbf{g}, \mathbf{F}, \dot{\mathbf{F}}, \boldsymbol{\sigma}) = \mathbf{I}\left(\theta, \mathbf{g}, \mathbf{F}, \dot{\mathbf{F}}, \hat{\boldsymbol{\xi}}(\theta, \mathbf{F}, \boldsymbol{\sigma})\right) \tag{5.1.46}$$

Show that

$$\hat{\mathbf{G}}(\theta, \mathbf{0}, \mathbf{F}, \mathbf{0}, \mathbf{0}) = \rho \mathbf{F} \frac{\partial \hat{u}(\theta, \mathbf{F}, \mathbf{0})}{\partial \mathbf{F}}^T \tag{5.1.47}$$

and

$$\hat{\mathbf{I}}(\theta, \mathbf{0}, \mathbf{F}, \mathbf{0}, \mathbf{0}) = \mathbf{0} \tag{5.1.48}$$

Equations (5.1.43), (5.1.47), and (5.1.48) show that departures from thermodynamic equilibrium are measured by $\mathbf{g}, \dot{\mathbf{F}}$ and $\boldsymbol{\sigma}$. For a process with $\mathbf{g} = \mathbf{0}$ and $\dot{\mathbf{F}} = \mathbf{0}$, the relaxation process is driven by the affinity. This statement is often argued by analogy with heat conduction models where the heat flux is driven by the temperature gradient. In any case, it must be stressed that the inversion leading to (5.1.40) was essential to the argument leading to (5.1.43), (5.1.47), and (5.1.48).

The concept of *shifting equilibrium* is important in applications of internal state variable models. A thermodynamic process is in *shifting equilibrium* if $\boldsymbol{\sigma} = \mathbf{0}$ for all time t. In this case the state variable is determined from $(\theta, \mathbf{F})$ by

$$\boldsymbol{\xi} = \hat{\boldsymbol{\xi}}(\theta, \mathbf{F}, \mathbf{0}) \tag{5.1.49}$$

a result which follows from (5.1.40). In certain special cases, the shifting equilibrium case arises as a consequence of relaxation processes which occur very fast relative to other characteristic times in the model. A material with internal state variables is *frozen* if $\boldsymbol{\xi}$ does not change in time. This case is achieved formally by taking the response function $\boldsymbol{\omega}$ in (5.1.4) to be zero. The frozen assumption simply reduces the model defined by (5.1.3) to the one analyzed in Sections 4.6 through 4.12. The frozen case can be viewed as arising from the assumption that the relaxation process is infinitely slow. In this sense, it is the opposite extreme from the shifting equilibrium case discussed above.

Exercise 5.1.6

Show that

$$\frac{\partial \hat{u}(\theta,\mathbf{F},\boldsymbol{\sigma})}{\partial \theta}=\frac{\partial u(\theta,\mathbf{F},\boldsymbol{\xi})}{\partial \theta}+\boldsymbol{\sigma}(\theta,\mathbf{F},\boldsymbol{\xi})\cdot\left(\frac{\partial \boldsymbol{\sigma}(\theta,\mathbf{F},\boldsymbol{\xi})}{\partial \boldsymbol{\xi}}\right)^{-1}\frac{\partial \boldsymbol{\sigma}(\theta,\mathbf{F},\boldsymbol{\xi})}{\partial \theta} \tag{5.1.50}$$

$$\frac{\partial \hat{u}(\theta,\mathbf{F},\boldsymbol{\sigma})}{\partial \mathbf{F}}=\frac{\partial u(\theta,\mathbf{F},\boldsymbol{\xi})}{\partial \mathbf{F}}+\boldsymbol{\sigma}(\theta,\mathbf{F},\boldsymbol{\xi})\cdot\left(\frac{\partial \boldsymbol{\sigma}(\theta,\mathbf{F},\boldsymbol{\xi})}{\partial \boldsymbol{\xi}}\right)^{-1}\frac{\partial \boldsymbol{\sigma}(\theta,\mathbf{F},\boldsymbol{\xi})}{\partial \mathbf{F}} \tag{5.1.51}$$

and

$$\frac{\partial \hat{u}(\theta,\mathbf{F},\boldsymbol{\sigma})}{\partial \boldsymbol{\sigma}}=\left(\frac{\partial \boldsymbol{\sigma}(\theta,\mathbf{F},\boldsymbol{\xi})}{\partial \boldsymbol{\xi}}\right)^{-1}\boldsymbol{\sigma}(\theta,\mathbf{F},\boldsymbol{\xi}) \tag{5.1.52}$$

Exercise 5.1.7

Given (5.1.40) and its implications (5.1.43), (5.1.47), and (5.1.48), show that $\hat{\boldsymbol{\omega}}\left(\theta,\mathbf{g},\mathbf{F},\dot{\mathbf{F}},\boldsymbol{\sigma}\right)$ must obey

$$\boldsymbol{\eta}\cdot\left(\frac{\partial \hat{\boldsymbol{\omega}}(\theta,\mathbf{0},\mathbf{F},\mathbf{0},\mathbf{0})}{\partial \boldsymbol{\sigma}}\boldsymbol{\eta}\right)\geq 0 \tag{5.1.53}$$

for all R-tuples $\boldsymbol{\eta}$.

Exercise 5.1.8

Consider the case where the response functions are independent of $\dot{\mathbf{F}}$. (See Exercise 5.1.1). Show that the Gibbs relation is

$$\dot{\psi}=-\eta\dot{\theta}+\frac{1}{\rho}\operatorname{tr}\mathbf{TL}-\boldsymbol{\sigma}\cdot\dot{\boldsymbol{\xi}} \tag{5.1.54}$$

and that balance of thermodynamic energy takes the form

$$\rho\theta\dot{\eta}=-\operatorname{div}\mathbf{q}+\rho\boldsymbol{\sigma}\cdot\dot{\boldsymbol{\xi}}+\rho r \tag{5.1.55}$$

Exercise 5.1.9

In the case where the response functions are independent of $\mathbf{g}$ and $\dot{\mathbf{F}}$, the results of Exercise 5.1.2 are valid. A special version of this model is used to describe relaxation in gases. It is convenient in this application to utilize $(\eta,\mathbf{F},\xi)$ as the independent variables rather than $(\theta,\mathbf{F},\xi)$. Follow the same line of reasoning utilized in Exercise 4.8.6 of Section 4.8 and show that

$$\varepsilon=e(\eta,\mathbf{F},\xi) \tag{5.1.56}$$

$$\theta=\frac{\partial e(\eta,\mathbf{F},\xi)}{\partial\eta} \tag{5.1.57}$$

$$\mathbf{T}=\rho\mathbf{F}\frac{\partial e(\eta,\mathbf{F},\xi)}{\partial\mathbf{F}}^{T} \tag{5.1.58}$$

and

$$\boldsymbol{\sigma}=-\frac{\partial e(\eta,\mathbf{F},\xi)}{\partial\xi} \tag{5.1.59}$$

Exercise 5.1.10

In problems where the shifting equilibrium is assumed, it is often convenient to change variables from $(\eta,\mathbf{F},\xi)$ to $(\eta,\mathbf{F},\boldsymbol{\sigma})$. The formal argument is similar to that used earlier in this section to change variables from $(\theta,\mathbf{F},\xi)$ to $(\theta,\mathbf{F},\boldsymbol{\sigma})$. Carry out this change and show that

$$\varepsilon=\bar{e}(\eta,\mathbf{F},\boldsymbol{\sigma}) \tag{5.1.60}$$

$$\theta=\frac{\partial\bar{e}(\eta,\mathbf{F},\boldsymbol{\sigma})}{\partial\eta}+\boldsymbol{\sigma}\cdot\frac{\partial\bar{\xi}(\eta,\mathbf{F},\boldsymbol{\sigma})}{\partial\eta} \tag{5.1.61}$$

and

$$\mathbf{T}=\rho\mathbf{F}\left(\frac{\partial\bar{e}(\eta,\mathbf{F},\boldsymbol{\sigma})}{\partial\mathbf{F}}+\boldsymbol{\sigma}\cdot\frac{\partial\bar{\xi}(\eta,\mathbf{F},\boldsymbol{\sigma})}{\partial\mathbf{F}}\right)^{T} \tag{5.1.62}$$

where

$$\xi=\bar{\xi}(\eta,\mathbf{F},\boldsymbol{\sigma}) \tag{5.1.63}$$

is the inverse of (5.1.59).

The restrictions on the constitutive equations of this section implied by material frame indifference can be read off from the results of Section 4.9 for those cases where the state variable is invariant under the transformation (4.9.1). The chemical reaction model is an example of this case. The restrictions implied by material symmetry follow the same formal development as presented in Section 4.10. For the sake of brevity, we shall not record material frame indifference and material symmetry restrictions in this section. Reference 5 does contain a discussion of these two concepts for the chemical reaction model.

As an illustration of a model of a material with internal state variables, consider an isotropic linear thermoelastic solid with heat conduction and a *single* internal state variable. This example is the one developed in Section 4.12 except that an internal state variable is present. The linearization is about a state of uniform temperature θ_o, no deformation, i.e., $\mathbf{x} = \mathbf{X}$ *and* zero affinity. It is readily shown that such a state is a solution of the field equations if we take

$$\mathbf{b} = \mathbf{0} \tag{5.1.64}$$

and

$$r = 0 \tag{5.1.65}$$

The equation which replaces (4.12.31) is

$$\begin{aligned}\rho_R\psi = {} & \rho_R u(\theta_o,\mathbf{I},\xi_o) - \rho_R\eta_o(\theta-\theta_o) - p\,\mathrm{tr}\,\tilde{\mathbf{E}} - \frac{\rho_R c_v}{2\theta_o}(\theta-\theta_o)^2 \\ & -\beta(\mathrm{tr}\,\tilde{\mathbf{E}})(\theta-\theta_o) + \frac{\lambda}{2}(\mathrm{tr}\,\tilde{\mathbf{E}})^2 + \mu\,\mathrm{tr}(\tilde{\mathbf{E}})^2 + \tfrac{1}{2}\rho_R\Phi(\xi-\xi_o)^2 \\ & +\delta(\xi-\xi_o)(\theta-\theta_o) + \alpha(\xi-\xi_o)\,\mathrm{tr}\,\tilde{\mathbf{E}}\end{aligned} \tag{5.1.66}$$

where, for simplicity, the order term has been omitted. Of course the symbol Φ in (5.1.66)(5.1.66) should not be confused with the function Φ introduced in (5.1.19). In (5.1.66) the coefficients, Φ, δ, and α arise from the dependence of ψ on the single state variable ξ. The quantity ξ_o is the value of the internal state variable in the reference state. The reference state has a zero affinity. Therefore, from (5.1.49),

$$\xi_o = \xi(\theta_o,\mathbf{I},0) \tag{5.1.67}$$

Given (5.1.66), it follows from (5.1.13), (5.1.15) and (5.1.9) that the entropy density, the stress and the affinity are given by

$$\rho_R\eta = \rho_R\eta_o + \frac{\theta-\theta_o}{\theta_o} + \beta\,\mathrm{tr}\,\tilde{\mathbf{E}} - \delta(\xi-\xi_o) \tag{5.1.68}$$

$$\mathbf{T}_R = -p(\mathbf{I}-\mathbf{H}^T) + \lambda(\mathrm{tr}\,\tilde{\mathbf{E}})\mathbf{I} + 2(\mu-p)\tilde{\mathbf{E}} - \beta(\theta-\theta_0)\mathbf{I} + \alpha(\xi-\xi_o)\mathbf{I} \tag{5.1.69}$$

and

$$\sigma = -\Phi(\xi - \xi_o) - \frac{\delta}{\rho_R}(\theta - \theta_o) - \frac{\alpha}{\rho_R} \operatorname{tr} \tilde{\mathbf{E}} \tag{5.1.70}$$

Because we are assuming (5.1.9) can be inverted to obtain (5.1.40), the linearized expression (5.1.70) must be invertible for $\xi - \xi_o$. Therefore, the material constant Φmust obey

$$\Phi \neq 0 \tag{5.1.71}$$

Below, we shall see reasons to require that Φ be positive.

The heat flux vector and the relaxation rate are taken to be

$$\mathbf{q}_R = -\kappa \text{ GRAD } \theta \tag{5.1.72}$$

and

$$\dot{\xi} = \frac{1}{\tau\Phi}\sigma \tag{5.1.73}$$

With the exception of the order term, equation (5.1.72) is formally the same as (4.12.40). In a linearized isotopic model a dependence of the heat flux vector on a scalar such as ξ is not allowed. Equation (5.1.73) is our special case of (5.1.41). It follows from the assumption that the departure from thermodynamic equilibrium is small and that the material is isotropic. In our case, where we have a single state variable and a dependence on $\dot{\mathbf{F}}$is not allowed, (5.1.41) reduces to

$$\dot{\xi} = \hat{\omega}(\theta, \mathbf{g}, \mathbf{F}, \sigma) \tag{5.1.74}$$

The constant $1/\tau\Phi$ in (5.1.73) is related to $\hat{\omega}$by the formula

$$\frac{1}{\tau\Phi} = \frac{\partial \hat{\omega}(\theta_o, \mathbf{0}, \mathbf{I}, 0)}{\partial \sigma} \tag{5.1.75}$$

Given (5.1.75), equation (5.1.53) tells us that

$$\frac{1}{\tau\Phi} \geq 0 \tag{5.1.76}$$

Equation (5.1.76) is one of the implications of the entropy inequality (5.1.16). The other implication is, as in Section 4.12,

$$\kappa \geq 0 \tag{5.1.77}$$

Because, from (5.1.70),

$$\Phi = \frac{\partial \sigma\left(\theta_o, \mathbf{I}, \xi_o\right)}{\partial \xi} \tag{5.1.78}$$

it is easily shown that the coefficient τ in (5.1.73) is given by

$$\frac{1}{\tau} = -\frac{\partial \omega\left(\theta_o, \mathbf{0}, I, \xi_o\right)}{\partial \xi} \tag{5.1.79}$$

The physical dimension of τ is time and it represents a characteristic relaxation time in our model. After we establish that Φ is positive, (5.1.76) will tell us that $1/\tau$ cannot be negative.

Exercise 5.1.11

Show that the linearized field equations which replace (4.12.45) and (4.12.46) are

$$\rho_R \ddot{\mathbf{w}} = (\lambda + \mu)\text{GRAD}(\text{Div}\,\mathbf{w}) + (\mu - p)\Delta\mathbf{w} - \beta\,\text{GRAD}\,\theta + \alpha\,\text{GRAD}\,\xi \tag{5.1.80}$$

$$\rho_R c_v \dot{\theta} + \beta\theta_o\,\text{Div}\,\dot{\mathbf{w}} - \delta\dot{\xi} = \kappa\Delta\theta \tag{5.1.81}$$

and

$$\tau\dot{\xi} = -(\xi - \xi_o) - \frac{\delta}{\rho_R\Phi}(\theta - \theta_o) - \frac{\alpha}{\rho_R\Phi}\text{Div}\,\mathbf{w} \tag{5.1.82}$$

Exercise 5.1.12

Show that (4.12.47) is replaced by

$$\begin{aligned} \rho_R\psi + \eta(\theta - \theta_o) &= \rho_R u(\theta_o, \mathbf{I}, \xi) - p\,\text{tr}\,\tilde{\mathbf{E}} + \frac{\rho_R c_v}{2\theta_o}(\theta - \theta_o)^2 \\ &+ \frac{\lambda}{2}\left(\text{tr}\,\tilde{\mathbf{E}}\right)^2 + \mu\,\text{tr}\left(\tilde{\mathbf{E}}\right)^2 + \tfrac{1}{2}\rho_R\Phi(\xi - \xi_o)^2 + \alpha(\xi - \xi_o)\,\text{tr}\,\tilde{\mathbf{E}} \end{aligned} \tag{5.1.83}$$

Exercise 5.1.13

Show that (4.12.48) is replaced by

$$\frac{\partial}{\partial t}\int_{\kappa(\mathscr{P})}\left(\Sigma + \tfrac{1}{2}\dot{\mathbf{w}}^2\right)dV \le \oint_{\partial\kappa(\mathscr{P})}\dot{\mathbf{w}}\cdot\left(\left(\mathbf{T}_R + p\mathbf{I}\right)d\mathbf{S}\right) - \oint_{\partial\kappa(\mathscr{P})}\frac{\theta - \theta_o}{\theta_o}\mathbf{q}_R \cdot d\mathbf{S} \tag{5.1.84}$$

where

$$\Sigma = \frac{\rho_R c_v}{2\theta_o}(\theta - \theta_o)^2 + \frac{\lambda}{2}\left(\text{tr}\,\tilde{\mathbf{E}}\right)^2 + \mu\,\text{tr}\left(\tilde{\mathbf{E}}\right)^2 + \tfrac{1}{2}\rho_R\Phi(\xi - \xi_o)^2 + \alpha(\xi - \xi_o)\,\text{tr}\tilde{\mathbf{E}} \tag{5.1.85}$$

Exercise 5.1.14

Show that Σ is positive definite if and only if

$$c_v > 0 \tag{5.1.86}$$

$$\Phi > 0 \tag{5.1.87}$$

$$\mu > 0 \tag{5.1.88}$$

and

$$\lambda - \frac{\alpha^2}{\rho_R \Phi} + \frac{2}{3}\mu > 0 \tag{5.1.89}$$

Equation (5.1.87) is the result anticipated above. Given (5.1.87) and (5.1.76), we see that

$$\frac{1}{\tau} \geq 0 \tag{5.1.90}$$

Also, given (5.1.89), it is true that

$$\lambda + \tfrac{2}{3}\mu > 0 \tag{5.1.91}$$

The material properties $c_v, \beta, \lambda,$ and μ arise in the representation (5.1.66) and are thus partial derivatives of ψ computed at constant ξ. For example,

$$c_v = \theta_o \frac{\partial^2 u(\theta, \mathbf{I}, \xi_o)}{\partial \theta^2} \tag{5.1.92}$$

It is convenient to refer to these properties as *frozen* properties. In the limit where the characteristic time τ approaches infinity, (5.1.82) reduces to $\dot{\xi} = \dot{\xi}_o$ and, thus, $\xi = \xi_o$. This limit is the *frozen* limit for our model. The limit $\tau \rightarrow 0$ yields, from (5.1.73), $\sigma = 0$. This limit is the *shifting equilibrium* limit for our model.

Exercise 5.1.15

Show that in the shifting equilibrium case

$$\rho_R \eta = \rho_R \eta_o + \rho_R \left(c_v + \theta_o \frac{\delta^2}{\rho_R^2 \Phi} \right) \left(\frac{\theta - \theta_o}{\theta_o} \right) + \left(\beta + \frac{\delta \alpha}{\rho_R \Phi} \right) \operatorname{tr} \tilde{\mathbf{E}} \tag{5.1.93}$$

and

$$\mathbf{T}_R = -p\left(\mathbf{I}-\mathbf{H}^T\right)+\left(\lambda-\frac{\alpha^2}{\rho_R\Phi}\right)\left(\operatorname{tr}\tilde{\mathbf{E}}\right)\mathbf{I}+2(\mu-p)\tilde{\mathbf{E}}-\left(\beta+\frac{\delta\alpha}{\rho_R\Phi}\right)(\theta-\theta_o)\mathbf{I} \tag{5.1.94}$$

These results enable us to identify $c_v+\theta_o\delta^2/\rho_R^2\Phi$ as the specific heat at constant volume for the shifting equilibrium case. In addition $\beta+\delta\alpha/\rho_R\Phi$ and $\lambda-\alpha^2/\rho_R\Phi$ are the shifting equilibrium properties which replace the frozen properties β and λ. Equation (5.1.89) asserts that the shifting equilibrium bulk modulus must be positive.

Exercise 5.1.16

Take $p=0$ and decouple (5.1.80), (5.1.81) and (5.1.82). The results are that θ,ξ and $\mathbf{w}$ obey

$$L\theta = L\xi = 0 \tag{5.1.95}$$

and

$$L\left(\frac{\partial^2}{\partial t^2}-\frac{\mu}{\rho_R}\Delta\right)\mathbf{w}=0 \tag{5.1.96}$$

where L is the differential operator

$$\begin{aligned} L = {} & \tau\frac{\partial}{\partial t}\left[\frac{\partial}{\partial t}\left(\frac{\partial^2}{\partial t^2}-a^{*2}\Delta\right)-\frac{\kappa}{\rho_R c_v}\Delta\left(\frac{\partial^2}{\partial t^2}-a^2\Delta\right)\right] \\ & +\left(1+\frac{\theta_o}{\rho_R^2}\frac{\delta^2}{c_v\Phi}\right)\frac{\partial}{\partial t}\left(\frac{\partial^2}{\partial t^2}-s^{*2}\Delta\right)-\frac{\kappa}{\rho_R c_v}\Delta\left(\frac{\partial^2}{\partial t^2}-s^2\Delta\right) \end{aligned} \tag{5.1.97}$$

The four constants $a, a^*, s,$ and s^* are longitudinal wave speeds defined by

$$a^2 = \frac{\lambda+2\mu}{\rho_R} \tag{5.1.98}$$

$$a^{*2} = \frac{1}{\rho_R}\left(\lambda+\frac{\theta_o}{\rho_R c_v}\beta^2+2\mu\right) \tag{5.1.99}$$

$$s^2 = \frac{1}{\rho_R}\left(\lambda-\frac{\alpha^2}{\rho_R\Phi}+2\mu\right) \tag{5.1.100}$$

and

$$s^{*2} = \frac{1}{\rho_R}\left(\lambda - \frac{\alpha^2}{\rho_R\Phi} + \frac{\theta_o}{\rho_R}\frac{\left(\beta + \frac{\alpha\delta}{\rho_R\Phi}\right)^2}{\left(c_v + \theta_o\frac{\delta^2}{\rho_R^2\Phi}\right)} + 2\mu\right) \tag{5.1.101}$$

The speeds a and a^* are *frozen isothermal* and *frozen isentropic* speeds, respectively. Likewise s and s* are equilibrium isothermal and equilibrium *isentropic* speeds, respectively. It is possible to show that

$$a^{*2} > s^{*2} > s^2 \tag{5.1.102}$$

and

$$a^{*2} > a^2 > s^2 \tag{5.1.103}$$

There are many aspects of the model with internal state variables which have not been touched on in this chapter. In particular, the stability theorems established by Coleman and Gurtin should be of interest to the reader.[Ref. 4]

5.2. Maxwell -Cattaneo Heat Conductor

As an illustration of a material with internal state variables which is not a reacting mixture or something similar, in this section we shall formulate the model of a *Maxwell-Cattaneo heat conductor*. The fundamental constitutive equation for this model is, in one dimensional form, (1.7.16). Therefore, we shall present an argument leading to a constitutive equation for the heat flux of the form

$$\tau\dot{\mathbf{q}} + \mathbf{q} = -\kappa\mathbf{g} \tag{5.2.1}$$

For simplicity, in this section we shall deal with the case of a rigid heat conductor. This assumption allows us to avoid a discussion of material frame indifference. A rigid material at rest is a material *constrained* by the deformation

$$\mathbf{x} = \mathbf{X} \tag{5.2.2}$$

Given (5.2.1), the entropy inequality (3.5.15) (or (3.5.16)) reduces to

$$-\rho\left(\dot{\psi} + \eta\dot{\theta}\right) - \mathbf{q}\cdot\mathbf{g}/\theta \geq 0 \tag{5.2.3}$$

and the stress $\mathbf{T}$ is completely indeterminate in so far as the entropy inequality is concerned. Balance of mass in the form (3.1.12) reduces to

$$\rho = \rho_R \tag{5.2.4}$$

and we shall take ρ_R to be a constant. The thermodynamic energy equation (3.5.20) reduces to

$$\rho\theta\dot{\eta}=-\rho\left(\dot{\psi}+\eta\dot{\theta}\right)-\operatorname{div}\mathbf{q}+\rho r \tag{5.2.5}$$

in the rigid case.

The constitutive equations to be investigated are as follows:

$$\left(\psi,\eta,\dot{\mathbf{q}}\right)=f\left(\theta,\mathbf{g},\mathbf{q}\right) \tag{5.2.6}$$

Relative to (5.1.3), it can be seen that we have taken the state variable to be the heat flux vector $\mathbf{q}$. Because of this choice, we have in effect taken the function $\mathbf{l}$ in (5.1.5) to be the identity function.

Given (5.2.6), we can write

$$\psi=u\left(\theta,\mathbf{g},\mathbf{q}\right) \tag{5.2.7}$$

and, thus,

$$\dot{\psi}=\frac{\partial u}{\partial\theta}\dot{\theta}+\frac{\partial u}{\partial\mathbf{g}}\cdot\dot{g}-\boldsymbol{\sigma}\cdot\dot{\mathbf{q}} \tag{5.2.8}$$

where the affinity $\boldsymbol{\sigma}$ is the *vector* defined by

$$\boldsymbol{\sigma}=-\frac{\partial u\left(\theta,\mathbf{g},\mathbf{q}\right)}{\partial\mathbf{q}} \tag{5.2.9}$$

If (5.2.8) is combined with (5.2.3), the same argument used in Section 5.2 yields

$$\psi=u\left(\theta,\mathbf{q}\right) \tag{5.2.10}$$

$$\eta=h\left(\theta,\mathbf{q}\right)=-\frac{\partial u\left(\theta,\mathbf{q}\right)}{\partial\theta} \tag{5.2.11}$$

And

$$\rho\boldsymbol{\sigma}\cdot\boldsymbol{\omega}-\mathbf{q}\cdot\mathbf{g}/\theta\geq 0 \tag{5.2.12}$$

where the relaxation rate law has been written

$$\dot{\mathbf{q}}=\boldsymbol{\omega}\left(\theta,\mathbf{g},\mathbf{q}\right) \tag{5.2.13}$$

For simplicity, we shall consider the equilibrium state defined by $\mathbf{g}=\boldsymbol{\sigma}=\mathbf{0}$ and assume (5.2.9) can be inverted to obtain

$$\mathbf{q}=\hat{\mathbf{q}}(\theta,\boldsymbol{\sigma}) \tag{5.2.14}$$

where (5.2.10) has been used to justify the omission of the argument **g** from (5.2.9). As in Section 5.1, the assumed inversion leading to (5.2.14) implies that the linear transformation $\partial\boldsymbol{\sigma}(\theta,\mathbf{q})/\partial\mathbf{q}$ is regular. Given (5.2.14), we can replace (5.2.13) by

$$\dot{\mathbf{q}}=\hat{\boldsymbol{\omega}}(\theta,\mathbf{g},\boldsymbol{\sigma}) \tag{5.2.15}$$

where

$$\hat{\boldsymbol{\omega}}(\theta,\mathbf{g},\boldsymbol{\sigma})=\boldsymbol{\omega}(\theta,\mathbf{g},\hat{\mathbf{q}}(\theta,\boldsymbol{\sigma})) \tag{5.2.16}$$

Exercise 5.2.1

Show that in thermodynamic equilibrium the following results must hold:

$$\hat{\boldsymbol{\omega}}(\theta,\mathbf{0},\mathbf{0})=\mathbf{0} \tag{5.2.17}$$

$$\hat{\mathbf{q}}(\theta,\mathbf{0})=\mathbf{0} \tag{5.2.18}$$

$$\frac{\partial\hat{\mathbf{q}}(\theta,\mathbf{0})}{\partial\boldsymbol{\sigma}}=\rho\theta\frac{\partial\hat{\boldsymbol{\omega}}(\theta,\mathbf{0},\mathbf{0})}{\partial\mathbf{g}} \tag{5.2.19}$$

and

$$\mathbf{a}\cdot\left(\frac{\partial\hat{\boldsymbol{\omega}}(\theta,\mathbf{0},\mathbf{0})}{\partial\boldsymbol{\sigma}}\mathbf{a}\right)\geq 0 \tag{5.2.20}$$

for all vectors **a**.

In order to reach a linear isotropic model, we take

$$\psi=u(\theta,\mathbf{q})=u(\theta_o,\mathbf{0})-\eta_o(\theta-\theta_o)-\frac{c_v}{2\theta_o}(\theta-\theta_o)^2+\frac{1}{2}\Phi\mathbf{q}\cdot\mathbf{q} \tag{5.2.21}$$

Essentially, (5.2.21) is a special case of (5.1.66). Relative to (5.1.66) we see that the terms arising from the deformation have been omitted and, in addition, a term like $\delta(\xi-\xi_o)(\theta-\theta_o)$ has been omitted. Such a term is not allowed by material symmetry where ξ is the vector **q**. As in Section 5.1, (5.2.21) is proposed as being a valid approximation about a state of constant temperature θ_o and zero affinity. Given (5.2.21), it then follows from (5.2.11) and (5.2.9) that

$$\eta=\eta_{o}+c_{v}\frac{(\theta-\theta_{o})}{\theta_{o}} \tag{5.2.22}$$

and

$$\boldsymbol{\sigma}=-\Phi\mathbf{q} \tag{5.2.23}$$

Of course, the coefficient Φ must be nonzero in order to construct the inversion (5.2.14). Also, given (5.2.23), the thermodynamic restriction (5.2.18) is satisfied. It follows from (5.2.21) and (5.2.22) that

$$\psi+\eta(\theta-\theta_{o})=u(\theta_{o},\mathbf{0})+\frac{c_{v}}{2\theta_{o}}(\theta-\theta_{o})^{2}+\frac{1}{2}\Phi\mathbf{q}\cdot\mathbf{q} \tag{5.2.24}$$

Roughly speaking, equation (5.2.24) is a special case of (5.1.83). The corresponding special case of (5.1.84) is

$$\frac{\partial}{\partial t}\int_{\kappa(\mathscr{P})}\Sigma dV\leq-\oint_{\partial\kappa(\mathscr{P})}\frac{(\theta-\theta_{o})}{\theta_{o}}\mathbf{q}\cdot d\mathbf{S} \tag{5.2.25}$$

where

$$\Sigma=\frac{c_{v}}{2\theta_{o}}(\theta-\theta_{o})^{2}+\frac{1}{2}\Phi\mathbf{q}\cdot\mathbf{q} \tag{5.2.26}$$

The usual requirement that Σ be positive definite yields

$$c_{v}>0 \tag{5.2.27}$$

and

$$\Phi>0 \tag{5.2.28}$$

For small departures from the state $\theta=\theta_{o}$ and $\boldsymbol{\sigma}=\mathbf{0}$ (5.2.15) is replaced by

$$\dot{\mathbf{q}}=\frac{1}{\tau\Phi}\boldsymbol{\sigma}-\alpha\mathbf{g} \tag{5.2.29}$$

where (5.2.17) has been used. It follows from (5.2.28) and (5.2.20) that

$$\frac{1}{\tau}\geq 0 \tag{5.2.30}$$

Based upon the results in Section 5.1, the reader will recognize τ as the characteristic relaxation time. It follows from (5.2.19), (5.2.29) and (5.2.23) that

$$\alpha=\frac{1}{\rho\theta_o\Phi} \tag{5.2.31}$$

Given (5.2.28), it follows from (5.2.31) that

$$\alpha>0 \tag{5.2.32}$$

If we define a coefficient κ by

$$\kappa=\tau\alpha \tag{5.2.33}$$

then (5.2.29) reduces to

$$\tau\dot{\mathbf{q}}+\mathbf{q}=-\kappa\mathbf{g} \tag{5.2.34}$$

where (5.2.23) has been used. Equation (5.2.34) is the Maxwell-Cattaneo constitutive equation (5.2.1).

Exercise 5.2.2

Show that the temperature is a solution of the following partial differential equation:

$$\tau\ddot{\theta}+\dot{\theta}=\frac{\kappa}{\rho c_v}\Delta\theta \tag{5.2.35}$$

For $\tau = 0$, (5.2.35) reduces to the parabolic diffusion equation. For $\tau \neq 0$, (5.2.35) is the hyperbolic damped wave equation. The wave-like solutions of (5.2.35) are the advantage of the Maxwell-Cattaneo constitutive equation over the Fourier equation (1.7.15). The Fourier constitutive equation implies that thermal disturbances propagate with infinite velocity.

The reader interested in additional details about the thermodynamic basis of the Maxwell-Cattaneo model should consult reference 8.

5.3. Maxwellian Materials

As an additional illustration of a material with internal state variables which is not a reacting mixture or something similar, in this section we shall formulate the model of what is known as a *Maxwellian* material. The arguments in this section should be viewed as a generalization of the model discussed in Section 4.2. As in Section 4.2, all thermodynamic effects will be suppressed by forcing the temperature to be constant. Unlike Section 4.2, we will allow the constitutive equations to depend upon $\dot{\mathbf{F}}$. In this way, viscous dissipation is included in the model. The constitutive equations which define the Maxwellian material are as follows:

$$\left(\psi,\dot{\mathbf{T}}\right)=f\left(\mathbf{F},\dot{\mathbf{F}},\mathbf{T}\right) \tag{5.3.1}$$

As should be apparent to the reader, we have taken the internal state variable to be the Cauchy stress tensor $\mathbf{T}$. Because of this choice, we have in effect taken the response function $\mathbf{G}$ in (5.1.5) to be the identity function. Of course, equation (1.7.12) is a one dimensional special case of the constitutive equation for $\dot{\mathbf{T}}$ implied by (5.3.1).

Given (5.3.1), we can write

$$\psi = u\left(\mathbf{F},\dot{\mathbf{F}},\mathbf{T}\right) \tag{5.3.2}$$

and

$$\dot{\mathbf{T}} = \Omega\left(\mathbf{F},\dot{\mathbf{F}},\mathbf{T}\right) \tag{5.3.3}$$

Exercise 5.3.1

Show that the axiom of material frame indifference forces the functions u and Ω to obey the following restrictions:

$$u\left(\mathbf{F},\dot{\mathbf{F}},\mathbf{T}\right) = u\left(\mathbf{QF},\dot{\mathbf{Q}}\mathbf{F}+\mathbf{Q}\dot{\mathbf{F}},\mathbf{QTQ}^T\right) \tag{5.3.4}$$

and

$$\mathbf{Q}\Omega\left(\mathbf{F},\dot{\mathbf{F}},\mathbf{T}\right)\mathbf{Q}^T + \dot{\mathbf{Q}}\mathbf{TQ}^T + \mathbf{QT}\dot{\mathbf{Q}}^T = \mathbf{\Omega}\left(\mathbf{QF},\dot{\mathbf{Q}}\mathbf{F}+\mathbf{Q}\dot{\mathbf{F}},\mathbf{QTQ}^T\right) \tag{5.3.5}$$

for all $\mathbf{Q}(t)$ in $\mathscr{Q}(\mathscr{V})$ and $\dot{\mathbf{Q}}(t)$ in $\mathscr{L}(\mathscr{V};\mathscr{V})$ with $\dot{\mathbf{Q}}(t)\mathbf{Q}(t)^T = -(\dot{\mathbf{Q}}(t)\mathbf{Q}(t)^T)^T$ such that $\left(\mathbf{QF},\dot{\mathbf{Q}}\mathbf{F}+\mathbf{Q}\dot{\mathbf{F}},\mathbf{QTQ}^T\right)$ is in the domain of u and $\mathbf{\Omega}$.

Exercise 5.3.2

Show that (5.3.4) and (5.3.5) force the two response functions to obey the following relationships:

$$u\left(\mathbf{F},\dot{\mathbf{F}},\mathbf{T}\right) = u\left(\mathbf{U},\mathbf{R}^T\mathbf{DRU},\mathbf{R}^T\mathbf{TR}\right) \tag{5.3.6}$$

and

$$\mathbf{\Omega}\left(\mathbf{F},\dot{\mathbf{F}},\mathbf{T}\right) = \mathbf{R\Omega}\left(\mathbf{U},\mathbf{R}^T\mathbf{DRU},\mathbf{R}^T\mathbf{TR}\right)\mathbf{R}^T + \mathbf{WT}-\mathbf{TW} \tag{5.3.7}$$

Note that (5.3.7) and (5.3.3) can be combined to obtain

$$\dot{\mathbf{T}} - \mathbf{WT}+\mathbf{TW} = \mathbf{R\Omega}\left(\mathbf{U},\mathbf{R}^T\mathbf{DRU},\mathbf{R}^T\mathbf{TR}\right)\mathbf{R}^T \tag{5.3.8}$$

The left side of (5.3.8) is the co-rotational stress rate introduced in Exercise 4.9.8.

Exercise 5.3.3

By an argument entirely similar to one given in Section 4.9, constitutive functions with argument $\left(\mathbf{U},\mathbf{R}^T\mathbf{DRU},\mathbf{R}^T\mathbf{TR}\right)$ can be replaced by functions of the argument $\left(\mathbf{C},\mathbf{F}^T\mathbf{DF},\mathbf{F}^T\mathbf{TF}\right)$. Provide the details of this argument and show that one can define functions $\hat{u}$ and $\hat{\boldsymbol{\Omega}}$ such that (5.3.2) and (5.3.3) can be replaced by

$$\psi=\hat{u}\left(\mathbf{C},\mathbf{F}^T\mathbf{DF},\mathbf{F}^T\mathbf{TF}\right) \tag{5.3.9}$$

and

$$\dot{\mathbf{T}}+\mathbf{L}^T\mathbf{T}+\mathbf{TL}=\mathbf{F}\hat{\boldsymbol{\Omega}}\left(\mathbf{C},\mathbf{F}^T\mathbf{DF},\mathbf{F}^T\mathbf{TF}\right)\mathbf{F}^T \tag{5.3.10}$$

The left side of (5.3.10)(5.3.10) is the convected stress rate introduced in Exercise 4.9.8.

Exercise 5.3.4

Show that (5.3.10) is equivalent to

$$\overline{\left(\mathbf{F}^T\mathbf{TF}\right)}^{\cdot}=\mathbf{C}\hat{\boldsymbol{\Omega}}\left(\mathbf{C},\mathbf{F}^T\mathbf{DF},\mathbf{F}^T\mathbf{TF}\right)\mathbf{C} \tag{5.3.11}$$

Equation (5.3.11) suggests an alternate formulation of the results of this section by taking the internal state variable to be the tensor $\mathbf{F}^T\mathbf{TF}$.

Exercise 5.3.5

Adopt definitions like those given in Section 4.10 and show that for an isotropic solid in its undistorted reference configuration

$$\psi=\overline{u}\left(\mathbf{B},\mathbf{D},\mathbf{T}\right) \tag{5.3.12}$$

and

$$\dot{\mathbf{T}}+\mathbf{L}^T\mathbf{T}+\mathbf{TL}=\overline{\boldsymbol{\Omega}}\left(\mathbf{B},\mathbf{D},\mathbf{T}\right) \tag{5.3.13}$$

where the functions $\overline{u}$ and $\overline{\boldsymbol{\Omega}}$ are defined analogous to the definition (4.10.37) and they must obey the following restrictions:

$$\overline{u}\left(\mathbf{B},\mathbf{D},\mathbf{T}\right)=\overline{u}\left(\mathbf{QBQ}^T,\mathbf{QDQ}^T,\mathbf{QTQ}^T\right) \tag{5.3.14}$$

and

$$\mathbf{Q}\overline{\boldsymbol{\Omega}}\left(\mathbf{B},\mathbf{D},\mathbf{T}\right)\mathbf{Q}^T=\overline{\boldsymbol{\Omega}}\left(\mathbf{QBQ}^T,\mathbf{QDQ}^T,\mathbf{QTQ}^T\right) \tag{5.3.15}$$

for all orthogonal linear transformations $\mathbf{Q}$ in $\mathscr{Q}(\mathscr{V})^{+}$.

Exercise 5.3.6

Show that for a fluid the constitutive equations (5.3.2) and (5.3.3) reduce to

$$\psi = \tilde{u}(\rho, \mathbf{D}, \mathbf{T}) \tag{5.3.16}$$

and

$$\dot{\mathbf{T}} + \mathbf{L}^{T}\mathbf{T} + \mathbf{T}\mathbf{L} = \tilde{\mathbf{\Omega}}(\rho, \mathbf{D}, \mathbf{T}) \tag{5.3.17}$$

where $\tilde{u}$ and $\tilde{\mathbf{\Omega}}$ must obey the following restrictions:

$$\tilde{u}(\rho, \mathbf{D}, \mathbf{T}) = \tilde{u}(\rho, \mathbf{Q}\mathbf{D}\mathbf{Q}^{T}, \mathbf{Q}\mathbf{T}\mathbf{Q}^{T}) \tag{5.3.18}$$

and

$$\mathbf{Q}\tilde{\mathbf{\Omega}}(\rho, \mathbf{D}, \mathbf{T})\mathbf{Q}^{T} = \tilde{\mathbf{\Omega}}(\rho, \mathbf{Q}\mathbf{D}\mathbf{Q}^{T}, \mathbf{Q}\mathbf{T}\mathbf{Q}^{T}) \tag{5.3.19}$$

for all orthogonal linear transformations $\mathbf{Q}$ in $\mathscr{Q}(\mathscr{V})^{+}$.

Next, we wish to investigate the thermodynamic restrictions implied by the entropy inequality on the constitutive equations (5.3.1). It follows from (5.3.2) that

$$\dot{\psi} = \operatorname{tr}\left(\frac{\partial u}{\partial \mathbf{F}}^{T}\dot{\mathbf{F}}\right) + \operatorname{tr}\left(\frac{\partial u}{\partial \dot{\mathbf{F}}}^{T}\ddot{\mathbf{F}}\right) + \operatorname{tr}\left(\frac{\partial u}{\partial \mathbf{T}}\dot{\mathbf{T}}\right) \tag{5.3.20}$$

If (5.3.2), (5.3.3) and (5.3.20) are substituted into the isothermal form of the entropy inequality, (4.2.1), it is readily shown that

$$\psi = u(\mathbf{F}, \mathbf{T}) \tag{5.3.21}$$

and

$$\operatorname{tr}\left(\mathbf{F}^{-1}\left(\mathbf{T} - \rho\mathbf{F}\frac{\partial u}{\partial \mathbf{F}}^{T}\right)\dot{\mathbf{F}}\right) + \rho \operatorname{tr}\left(\mathbf{\Sigma}(\mathbf{F}, \mathbf{T})\mathbf{\Omega}(\mathbf{F}, \dot{\mathbf{F}}, \mathbf{T})\right) \geq 0 \tag{5.3.22}$$

where the affinity $\mathbf{\Sigma}$ is a symmetric linear transformation defined by

$$\mathbf{\Sigma} = \frac{\partial u}{\partial \mathbf{T}} \tag{5.3.23}$$

We shall restrict ourselves to the case where the equilibrium state is defined by $\dot{\mathbf{F}}=\Sigma=\mathbf{0}$ and assume (5.3.23) can be inverted to yield

$$\mathbf{T}=\hat{\hat{\mathbf{T}}}(\mathbf{F},\boldsymbol{\Sigma}) \tag{5.3.24}$$

As in Section 5.1, the assumed inversion leading to (5.3.24) implies that the symmetric fourth order tensor $\partial\boldsymbol{\Sigma}(\mathbf{F},\mathbf{T})/\partial\mathbf{T}$ is regular as a linear transformation from $\mathscr{L}(\mathscr{V};\mathscr{V})$ to $\mathscr{L}(\mathscr{V};\mathscr{V})$. Given (5.3.24), we can replace (5.3.3) by

$$\dot{\mathbf{T}}=\hat{\hat{\boldsymbol{\Omega}}}(\mathbf{F},\dot{\mathbf{F}},\boldsymbol{\Sigma}) \tag{5.3.25}$$

where

$$\hat{\hat{\boldsymbol{\Omega}}}(\mathbf{F},\dot{\mathbf{F}},\boldsymbol{\Sigma})=\boldsymbol{\Omega}\left(\mathbf{F},\dot{\mathbf{F}},\hat{\hat{\mathbf{T}}}(\mathbf{F},\boldsymbol{\Sigma})\right) \tag{5.3.26}$$

By the same type of argument, we can replace (5.3.21) by

$$\psi=\hat{\hat{u}}(\mathbf{F},\boldsymbol{\Sigma}) \tag{5.3.27}$$

where

$$\hat{\hat{u}}(\mathbf{F},\boldsymbol{\Sigma})=u\left(\mathbf{F},\hat{\hat{\mathbf{T}}}(\mathbf{F},\boldsymbol{\Sigma})\right) \tag{5.3.28}$$

Exercise 5.3.7

Show that

$$\frac{\partial\hat{\hat{u}}(\mathbf{F},\boldsymbol{\Sigma})}{\partial F_{jJ}}=\frac{\partial u(\mathbf{F},\mathbf{T})}{\partial F_{jJ}}-\Sigma_{rk}(\mathbf{F},\mathbf{T})\frac{\partial\hat{\hat{T}}_{rk}(\mathbf{F},\boldsymbol{\Sigma})}{\partial F_{jJ}} \tag{5.3.29}$$

Exercise 5.3.8

Show that in thermodynamic equilibrium the following results must hold:

$$\hat{\hat{\boldsymbol{\Omega}}}(\mathbf{F},\mathbf{0},\mathbf{0})=\mathbf{0} \tag{5.3.30}$$

$$\hat{\hat{\mathbf{T}}}(\mathbf{F},\mathbf{0})=\rho\mathbf{F}\frac{\partial\hat{\hat{u}}(\mathbf{F},\mathbf{0})}{\partial\mathbf{F}}^{T} \tag{5.3.31}$$

$$\operatorname{tr}\left(\mathbf{F}^{-1}\left(\frac{\partial \mathbf{K}(\mathbf{F},\mathbf{0})}{\partial \boldsymbol{\Sigma}}[\mathbf{B}]\right)\mathbf{A}\right)+\rho \operatorname{tr}\left(\mathbf{B}\left(\frac{\partial \hat{\hat{\Omega}}(\mathbf{F},\mathbf{0},\mathbf{0})}{\partial \dot{\mathbf{F}}}[\mathbf{A}]\right)\right)=0 \tag{5.3.32}$$

and

$$\operatorname{tr}\left(\mathbf{B}\left(\frac{\partial \hat{\hat{\Omega}}(\mathbf{F},\mathbf{0},\mathbf{0})}{\partial \boldsymbol{\Sigma}}[\mathbf{B}]\right)\right)\geq 0 \tag{5.3.33}$$

for all linear transformations $\mathbf{A}$ and symmetric linear transformations $\mathbf{B}$. The quantity $\mathbf{K}(\mathbf{F},\boldsymbol{\Sigma})$ in (5.3.31) is defined by

$$\mathbf{K}(\mathbf{F},\boldsymbol{\Sigma})=\hat{\hat{\mathbf{T}}}(\mathbf{F},\boldsymbol{\Sigma})-\rho \mathbf{F}\left.\frac{\partial u(\mathbf{F},\mathbf{T})}{\partial \mathbf{F}}^{T}\right|_{\mathbf{T}=\hat{\hat{\mathbf{T}}}(\mathbf{F},\boldsymbol{\Sigma})} \tag{5.3.34}$$

The formal similarity of the results in this section with those in Section 5.2 should be clear to the reader. The results (5.3.29), (5.3.30), (5.3.31) and (5.3.32) are similar in form to (5.2.17), (5.2.18), (5.2.19) and (5.2.20), respectively. The reduction of the results in this section to various linear isotropic models can also be carried out. The results are formally very complicated and will not be developed here. The reader interested in additional information on the thermodynamics of Maxwellian materials should consult Reference 9.

The literature on continuum mechanics is full of models which utilize internal state variables. The example contained in this section and in Sections 5.1 and 5.2 represent a short introduction to an extensive subject. The reader interested in additional references on models which utilize internal state variables might wish to consult References 10 through 13.

5.4. Closing Remarks-Alternate Forms of the Entropy Inequality

As the results in Chapter 4 and 5 illustrate, one can make productive use of the entropy inequality in the formulation of constitutive equations for materials. The entropy inequality we have adopted was introduced in Sections 1.6 and 3.5. In this closing section of this textbook, it is useful to comment on other forms of the entropy inequality which have been introduced in the continuum mechanics literature as generalizations of the one given here.

Müller, in Ref. 14, proposed that (3.5.8) be replaced by

$$\dot{\overline{\int_{\chi(\mathscr{P},t)} \rho\eta\, dv}} \geq -\oint_{\partial\chi(\mathscr{P},t)} \mathbf{h}\cdot d\mathbf{s}+\int_{\chi(\mathscr{P},t)} \frac{\rho r}{\theta} dv \tag{5.4.1}$$

where $\mathbf{h}$ is the *entropy flux vector*. In the formulation proposed by Müller, $\mathbf{h}$, like $\psi,\eta,\mathbf{T}$, and $\mathbf{q}$, is prescribed by a constitutive equation. In reference 14, Müller discusses cases where the restrictions implied by his entropy inequality, combined with material frame indifference and material symmetry considerations, implies that **h** is given by

$$\mathbf{h}=\frac{\mathbf{q}}{\theta} \tag{5.4.2}$$

The motivation for replacing (3.5.8) with the more general statement (5.4.1) can be found in Reference 15. In this work, Müller formulated a model of a Maxwellian gas for the purpose of obtaining a model which did not imply that temperature disturbances propagate with infinite velocity. In effect, his results show that if one does not assume in advance the special relationship (5.4.2), a result could be obtained which was sufficiently general to allow temperature disturbances to propagate with finite velocity. More importantly, results were obtained by Müller in Ref. 15 which were self consistent with predictions of the kinetic theory of gases.

Exercise 5.4.1

Given (5.4.1), show that (3.5.15) must be replaced by

$$-\rho\left(\dot{\psi}+\eta\dot{\theta}\right)+\operatorname{tr}\mathbf{TL}+\theta\operatorname{div}\mathbf{p}-\frac{\mathbf{q}}{\theta}\cdot\operatorname{grad}\theta\geq 0 \tag{5.4.3}$$

where the vector **p** is given by

$$\mathbf{p}=\mathbf{k}-\frac{\mathbf{q}}{\theta} \tag{5.4.4}$$

Exercise 5.4.2

Consider a material defined by the following special constitutive equations:

$$\left(\psi,\eta,\mathbf{T},\mathbf{q},\mathbf{p}\right)=f\left(\theta,\mathbf{g},\mathbf{F}\right) \tag{5.4.5}$$

Derive all of the restrictions implied by (5.4.3) on these constitutive equations. Among the results you should obtain is that the vector **p** must take the special form

$$\mathbf{p}_R\equiv\left|\det\mathbf{F}\right|\mathbf{F}^{-1}\mathbf{p}=\mathbf{\Omega}\left(\theta\right)\operatorname{GRAD}\theta+\boldsymbol{\omega}\left(\theta\right) \tag{5.4.6}$$

where $\mathbf{\Omega}(\theta)$ is an arbitrary skew-symmetric linear transformation and $\boldsymbol{\omega}\left(\theta\right)$ is an arbitrary vector. Therefore, in this case **p** is not necessarily zero.

Exercise 5.4.3

Show that if the material defined by (5.4.5) is an isotropic material, then it follows from (5.4.6) that $\mathbf{p}$ is zero, and thus (5.4.2) is obeyed.

Müller has also formulated a procedure for exploiting the entropy inequality (5.4.1) by taking the external heat supply density r and the body force density $\mathbf{b}$ to be zero [Ref. 16].

Gurtin and Williams, in Ref. 17, proposed an entropy inequality of the form

$$\dot{\overline{\int_{\chi(\mathscr{P},t)} \rho\eta\, dv}} \geq -\oint_{\partial\chi(\mathscr{P},t)} \frac{\mathbf{q}}{\varphi}\cdot d\mathbf{s} + \int_{\chi(\mathscr{P},t)} \frac{\rho r}{\theta} dv \tag{5.4.7}$$

where φ is a *surface-relevant temperature* and θ is a *volume-relevant temperature.*

Exercise 5.4.4

Given (5.4.7) show that (3.5.15) must be replaced by

$$-\rho\left(\dot{\psi} + \eta\dot{\theta}\right) + \operatorname{tr}\mathbf{TL} + \left(\frac{\theta}{\varphi} - 1\right)\operatorname{div}\mathbf{q} - \frac{\theta}{\varphi^2}\mathbf{q}\cdot\operatorname{grad}\varphi \geq 0 \tag{5.4.8}$$

Exercise 5.4.5

Consider a material defined by the following special constitutive equations:

$$\left(\psi, \eta, \mathbf{T}, \mathbf{q}, \theta\right) = f\left(\varphi, \operatorname{grad}\varphi, \mathbf{F}\right) \tag{5.4.9}$$

Derive all of the restrictions implied by (5.4.9) on these constitutive equations. In particular, determine for the material defined by (5.4.9) whether or not one can prove that $\theta = \varphi$. Additional discussion of the Gurtin and Williams proposal can be found in references 18 through 22.

Green and Laws, in reference 23, proposed an entropy inequality in the form

$$\dot{\overline{\int_{\chi(\mathscr{P},t)} \rho\eta\, dv}} \geq -\oint_{\partial\chi(\mathscr{P},t)} \frac{\mathbf{q}}{\varphi}\cdot d\mathbf{s} + \int_{\chi(\mathscr{P},t)} \frac{\rho r}{\varphi} dv \tag{5.4.10}$$

where φ is positive valued function that is prescribed by a constitutive equation. In the Green and Laws formulation the temperature θ is regarded as a primitive quantity which need not necessarily equal the quantity φ. They do require that φ equals θ in equilibrium.

Exercise 5.4.6

Given (5.4.10) show that (3.5.15) must be replaced by

$$-\rho\left(\dot{\psi}+\eta\dot{\varphi}\right)+\mathbf{tr\,TL}-\frac{1}{\varphi}\mathbf{q}\cdot\operatorname{grad}\varphi\geq 0 \tag{5.4.11}$$

when ψ is defined by

$$\psi=\varepsilon-\eta\varphi \tag{5.4.12}$$

rather than by (3.5.14).

Exercise 5.4.7

Green and Laws considered the case where the material is at rest and proposed the following constitutive equations:

$$\left(\psi,\eta,\mathbf{q},\varphi\right)=f\left(\theta,\dot{\theta},\operatorname{grad}\theta\right) \tag{5.4.13}$$

Substitute (5.4.13) into (5.4.11) and show that

$$\frac{\partial\psi}{\partial\dot{\theta}}+\eta\frac{\partial\varphi}{\partial\dot{\theta}}=0 \tag{5.4.14}$$

$$\rho\left(\frac{\partial\psi}{\partial\operatorname{grad}\theta}+\eta\frac{\partial\varphi}{\partial\operatorname{grad}\theta}\right)+\frac{\mathbf{q}}{\varphi}\frac{\partial\varphi}{\partial\dot{\theta}}=\mathbf{0} \tag{5.4.15}$$

$$\mathbf{q}\otimes\frac{\partial\varphi}{\partial\, grad\,\theta}+\frac{\partial\varphi}{\partial\, grad\,\theta}\otimes\mathbf{q}=\mathbf{0} \tag{5.4.16}$$

and

$$-\rho\left(\frac{\partial\psi}{\partial\theta}+\eta\frac{\partial\varphi}{\partial\theta}\right)\dot{\theta}-\frac{\partial\varphi}{\partial\theta}\frac{\mathbf{q}}{\varphi}\cdot\operatorname{grad}\theta\geq 0 \tag{5.4.17}$$

Exercise 5.4.8

Assume that the heat flux vector $\mathbf{q}$ is not zero and prove from the above results that

$$\varphi=\varphi\left(\theta,\dot{\theta}\right) \tag{5.4.18}$$

Exercise 5.4.9

On the assumption that $\partial\varphi/\partial\dot{\theta}\neq 0$ and $\partial\psi/\partial\operatorname{grad}\theta\neq\mathbf{0}$, show that $\mathbf{q}$ must obey

$$\frac{\partial \mathbf{q}}{\partial \operatorname{grad} \theta} = \left(\frac{\partial \mathbf{q}}{\partial \operatorname{grad} \theta} \right)^T \tag{5.4.19}$$

Green and Laws pointed out that for a material which conducts heat according to a Fourier law of the form

$$\mathbf{q} = -\mathbf{K}\left(\theta, \dot{\theta}\right) \operatorname{grad} \theta \tag{5.4.20}$$

equation (5.4.19) shows that the conductivity tensor **K** must be symmetric. The formulation of Green and Laws has been used by Green and Lindsay to formulate a thermoelasticity theory. [Ref. 24]. This theory has been adopted by Prevost and Tao to numerically analyze transient phenomena in thermoelastic solids.[Ref. 25]

References

1. VINCENTI, W.G., and C.H. KRUGER, JR., *Introduction to Physical Gas Dynamics*, John Wiley and Sons, New York (1965).
2. BOWEN, R.M., On the Stoichiometry of Chemically Reacting Materials, *Arch. Rational Mech. Analysis*, **29**, 114-124 (1963).
3. BOWEN, R.M., Theory of Mixtures, *Continuum Physics*, Vol. III (ed. A.C. Eringen), Academic Press, New York (1976).
4. COLEMAN, B.D., and M.E. GURTIN, Thermodynamics with Internal State Variables, *J. Chem. Phys.*, **47**, 597-613 (1967).
5. BOWEN, R.M., Thermochemistry of Reacting Materials, *J.Chem.Phys.*, 49, 1625-1637 (1968), **50**, 4601-4602 (1969).
6. TRUESDELL, C., *Rational Thermodynamics*, Second Edition, Springer-Verlag, New York (1984).
7. PRIGOGINE, I. and R. DEFAY, *Chemical Thermodynamics* (trans. by D.H. Everett), Longmans Green and Co., New York (1954).
8. NUNZIATO, J.W., and D.S. DRUMHELLER, The Thermodynamics of Maxwellian Materials, *Int.J.Solid Strs.*, **14**, 545-588 (1978).
9. COLEMAN, B. D., FABRIZIO, M., and D. R. OWEN, On the Thermodynamics of Second Sound in Dielectric Crystals, *Arch. Rational Mech. Anal.*, **80**, 135-158 (1982).
10. KRATOCHVIL, J., and O. W. DILLON, JR., Thermodynamics of Elastic-Plastic Materials as a Theory with Internal State Variables, *J. Applied Phys.*, **40**, 3207-3218 (1969).
11. DASHNER, P. A., and W. E. VANARSDALE, A Phenomenological Theory for Elastic Fluids, *J. Non-Newtonian Fluid Mechanics*, **8**, 59-67 (1981).
12. WRIGHT, THOMAS W., *Some Aspects of Adiabatic Shear Bands, Metastability and Partial Differential Equations*, (ed. S. Antman, J. L. Ericksen, D. KindeSpringer-Verlag, New York (1987).
13. DESAI, C. S., and R. H. GALLAHER, (ed.), *Proceedings International Conference on Constitutive Laws for Engineering Materials*, Tucson (1983).
14. MÜLLER, I., On the Entropy Inequality, *Arch. Rational Mech. Anal.*, **26**, 118-141 (1967).
15. MÜLLER, I., Zum Paradoxon der Warmeleitungstheorie, *Zeitschrift fur Physik*, **198**, 329-344 (1967).

16. MÜLLER, I., The Coldness, a Universal Function in Thermoelastic Bodies, *Arch Rational Mech. Anal.*, **41**, 319-332 (1971).
17. GURTIN, M. E., and W. O. WILLIAMS, On the Clausius-Duhem Inequality, *Z. Angew. Math. Physik*, **17**, 626-633 (1966).
18. GURTIN, M. E., and W. O. WILLIAMS, An Axiomatic Foundation for Continuum Thermodynamics, *Arch. Rational Mech. Anal.*, **26,** 83-117 (1967).
19. CHEN, P. J., and M. E. GURTIN, On a Theory of Heat Conduction Involving Two Temperatures, *Z. Angew. Math. Physik*, **19**, 614-627 (1968).
20. CHEN, P. J., GURTIN, M. E., and W. O. WILLIAMS, A Note on Non-simple Heat Conduction, *Z. Angew. Math. Physik*, **19**, 969-970 (1968).
21. CHEN, P. J., GURTIN, M. E., and W. O. WILLIAMS, On the Thermodynamics of Non-simple Elastic Materials with Two Temperatures, *Z. Angew. Math. Physik*, **20**, 107-112 (1969).
22. GURTIN, M. E., On the Existence of a Single Temperature in Continuum Thermodynamics, *Z. Angew. Math. Physik*, **27**, 775-779 (1976).
23. GREEN, A. E., and N. LAWS, On the Entropy Production Inequality, *Arch. Rational Mech. Anal.*, **45**, 47-53 (1972).
24. GREEN, A. E., and K. A. LINDSAY, Thermoelasticity, *J. of Elasticity*, **2**, 1-7 (1972).
25. PREVOST, J. H., and D. TAO, Finite Element Analysis of Dynamic Coupled Thermoelasticity Problems With Relaxation Times, *J. Appl. Mech.*, **50**, 817-822 (1983).

Bibliography

26. LEBON, G., A Thermodynamic Analysis of Rigid Heat Conductors, *Intl. J. Engr.Sci.*, **18**, 727-739 (1980).
27. KRANYS, M., Hyperbolic Elasticity of Dissipative Media and its Wave Propagation Modes, *J. Phys. A: Math.* Gen., **10**, 689-709 (1977).
28. STAVERMAN, A.J. and F. SCHWARZL, Linear Deformation Behavior of High Polymers, *Physik der Hochpolymeren*, Vol. 4 (ed. M. Stuart) Springer-Verlag, Berlin (1956).
29. MEIXNER, J. and H.G. REIK, *Thermodynamik der Irreversiblen Processe*, Handbuch der Physik, Vol III/2 (ed. S. Flugge), Springer-Verlag, Berlin (1959).
30. GURTIN, M.E. and A.C. PIPKIN, A General Theory of Heat Conduction Finite Wave Speeds, *Arch. Rational Mech.* Anal., **31**, 113-126 (1968).
31. LORD, H.W. and Y. SHULMAN, A Generalized Dynamical Theory of Thermoelasticity, *J. Mech. Phys. Solids*, **15**, 299-309 (1967).
32. PARKER, D.F., Non-Linearity, *Relaxation and Diffusion in Acoustics and Ultrasonics, J. Fluid Mech.*, **39**, 793-815 (1969).
33. DILLON, O. W., C. T. TSAI and R. J. DEANGELIS, Dislocation Dynamics During the Growth of Silicon Ribbon, *J. Appl. Phys.*, **60**, 1784-1792 (1986).
34. DILLON, O. W., C. T. TSAI and R. J. DEANGELIS, Dislocation Dynamics of Web Type Silicon Rubber, *J. of Crystal Growth*, **82**, 50-59 (1987

Appendix A

Mathematical Preliminaries

The purpose of this appendix is to present certain of the mathematical concepts which are necessary for the study of three dimensional continuum mechanics. This appendix contains a brief discussion of vector spaces, linear transformations, inner product spaces, Euclidean point spaces, cross products, determinants, tensor algebra and vector calculus.

This appendix is not self contained. The reader is assumed to be familiar with the concept of a set, with the definition of the Cartesian product of a finite number of sets and with the definition of a function. In addition, the reader is assumed to be familiar with the algebraic concept of a group. These concepts, as well as the others discussed below, can be found in any linear algebra textbook.

A.1. Vector Spaces

In this appendix, the symbol $\mathscr{R}$ denotes the set of real numbers. Subsets of $\mathscr{R}$ are defined as follows:

1. $\mathscr{R}^{+} = \{x \mid x \text{ is in } \mathscr{R} \text{ and } 0 < x < \infty\}$,
2. $(a,b) = \{x \mid x \text{ is in } \mathscr{R} \text{ and } a < x < b\}$,
3. $[a,b] = \{x \mid x \text{ is in } \mathscr{R} \text{ and } a \le x \le b\}$,
4. $[a,b) = \{x \mid x \text{ is in } \mathscr{R} \text{ and } a \le x < b\}$

and

5. $(a,b] = \{x \mid x \text{ is in } \mathscr{R} \text{ and } a < x \le b\}$

Definition. Let $\mathscr{V}$ be a set and $\mathscr{R}$ the set of real numbers. $\mathscr{V}$ is a real *vector space* if it satisfies the following rules:

(a) There exists a binary operation in $\mathscr{V}$ called *addition* and denoted by $+$ such that
   1. $(\mathbf{u}+\mathbf{v})+\mathbf{w} = \mathbf{u}+(\mathbf{v}+\mathbf{w})$, for all $\mathbf{u},\mathbf{v},\mathbf{w}$ in $\mathscr{V}$.
   2. $\mathbf{u}+\mathbf{v} = \mathbf{v}+\mathbf{u}$, all $\mathbf{u},\mathbf{v}$ in $\mathscr{V}$.
   3. There exists an element $\mathbf{0}$ in $\mathscr{V}$ such that $\mathbf{u}+\mathbf{0} = \mathbf{u}$, for all $\mathbf{u}$ in $\mathscr{V}$.
   4. For every $\mathbf{u}$ in $\mathscr{V}$, there exists an element $-\mathbf{u}$ in $\mathscr{V}$ such that $\mathbf{u}+(-\mathbf{u}) = \mathbf{0}$.

(b) There exists an operation called *scalar multiplication* in which every real number λ in $\mathscr{R}$ can be combined with every $\mathbf{u}$ in $\mathscr{V}$ to give an element $\lambda\mathbf{u}$ in $\mathscr{V}$ such that
   1. $\lambda(\mu\mathbf{u}) = (\lambda\mu)\mathbf{u}$.
   2. $(\lambda+\mu)\mathbf{u} = \lambda\mathbf{u}+\mu\mathbf{u}$.

3. $\lambda(\mathbf{u}+\mathbf{v}) = \lambda\mathbf{u} + \lambda\mathbf{v}$.
4. $1\mathbf{u} = \mathbf{u}$, for all λ, μ in $\mathscr{R}$ and all $\mathbf{u}, \mathbf{v}$ in $\mathscr{V}$.

Since we shall always deal with vector spaces over the real numbers, we shall simply call $\mathscr{V}$ a vector space. The elements of $\mathscr{V}$ are called *vectors* and the elements of $\mathscr{R}$ are called *scalars*.

Example. A common example of a vector space is the set of all *ordered* n-tuples of real numbers

$$\mathbf{u} = (u_1, u_2, u_3, \ldots, u_n) \tag{A.1.1}$$

where n is a fixed integer. Addition of two n-tuples is defined by

$$\begin{aligned} \mathbf{u} + \mathbf{v} &= (u_1, u_2, \ldots, u_n) + (\upsilon_1, \upsilon_2, \ldots, \upsilon_n) \\ &= (u_1 + \upsilon_1, u_2 + \upsilon_2, \ldots, u_n + \upsilon_n) \end{aligned} \tag{A.1.2}$$

and multiplication by a scalar is defined by

$$\lambda\mathbf{u} = (\lambda u_1, \lambda u_2, \lambda u_3, \ldots, \lambda u_n) \tag{A.1.3}$$

With these definitions, it is easy to prove that the set of all n-tuples form a vector space. We shall always denote this vector space by $\mathscr{R}^n$.

Theorem A.1.1. $\lambda\mathbf{v} = \mathbf{0}$ if and only if $\lambda = 0$ or $\mathbf{v} = \mathbf{0}$.

Theorem A.1.2. $(-\lambda)\mathbf{v} = -\lambda\mathbf{v}$.

Theorem A.1.3. $\lambda(-\mathbf{v}) = -\lambda\mathbf{v}$.

The proofs of these theorems can be found in any linear algebra text.

It is easily shown by induction that the distributive laws in the definition of a vector space imply

$$\left(\sum_{i=1}^{k} \lambda_k\right)\mathbf{v} = \sum_{j=1}^{k} (\lambda_j \mathbf{v}) \tag{A.1.4}$$

and

$$\lambda \sum_{j=1}^{k} \mathbf{v}_j = \sum_{j=1}^{k} \lambda \mathbf{v}_j \tag{A.1.5}$$

for k equal to a finite integer.

Definition. A set of p $(p \geq 1)$ vectors $\{\mathbf{v}_1, \mathbf{v}_2, ..., \mathbf{v}_p\}$ of a vector space $\mathscr{V}$ is said to be *linearly dependent* if there exist coefficients $\{\lambda_1, ..., \lambda_p\}$ *not all zero* such that

$$\sum_{j=1}^{p} \lambda_j \mathbf{v}_j = \mathbf{0}$$

Definition. A set of p $(p \geq 1)$ vectors $\{\mathbf{v}_1, \mathbf{v}_2, ..., \mathbf{v}_p\}$ of a vector space $\mathscr{V}$ is said to be *linearly independent* if they are not linearly dependent..

Therefore, for a set of p linearly independent vectors, the sum

$$\sum_{j=1}^{p} \lambda_j \mathbf{v}_j = \mathbf{0}$$

implies

$$\lambda_1 = \lambda_2 = \cdots = \lambda_p = 0$$

Definition. A vector space $\mathscr{V}$ is said to have *dimension n* (*n* a positive *integer*) if there exists a set of *n* linearly independent vectors and if there exists no set of more than *n* linearly independent vectors.

In this discussion we are only interested in vector spaces which have a finite number of independent vectors, i.e., *finite dimensional vector spaces*. We shall write $\dim \mathscr{V}$ to denote the dimension of a vector space $\mathscr{V}$.

Definition. A set of *n* linearly independent vectors in $\mathscr{V}$ is said to be a *basis* of $\mathscr{V}$ when $n = \dim \mathscr{V}$.

Theorem A.1.4, If $\{\mathbf{e}_1, ..., \mathbf{e}_n\}$ is a basis for $\mathscr{V}$, then every vector $\mathbf{v}$ in $\mathscr{V}$ can be written

$$\mathbf{v} = \sum_{j=1}^{n} \xi_j \mathbf{e}_j \tag{A.1.6}$$

Proof. Since $n = \dim \mathscr{V}$, it follows that the $n + 1$ vectors $\{\mathbf{v}, \mathbf{e}_1, ..., \mathbf{e}_n\}$ are linearly dependent. Therefore

$$\lambda \mathbf{v} + \sum_{j=1}^{n} \lambda_j \mathbf{e}_j = \mathbf{0}$$

where at least one coefficient is not zero. In particular $\lambda \neq 0$, because $\lambda = 0$ would require that $\lambda_1 = \lambda_2 = \cdots = \lambda_n = 0$, and then we would be forced to conclude that the $n + 1$ vectors $\{\mathbf{v}, \mathbf{e}_1, ..., \mathbf{e}_n\}$ are linearly independent. Therefore,

$$\mathbf{v} = \sum_{j=1}^{n} (-\lambda_j / \lambda)\mathbf{e}_j \equiv \sum_{j=1}^{n} \xi_j \mathbf{e}_j$$

Theorem A.1.5. The scalars $\{\xi_1, \xi_2, ..., \xi_n\}$ in the formula

$$\mathbf{v} = \sum_{j=1}^{n} \xi_j \mathbf{e}_j$$

are unique.

Proof. Let

$$\mathbf{v} = \sum_{j=1}^{n} \xi_j \mathbf{e}_j = \sum_{j=1}^{n} \xi'_j \mathbf{e}_j$$

Therefore,

$$\sum_{j=1}^{n} (\xi_j - \xi'_j)\mathbf{e}_j = \mathbf{0}$$

Since the vectors $\{\mathbf{e}_1, ..., \mathbf{e}_n\}$ are linearly independent, it immediately follows that

$$\xi_j = \xi'_j$$

for $j = 1, 2, ..., n$

Definition : The scalars $(\xi_1, ..., \xi_n)$ in the formula

$$\mathbf{v} = \sum_{j=1}^{n} \xi_j \mathbf{e}_j$$

are the *components* of $\mathbf{v}$ with respect to the basis $\{\mathbf{e}_1, ..., \mathbf{e}_n\}$.

Example. In $\mathscr{R}^n$ it is easily seen that the n vectors

$$
\begin{aligned}
\mathbf{i}_1 &= (1,0,\ldots,0),\\
\mathbf{i}_2 &= (0,1,\ldots,0),\\
&\cdot\\
&\cdot\\
&\cdot\\
\mathbf{i}_n &= (0,0,\ldots,1)
\end{aligned}
\tag{A.1.7}
$$

are linearly independent. Since any vector $\mathbf{v}$ in $\mathscr{R}^n$ can be written

$$\mathbf{v} = (\upsilon_1, \upsilon_2, \ldots, \upsilon_n) = \upsilon_1\mathbf{i}_1 + \upsilon_2\mathbf{i}_2 + \ldots + \upsilon_n\mathbf{i}_n \tag{A.1.8}$$

it follows that there exists no set of vectors in $\mathscr{R}^n$ which contains more than n linearly independent vectors. Consequently $\dim\mathscr{R}^n = n$ and $\{\mathbf{i}_1,\ldots,\mathbf{i}_n\}$ is a basis for $\mathscr{R}^n$.

It must be pointed out that a basis of a vector space is not unique. If $\{\mathbf{e}_1,\ldots,\mathbf{e}_n\}$ and $\{\overline{\mathbf{e}}_1,\ldots,\overline{\mathbf{e}}_n\}$ are both bases of $\mathscr{V}$, then by application of Theorem A.1.4 it follows that

$$\overline{\mathbf{e}}_k = \sum_{j=1}^{n} T_{jk}\mathbf{e}_j \tag{A.1.9}$$

and

$$\mathbf{e}_j = \sum_{q=1}^{n} \hat{T}_{qj}\overline{\mathbf{e}}_q \tag{A.1.10}$$

These formulas characterize the basis transformation from $\{\mathbf{e}_1,\ldots,\mathbf{e}_n\}$ to $\{\overline{\mathbf{e}}_1,\ldots,\overline{\mathbf{e}}_n\}$ and vice versa. It is elementary to use (A.1.9) and (A.1.10) to show that

$$\sum_{j=1}^{n} \hat{T}_{qj}T_{jk} = \delta_{qk} \tag{A.1.11}$$

and

$$\sum_{j=1}^{n} T_{qj}\hat{T}_{jk} = \delta_{qk} \tag{A.1.12}$$

where δ_{qk} denotes the *Kronecker delta* defined by

$$\delta_{qk} = \begin{cases} 1 & \text{if } q = k \\ 0 & \text{if } q \neq k \end{cases} \tag{A.1.13}$$

Equations (A.1.11) and (A.1.12) show that the basis transformation is characterized by a matrix $\left[T_{jk}\right]$ whose inverse is $\left[\hat{T}_{jk}\right]$.

Exercise A.1.1

If $\mathbf{v} = \sum_{j=1}^{n} \upsilon_j \mathbf{e}_j$ and $\mathbf{v} = \sum_{j=1}^{n} \bar{\upsilon}_j \bar{\mathbf{e}}_j$, show that the two sets of components are related by

$$\bar{\upsilon}_q = \sum_{j=1}^{n} \hat{T}_{qj} \upsilon_j \tag{A.1.14}$$

and

$$\upsilon_j = \sum_{k=1}^{n} T_{jk} \bar{\upsilon}_k \tag{A.1.15}$$

Equations (A.1.14) and (A.1.15) characterize the *transformation rule* for the components of a vector.

A.2. Linear Transformations

Definition. Let $\mathscr{U}$ and $\mathscr{V}$ denote vector spaces. A *linear transformation* is a function $\mathbf{A}: \mathscr{V} \to \mathscr{U}$ such that

$$\mathbf{A}(\mathbf{u}+\mathbf{v}) = \mathbf{A}(\mathbf{u}) + \mathbf{A}(\mathbf{v}) \tag{A.2.1}$$

and

$$\mathbf{A}(\lambda \mathbf{v}) = \lambda \mathbf{A}(\mathbf{v}) \tag{A.2.2}$$

for all $\mathbf{u}, \mathbf{v}$ in $\mathscr{V}$ and λ in $\mathscr{R}$.

It is obvious that (A.2.1) and (A.2.2)are equivalent to the single condition, $\mathbf{A}(\lambda \mathbf{u} + \mu \mathbf{v}) = \lambda \mathbf{A}(\mathbf{u}) + \mu \mathbf{A}(\mathbf{v})$. For simplicity of notation, it is often convenient to write

$$\mathbf{A}(\mathbf{v}) = \mathbf{A}\mathbf{v}$$

Prove the following theorems:

Exercise A.2.1

A linear transformation $\mathbf{A}:\mathscr{V} \rightarrow \mathscr{U}$ maps the zero vector of $\mathscr{V}$ into the zero vector of $\mathscr{U}$.

Exercise A.2.2

If $\{\mathbf{v}_1, \mathbf{v}_2, ..., \mathbf{v}_p\}$ is a system of linearly dependent vectors in $\mathscr{V}$, then $\{\mathbf{A}\mathbf{v}_1, \mathbf{A}\mathbf{v}_2, ..., \mathbf{A}\mathbf{v}_p\}$ is a system of linearly dependent vectors in $\mathscr{U}$.

Linearly independent vectors in $\mathscr{V}$ are not necessarily mapped into linearly independent vectors in $\mathscr{U}$. As an example, consider the linear transformation $\mathbf{A}$ that maps every vector in $\mathscr{V}$ into the zero vector in $\mathscr{U}$.

Definition. A nonempty subset $\mathscr{U}$ of a vector space $\mathscr{V}$ is a *subspace* if:

1. $\mathbf{u}, \mathbf{v}$ in $\mathscr{U}$ implies $\mathbf{u} + \mathbf{w}$ is in $\mathscr{U}$, for all $\mathbf{u}, \mathbf{v}$ in $\mathscr{U}$; and
2. $\mathbf{u}$ in $\mathscr{U}$ implies $\lambda \mathbf{u}$ is in $\mathscr{U}$; for all $\mathbf{u}$ in $\mathscr{U}$ and λ in $\mathscr{R}$

Definition. The *kernel*[1] of $\mathbf{A}$, written $\ker \mathbf{A}$ is defined by

$$\ker \mathbf{A} = \{\mathbf{v} \mid \mathbf{A}\mathbf{v} = \mathbf{0}, \mathbf{v} \text{ in } \mathscr{V}\}$$

It is easily shown that $\ker \mathbf{A}$ is a subspace of $\mathscr{V}$.

Definition. A linear transformation $\mathbf{A}:\mathscr{V} \rightarrow \mathscr{U}$ is *regular* if $\ker \mathbf{A} = \{\mathbf{0}\}$.

It is easily shown that $\ker \mathbf{A} = \{\mathbf{0}\}$ implies that $\mathbf{A}$ is one-to-one. Also, regular linear transformations map a system of linearly independent vectors into a system of linearly independent vectors.

Definition. The image space of $\mathbf{A}$, written $\operatorname{Im} \mathbf{A}$, is defined by

$$\operatorname{Im} \mathbf{A} = \{\mathbf{A}\mathbf{v} \mid \mathbf{v} \text{ in } \mathscr{V}\}$$

The image space is a subspace of $\mathscr{U}$.

Definition. If $\operatorname{Im} \mathbf{A} = \mathscr{U}$, $\mathbf{A}$ is a mapping of $\mathscr{V}$ onto $\mathscr{U}$.

Definition. The *rank* of $\mathbf{A}$ is $\dim \operatorname{Im} \mathbf{A}$.

---

[1] The subspace $\ker \mathbf{A}$ is also called the *nullity* of $\mathbf{A}$

Theorem A.2.1. $\dim \operatorname{Im} \mathbf{A} = \dim \mathscr{U}$ if and only if $\mathbf{A}$ is onto.

Proof. If $\dim \operatorname{Im} \mathbf{A} = \dim \mathscr{U}$, then $\operatorname{Im} \mathbf{A} = \mathscr{U}$ because $\operatorname{Im} \mathbf{A}$ is a subspace of $\mathscr{U}$. Conversely, if $\mathbf{A}$ is onto, $\operatorname{Im} \mathbf{A} = \mathscr{U}$, which implies that $\dim \operatorname{Im} \mathbf{A} = \dim \mathscr{U}$.

It is possible to show that

$$\dim \mathscr{V} = \dim \operatorname{Im} \mathbf{A} + \dim \ker \mathbf{A} \tag{A.2.3}$$

for a linear transformation $\mathbf{A}: \mathscr{V} \to \mathscr{U}$. Therefore,

$$\dim \operatorname{Im} \mathbf{A} \leq \dim \mathscr{V} \tag{A.2.4}$$

and

$$\dim \ker \mathbf{A} \leq \dim \mathscr{V} \tag{A.2.5}$$

In the special case where $\dim \mathscr{V} = \dim \mathscr{U}$, it is possible to state the following important theorem.

Theorem A.2.2. If $\mathbf{A}: \mathscr{V} \to \mathscr{U}$ is a linear transformation and if $\dim \mathscr{V} = \dim \mathscr{U}$, then $\mathbf{A}$ is a linear transformation onto $\mathscr{U}$ if and only if $\mathbf{A}$ is regular.

Proof. Assume that $\mathbf{A}: \mathscr{V} \to \mathscr{U}$ is onto $\mathscr{U}$, then (A.2.3) and Theorem A.2.1 show that

$$\dim \mathscr{V} = \dim \mathscr{U} = \dim \mathscr{U} + \dim \ker \mathbf{A}$$

Therefore $\dim \ker \mathbf{A} = 0$ and thus $\ker \mathbf{A} = \{\mathbf{0}\}$ and $\mathbf{A}$ is one to one. Next, assume $\mathbf{A}$ is one to one. Therefore, $\ker \mathbf{A} = \{\mathbf{0}\}$ and $\dim \ker \mathbf{A} = 0$. Equation (A.2.3) yields $\dim \mathscr{V} = \dim \operatorname{Im} \mathbf{A}$. Because $\dim \mathscr{V} = \dim \mathscr{U}$, it then follows that $\dim \mathscr{U} = \dim \operatorname{Im} \mathbf{A}$. This fact and Theorem A.2.1 show that $\mathbf{A}$ is onto.

Definition. A *regular onto* linear transformation $\mathbf{A}: \mathscr{V} \to \mathscr{U}$ is an *isomorphism*

Therefore, for an isomorphism $\mathbf{A}: \mathscr{V} \to \mathscr{U}$, there exists an inverse function $\mathbf{A}^{-1}: \mathscr{U} \to \mathscr{V}$.

Exercise A.2.3

Show that the inverse function $\mathbf{A}^{-1}: \mathscr{U} \to \mathscr{V}$ is a linear transformation and, thus, an isomorphism.

Definition. Two linear spaces, $\mathscr{V}$ and $\mathscr{U}$, are *isomorphic* if there exists an isomorphism of $\mathscr{V}$ onto $\mathscr{U}$.

Theorem A.2.3. If $\mathbf{A}: \mathscr{V} \to \mathscr{U}$ is an isomorphism, then $\dim \mathscr{V} = \dim \mathscr{U}$.

Proof. Since $\mathbf{A}$ is regular, $\dim \ker \mathbf{A} = 0$. Therefore, by this result, Theorem A.2.1 and (A.2.3), $\dim \mathscr{U} = \dim \operatorname{Im} \mathbf{A} = \dim \mathscr{V}$.

Definition. A linear transformation $\mathbf{A}$ of $\mathscr{V}$ into $\mathscr{V}$ is an *endomorphism*.

Definition. A regular linear transformation $\mathbf{A}$ of $\mathscr{V}$ into $\mathscr{V}$ is an *automorphism*.

Definition. Let $\mathscr{A}$, $\mathscr{B}$ and $\mathscr{C}$ be three linear spaces. If $\mathbf{A}:\mathscr{A} \to \mathscr{B}$ and $\mathbf{B}:\mathscr{B} \to \mathscr{C}$ are linear transformations, then the product of $\mathbf{B}$ and $\mathbf{A}$, written $\mathbf{BA}$ is a linear transformation from $\mathscr{A}$ to $\mathscr{C}$ defined by

$$(\mathbf{BA})(\mathbf{u}) = \mathbf{B}(\mathbf{A}(\mathbf{u}))$$

for all $\mathbf{u}$ in $\mathscr{A}$.

It follows from this definition that the product of two linear transformations is distributive and associative. In general, $\mathbf{AB} \neq \mathbf{BA}$. If $\mathbf{A}$ and $\mathbf{B}$ are isomorphisms, then $\mathbf{BA}$ is an isomorphism. Therefore, there exists an isomorphism $(\mathbf{BA})^{-1}$.

Theorem A.2.4. If $\mathbf{A}:\mathscr{A} \to \mathscr{B}$ and $\mathbf{B}:\mathscr{B} \to \mathscr{C}$ are isomorphisms, then

$$(\mathbf{BA})^{-1} = \mathbf{A}^{-1}\mathbf{B}^{-1} \tag{A.2.6}$$

Proof. If $\mathbf{u}$ is an arbitrary vector in $\mathscr{A}$, then $\mathbf{v} = \mathbf{Au}$ is in $\mathscr{B}$ and $\mathbf{c} = \mathbf{BAu}$ is in $\mathscr{C}$. Because $\mathbf{BA}$ is an isomorphism

$$\mathbf{u} = (\mathbf{BA})^{-1}\mathbf{c}$$

But

$$\mathbf{u} = \mathbf{A}^{-1}\mathbf{v} \qquad \text{and} \qquad \mathbf{v} = \mathbf{B}^{-1}\mathbf{c}$$

Therefore,

$$\mathbf{u} = \mathbf{A}^{-1}\mathbf{B}^{-1}\mathbf{c} = (\mathbf{BA})^{-1}\mathbf{c}$$

and, thus,

$$\mathbf{A}^{-1}\mathbf{B}^{-1} = (\mathbf{BA})^{-1}$$

Definition. The *identity* automorphism $\mathbf{I}$ is defined by

$$\mathbf{v} = \mathbf{Iv} \tag{A.2.7}$$

for all $\mathbf{v}$ in $\mathscr{V}$.

Theorem A.2.5. If $\mathbf{A}:\mathscr{V}\rightarrow\mathscr{U}$ is an isomorphism, then

$$\mathbf{A}\mathbf{A}^{-1}=\mathbf{A}^{-1}\mathbf{A}=\mathbf{I} \tag{A.2.8}$$

Proof. By definition,

$$\mathbf{u}=\mathbf{A}\mathbf{v}$$

for all $\mathbf{v}$ in $\mathscr{V}$, $\mathbf{u}$ in $\mathscr{U}$ and

$$\mathbf{v}=\mathbf{A}^{-1}\mathbf{u}$$

for all $\mathbf{v}$ in $\mathscr{V}$, $\mathbf{u}$ in $\mathscr{U}$. Therefore,

$$\mathbf{u}=\mathbf{A}\mathbf{A}^{-1}\mathbf{u} \qquad \text{and} \qquad \mathbf{v}=\mathbf{A}^{-1}\mathbf{A}\mathbf{v}$$

which is the desired result.

Given two vector spaces $\mathscr{V}$ and $\mathscr{U}$, consider the set $\mathscr{L}(\mathscr{V};\mathscr{U})$ of all linear transformations $\mathbf{A}:\mathscr{V}\rightarrow\mathscr{U}$. By the definitions

$$(\mathbf{A}_1+\mathbf{A}_2)\mathbf{v}=\mathbf{A}_1\mathbf{v}+\mathbf{A}_2\mathbf{v} \tag{A.2.9}$$

and

$$(\mu\mathbf{A})\mathbf{v}=\mu(\mathbf{A}\mathbf{v}) \tag{A.2.10}$$

the set $\mathscr{L}(\mathscr{V};\mathscr{U})$ is a *vector space*. The proof of this assertion is left as an exercise. Equation (A.2.9) defines the addition of two linear transformations while (A.2.10) defines the product of a linear transformation by a scalar. It is possible to show that $\dim\mathscr{L}(\mathscr{V};\mathscr{U})=\dim\mathscr{V}\dim\mathscr{U}$.

Definition. $\mathscr{GL}(\mathscr{V})=\{\mathbf{A}\,|\,\mathbf{A}\text{ in }\mathscr{L}(\mathscr{V};\mathscr{V})\text{ and }\mathbf{A}\text{ an automorphism}\}$.

It can be shown that $\mathscr{GL}(\mathscr{V})$ is a *group* under the operation of multiplication of linear transformations. This group is called the *general linear group*.

Consider a linear transformation $\mathbf{A}:\mathscr{V}\rightarrow\mathscr{U}$. Let $\{\mathbf{e}_1,...,\mathbf{e}_n\}$ denote the basis for $\mathscr{V}$ and $\{\mathbf{g}_1,...,\mathbf{g}_m\}$ denote the basis for $\mathscr{U}$. It follows that $\mathbf{A}\mathbf{e}_j$, since it is a vector of $\mathscr{U}$, can be expanded in the basis $\{\mathbf{g}_1,...,\mathbf{g}_m\}$. Therefore we can write

$$\mathbf{A}\mathbf{e}_j = \sum_{k=1}^{m} A_{kj}\mathbf{g}_k \tag{A.2.11}$$

for $j = 1,2,...,n$.

Definition. The *matrix* of $\mathbf{A}$, written $[\mathbf{A}]$, with respect to the bases $\{\mathbf{e}_1,...,\mathbf{e}_n\}$ and $\{\mathbf{g}_1,...,\mathbf{g}_m\}$ is the $m \times n$ array of scalars,

$$[\mathbf{A}] = \begin{bmatrix} A_{11} & A_{12} & A_{13} & \cdot & \cdot & \cdot & A_{1n} \\ A_{21} & A_{22} & A_{23} & & & & A_{2n} \\ A_{31} & & & & & & \\ \cdot & & & & & & \\ \cdot & & & & & & \\ A_{m1} & A_{m2} & A_{m3} & \cdot & \cdot & \cdot & A_{mn} \end{bmatrix} \tag{A.2.12}$$

In this work, the reader is assumed to be familiar with elementary matrix algebra.

Exercise A.2.4

Show that

$$[\mathbf{A}+\mathbf{B}] = [\mathbf{A}] + [\mathbf{B}] \tag{A.2.13}$$

and

$$[\lambda\mathbf{A}] = \lambda[\mathbf{A}] \tag{A.2.14}$$

Exercise A.2.5

If the basis of $\mathscr{V}$ is transformed by (A.1.9) and the basis of $\mathscr{U}$ is transformed by

$$\overline{\mathbf{g}}_k = \sum_{j=1}^{m} S_{jk}\mathbf{g}_j \tag{A.2.15}$$

for $k = 1,...,m$, show that the components of $\mathbf{A}$ transform according to the formula

$$\overline{A}_{sk} = \sum_{p=1}^{m}\sum_{j=1}^{n} \hat{S}_{sp} A_{pj} T_{jk} \tag{A.2.16}$$

where $\left[\hat{S}_{sp}\right] = \left[S_{sp}\right]^{-1}$, $j,k = 1,...,n$ and $s,p = 1,...,m$.

We denote by $\mathscr{M}^{m\times n}$ the set of matrices with m rows and n columns. It is routine to show that $\mathscr{M}^{m\times n}$ is a vector space and, in addition, is isomorphic to $\mathscr{L}(\mathscr{V};\mathscr{U})$.

A.3. Inner Product Spaces

From the definition of a real vector space, we see that such things as distance, angles, and length do not have any meaning unless we postulate that $\mathscr{V}$ has some additional structure. In order to introduce these concepts, we introduce the concept of an *inner produce space.*

Definition. An *inner product space* is a vector space $\mathscr{V}$ and a function from $\mathscr{V}\times\mathscr{V}\to\mathscr{R}$, written $\mathbf{u}\cdot\mathbf{v}$, having the following properties:

1. For all $\mathbf{u}$ and $\mathbf{v}$ in $\mathscr{V}$,

$$\mathbf{u}\cdot\mathbf{v}=\mathbf{v}\cdot\mathbf{u}$$

2. For all $\mathbf{u}$ and $\mathbf{v}$ in $\mathscr{V}$ and μ in $\mathscr{R}$,

$$(\mu\mathbf{u})\cdot\mathbf{v}=\mathbf{u}\cdot(\mu\mathbf{v})=\mu\mathbf{u}\cdot\mathbf{v}$$

3. For all $\mathbf{u},\mathbf{v}$ and $\mathbf{w}$ in $\mathscr{V}$,

$$(\mathbf{u}+\mathbf{v})\cdot\mathbf{w}=\mathbf{u}\cdot\mathbf{w}+\mathbf{v}\cdot\mathbf{w}$$
$$\mathbf{u}\cdot(\mathbf{v}+\mathbf{w})=\mathbf{u}\cdot\mathbf{v}+\mathbf{u}\cdot\mathbf{w}$$

4. For all $\mathbf{u}$ in $\mathscr{V}$,

$$\mathbf{u}\cdot\mathbf{u}\geq 0$$

5. For all $\mathbf{u}$ in $\mathscr{V}$,

$$\mathbf{u}\cdot\mathbf{u}=0$$

if and only if $\mathbf{u}=\mathbf{0}$.

The above axioms of an inner product can be paraphrased as follows: (1) asserts that $\mathbf{u}\cdot\mathbf{v}$ is *symmetric*, (2) and (3) assert that $\mathbf{u}\cdot\mathbf{v}$ is a *linear* function of each variable, and (4) and (5) assert that $\mathbf{u}\cdot\mathbf{u}$ is *positive definite.*

Definition. The *length* or *norm* of a vector $\mathbf{u}$ in $\mathscr{V}$ is

$$\|\mathbf{u}\|=(\mathbf{u}\cdot\mathbf{u})^{1/2} \tag{A.3.1}$$

Definition. The *distance* between two vectors $\mathbf{u}$ and $\mathbf{v}$ in $\mathscr{V}$ is denoted by $d(\mathbf{u},\mathbf{v})$ and is defined by

$$d(\mathbf{u},\mathbf{v}) = \|\mathbf{u}-\mathbf{v}\| = \{(\mathbf{u}-\mathbf{v})\cdot(\mathbf{u}-\mathbf{v})\}^{1/2} \tag{A.3.2}$$

Theorem A.3.1. The length function has the following properties:

1. For all $\mathbf{u}$ in $\mathscr{V}$,

$$\|\mathbf{u}\| \geq 0$$

2. If $\mathbf{u}$ is in $\mathscr{V}$, then

$$\|\mathbf{u}\| = 0$$

if and only if $\mathbf{u} = \mathbf{0}$.

3. For all $\mathbf{u}$ in $\mathscr{V}$ and μ in $\mathscr{R}$,

$$\|\mu\mathbf{u}\| = |\mu|\|\mathbf{u}\|$$

where $|\mu|$ is the absolute value of μ.

4. For all $\mathbf{u}$ and $\mathbf{v}$ in $\mathscr{V}$,

$$|\mathbf{u}\cdot\mathbf{v}| \leq \|\mathbf{u}\|\|\mathbf{v}\| \qquad \text{(Schwarz inequality)}$$

5. For all $\mathbf{u}$ and $\mathbf{v}$ in $\mathscr{V}$,

$$\|\mathbf{u}+\mathbf{v}\| \leq \|\mathbf{u}\| + \|\mathbf{v}\| \qquad \text{(Triangle inequality)}$$

Parts (1), (2), and (3) are immediate consequences of the definition of the inner product. The proof of (4) and (5) can be found in any linear algebra book. The essentials are straight forward. First, we note that that the Schwarz inequality is trivially true of $\mathbf{u}$ or $\mathbf{v}$ is zero. Next, assume neither $\mathbf{u}$ or $\mathbf{v}$ is zero and calculate the length equared of the vector $(\mathbf{u}\cdot\mathbf{u})\mathbf{v}-(\mathbf{v}\cdot\mathbf{u})\mathbf{u}$. The result of this calculation is

$$\begin{aligned}\|(\mathbf{u}\cdot\mathbf{u})\mathbf{v}-(\mathbf{v}\cdot\mathbf{u})\mathbf{u}\|^2 &= \left((\mathbf{u}\cdot\mathbf{u})\mathbf{v}-(\mathbf{v}\cdot\mathbf{u})\mathbf{u}\right)\cdot\left((\mathbf{u}\cdot\mathbf{u})\mathbf{v}-(\mathbf{v}\cdot\mathbf{u})\mathbf{u}\right)\\ &= \left(\|\mathbf{u}\|^2\|\mathbf{v}\|^2-(\mathbf{v}\cdot\mathbf{u})^2\right)\|\mathbf{u}\|^2\end{aligned} \tag{A.3.3}$$

Because of the property 1. above, and the fact that $\mathbf{u}$ is nonzero, (A.3.3) implies that

$$\|\mathbf{u}\|^2 \|\mathbf{v}\|^2 \geq (\mathbf{v}\cdot\mathbf{u})^2 = |\mathbf{v}\cdot\mathbf{u}|^2 \tag{A.3.4}$$

The positive square root of (A.3.4) is the Schwarz inequality. The triangle inequality follows from

$$\begin{aligned}\|\mathbf{u}+\mathbf{v}\|^2 &= (\mathbf{u}+\mathbf{v})\cdot(\mathbf{u}+\mathbf{v}) \\ &= \|\mathbf{u}\|^2 + \|\mathbf{v}\|^2 + 2\mathbf{v}\cdot\mathbf{u}\end{aligned} \tag{A.3.5}$$

If Schwartz's inequality is used, (A.3.5) yields

$$\|\mathbf{u}+\mathbf{v}\|^2 = \|\mathbf{u}\|^2 + \|\mathbf{v}\|^2 + 2\mathbf{v}\cdot\mathbf{u} \leq \left(\|\mathbf{u}\| + \|\mathbf{v}\|\right)^2 \tag{A.3.6}$$

The positive square root of this equation yields the triangle inequality.

Definition. The *angle* θ between $\mathbf{u}$ and $\mathbf{v}$ in $\mathscr{V}$ is defined by

$$\cos\theta = \frac{\mathbf{u}\cdot\mathbf{v}}{\|\mathbf{u}\|\|\mathbf{v}\|} \tag{A.3.7}$$

It follows from the Schwarz inequality that

$$-1 \leq \cos\theta \leq 1$$

Definition. Two vectors $\mathbf{u}$ and $\mathbf{v}$ in $\mathscr{V}$ are said to be *orthogonal* if $\mathbf{u}\cdot\mathbf{v} = 0$.

Example. In $\mathscr{R}^n$, the inner product of $\mathbf{u}, \mathbf{v}$ is defined to be

$$\mathbf{u}\cdot\mathbf{v} = \sum_{k=1}^{n} u_k \upsilon_k \tag{A.3.8}$$

It follows then from (A.1.7) that

$$\mathbf{i}_j \cdot \mathbf{i}_k = \begin{Bmatrix} 1 & \text{if } i = j \\ 0 & \text{if } i \neq j \end{Bmatrix} = \delta_{jk} \tag{A.3.9}$$

From (A.3.1) and (A.3.8), the norm of a vector $\mathbf{u}$ in $\mathscr{R}^n$ can be written

$$\|\mathbf{u}\| = \left(\sum_{j=1}^{n} u_j u_j\right)^{1/2} \tag{A.3.10}$$

From (A.3.10), the vector $\mathbf{i}_j$ has a unit norm. From (A.3.2), the distance between $\mathbf{u}$ and $\mathbf{v}$ in $\mathscr{R}^n$ can be written

$$d(\mathbf{u},\mathbf{v}) = \left(\sum_{j=1}^{n} (u_j - \upsilon_j)(u_j - \upsilon_j) \right)^{1/2} \tag{A.3.11}$$

and, from (A.3.7), $\cos\theta$ can be written

$$\cos\theta = \frac{\sum_{i=1}^{n} u_i \upsilon_i}{\sqrt{\left(\sum_{j=1}^{n} u_j u_j \right)\left(\sum_{k=1}^{n} \upsilon_k \upsilon_k \right)}} \tag{A.3.12}$$

We shall assume in the following that all vector spaces have the inner product structure.

Definition. A basis $\{\mathbf{i}_1,\ldots,\mathbf{i}_n\}$ for a vector space $\mathscr{V}$ is *orthonormal* if

$$\mathbf{i}_j \cdot \mathbf{i}_k = \delta_{jk} \tag{A.3.13}$$

It is possible to show that every inner product space has an orthonormal basis. Clearly, equations (A.3.8), (A.3.10), (A.3.11) and (A.3.12)are valid for the space $\mathscr{V}$ when the components are those with respect to the orthonormal basis.

Definition. If $\mathbf{A}$ is in $\mathscr{L}(\mathscr{V};\mathscr{U})$, the *transpose* of $\mathbf{A}$ is a linear transformation $\mathbf{A}^T$ in $\mathscr{L}(\mathscr{U};\mathscr{V})$ defined by

$$(\mathbf{A}^T\mathbf{u}) \cdot \mathbf{v} = \mathbf{u} \cdot (\mathbf{A}\mathbf{v}) \tag{A.3.14}$$

for all $\mathbf{v}$ in $\mathscr{V}$ and $\mathbf{u}$ in $\mathscr{U}$.

Exercise A.3.1

If $\mathbf{A}_1$ and $\mathbf{A}_2$ are in $\mathscr{L}(\mathscr{V};\mathscr{U})$, show that

$$(\lambda\mathbf{A}_1 + \mu\mathbf{A}_2)^T = \lambda\mathbf{A}_1^T + \mu\mathbf{A}_2^T \tag{A.3.15}$$

for all λ, μ in $\mathscr{R}$.

Exercise A.3.2

If $\mathbf{A}$ is in $\mathscr{L}(\mathscr{V};\mathscr{U})$, show that with respect to orthonormal bases for $\mathscr{V}$ and $\mathscr{U}$

$$\left[\mathbf{A}^T\right]=\left[\mathbf{A}\right]^T \tag{A.3.16}$$

Theorem A.3.2. If $\mathbf{A}:\mathscr{A}\rightarrow\mathscr{B}$ and $\mathbf{B}:\mathscr{B}\rightarrow\mathscr{C}$ are linear transformations, then

$$(\mathbf{BA})^T=\mathbf{A}^T\mathbf{B}^T \tag{A.3.17}$$

Proof. By definition $(\mathbf{BA})^T$ is defined by

$$\mathbf{a}\cdot\left((\mathbf{BA})^T\mathbf{c}\right)=(\mathbf{BAa})\cdot\mathbf{c}$$

But,

$$\begin{aligned}(\mathbf{BAa})\cdot\mathbf{c}&=\left(\mathbf{B}(\mathbf{Aa})\right)\cdot\mathbf{c}=(\mathbf{Aa})\cdot(\mathbf{B}^T\mathbf{c})\\&=\mathbf{a}\cdot\left(\mathbf{A}^T(\mathbf{B}^T\mathbf{c})\right)=\mathbf{a}\cdot\left(\mathbf{A}^T\mathbf{B}^T\mathbf{c}\right)\end{aligned}$$

Therefore,

$$\mathbf{a}\cdot\left((\mathbf{BA})^T\mathbf{c}\right)=\mathbf{a}\cdot\left(\mathbf{A}^T\mathbf{B}^T\mathbf{c}\right)$$

which is the desired result.

Exercise A.3.3

Show that $\mathbf{A}^{-1^T}=\mathbf{A}^{T^{-1}}$ if $\mathbf{A}$ is an isomorphism.

Definition. An endomorphism $\mathbf{A}:\mathscr{V}\rightarrow\mathscr{V}$ is said to be *symmetric* if $\mathbf{A}=\mathbf{A}^T$.

Definition. An endomorphism $\mathbf{A}:\mathscr{V}\rightarrow\mathscr{V}$ is said to be *skew - symmetric* if $\mathbf{A}=-\mathbf{A}^T$.

Theorem A.3.3. An endomorphism $\mathbf{A}:\mathscr{V}\rightarrow\mathscr{V}$ can be uniquely decomposed into the sum of a symmetric endomorphism and a skew-symmetric endomorphism.

It is easily seen that the decomposition,

$$\mathbf{A}=\frac{1}{2}(\mathbf{A}+\mathbf{A}^T)+\frac{1}{2}(\mathbf{A}-\mathbf{A}^T) \tag{A.3.18}$$

has the properties asserted in the theorem.

Definition. A linear transformation $\mathbf{A}:\mathscr{V}\to\mathscr{U}$ is said to be *orthogonal* if the inner product is preserved under $\mathbf{A}$, i.e.,

$$\mathbf{A}\mathbf{v}_1\cdot\mathbf{A}\mathbf{v}_2=\mathbf{v}_1\cdot\mathbf{v}_2 \tag{A.3.19}$$

for all $\mathbf{v}_1,\mathbf{v}_2$ in $\mathscr{V}$.

If we set $\mathbf{v}_1=\mathbf{v}_2$, it follows from (A.3.19) that

$$\|\mathbf{A}\mathbf{v}\|=\|\mathbf{v}\| \tag{A.3.20}$$

Thus, when $\mathbf{A}$ is orthogonal, the length of $\mathbf{v}$ in $\mathscr{V}$ equals the length of $\mathbf{A}\mathbf{v}$ in $\mathscr{U}$.

Theorem A.3.4. An orthogonal linear transformation is regular.

The proof of this result follows immediately from (A.3.20). Therefore, if $\dim\mathscr{V}=\dim\mathscr{U}$, an orthogonal mapping $\mathbf{A}:\mathscr{V}\to\mathscr{U}$ has an inverse. It follows from (A.3.19) and (A.3.14) that, in this case,

$$\mathbf{A}^T=\mathbf{A}^{-1} \tag{A.3.21}$$

Definition. The *orthogonal group* is a subgroup (a subset of a group that is also a group) of $\mathscr{GL}(\mathscr{V})$ defined by

$$\mathscr{O}(\mathscr{V})=\left\{\mathbf{Q}\,|\,\mathbf{Q}\text{ in }\mathscr{GL}(\mathscr{V})\text{ and }\mathbf{Q}^T\mathbf{Q}=\mathbf{I}\right\} \tag{A.3.22}$$

Exercise A.3.4

Prove that $\mathscr{O}(\mathscr{V})$ is indeed a subgroup of $\mathscr{GL}(\mathscr{V})$.

Definition. A symmetric endomorphism $\mathbf{A}:\mathscr{V}\to\mathscr{V}$ is said to be

$$\left\{\begin{matrix}\text{positive definite}\\ \text{positive semidefinite}\\ \text{negative definite}\\ \text{negative semidefinite}\end{matrix}\right\}\text{ if }\mathbf{v}\cdot\mathbf{A}\mathbf{v}\left\{\begin{matrix}>0\\ \geq 0\\ <0\\ \leq 0\end{matrix}\right., \tag{A.3.23}$$

for all nonzero vectors $\mathbf{v}$ in $\mathscr{V}$.

Exercise A.3.5

Show that the set of all symmetric linear transformations does not form a group under multiplication.

Definition. If $\mathbf{v}$ is in $\mathscr{V}$ and $\mathbf{u}$ is in $\mathscr{U}$, their *tensor product*, written $\mathbf{u}\otimes\mathbf{v}$, is a linear transformation in $\mathscr{L}(\mathscr{V};\mathscr{U})$ defined by

$$(\mathbf{u}\otimes\mathbf{v})\mathbf{w}=\mathbf{u}(\mathbf{v}\cdot\mathbf{w}) \tag{A.3.24}$$

for all $\mathbf{w}$ in $\mathscr{V}$.

Exercise A.3.6

Show that

$$(\lambda\mathbf{u}_1+\mu\mathbf{u}_2)\otimes\mathbf{v}=\lambda\mathbf{u}_1\otimes\mathbf{v}+\mu\mathbf{u}_2\otimes\mathbf{v} \tag{A.3.25}$$

Exercise A.3.7

Show that

$$\mathbf{u}\otimes(\lambda\mathbf{v}_1+\mu\mathbf{v}_2)=\lambda\mathbf{u}\otimes\mathbf{v}_1+\mu\mathbf{u}\otimes\mathbf{v}_2 \tag{A.3.26}$$

Exercise A.3.8

Show that

$$(\mathbf{u}\otimes\mathbf{v})^T=\mathbf{v}\otimes\mathbf{u} \tag{A.3.27}$$

Exercise A.3.9

Show that

$$(\mathbf{u}\otimes\mathbf{v})(\mathbf{a}\otimes\mathbf{b})=(\mathbf{v}\cdot\mathbf{a})(\mathbf{u}\otimes\mathbf{b}) \tag{A.3.28}$$

Exercise A.3.10

Show that

$$[\mathbf{u}\otimes\mathbf{v}]=\begin{bmatrix} u_1v_1 & u_1v_2 & u_1v_3 & \cdot & \cdot & \cdot & u_1v_n \\ u_2v_1 & u_2v_2 & u_2v_3 & \cdot & \cdot & \cdot & u_2v_n \\ u_3v_1 & & & & & & \cdot \\ \cdot & & & & & & \cdot \\ \cdot & & & & & & \cdot \\ u_mv_1 & u_mv_2 & u_mv_3 & \cdot & \cdot & \cdot & u_mv_n \end{bmatrix} \tag{A.3.29}$$

with respect to the orthonormal bases $\{\mathbf{i}_1,...,\mathbf{i}_n\}$ and $\{\mathbf{i}_1,...,\mathbf{i}_m\}$. It is important to note that $\mathbf{u}\otimes\mathbf{v}\neq\mathbf{v}\otimes\mathbf{u}$

A.4. Components of Vectors and Linear Transformations

It is necessary to be able to express some of the previous formulas in terms of components. For simplicity, we deal with a single vector space $\mathscr{V}$ and consider only linear transformations from $\mathscr{V}$ into $\mathscr{V}$. Recall that the vectors defined by (A.3.13) constitute a basis for $\mathscr{V}$. Therefore, every vector $\mathbf{v}$ of $\mathscr{V}$ can be written in the form

$$\mathbf{v}=\upsilon_j\mathbf{i}_j \tag{A.4.1}$$

where $\{\mathbf{i}_1,...,\mathbf{i}_n\}$ is an orthonormal basis. The reader should note that from this point on a *summation convention* is adopted. Rather than indicate a summation from 1 to n explicitly, the presence of the repeated index tells the reader to automatically sum the equation from 1 to $n=\dim\mathscr{V}$. It is easily seen that many of the previous equations in this appendix which involved summations could have been simplified in their formal appearance if this convention had been adopted.

Theorem A.4.1. If $\mathbf{A}$ is in $\mathscr{L}(\mathscr{V};\mathscr{V})$, then the components of $\mathbf{A}$ with respect to the basis $\{\mathbf{i}_1,...,\mathbf{i}_n\}$ are given by

$$A_{kj}=\mathbf{i}_k\cdot(\mathbf{A}\mathbf{i}_j) \tag{A.4.2}$$

Proof. From (A.2.11), it follows that

$$\mathbf{A}\mathbf{i}_j=A_{sj}\mathbf{i}_s$$

and, since $\mathbf{i}_s\cdot\mathbf{i}_k=\delta_{sk}$, we have

$$\mathbf{i}_k\cdot(\mathbf{A}\mathbf{i}_j)=A_{sj}\mathbf{i}_s\cdot\mathbf{i}_k=A_{sj}\delta_{sk}=A_{kj}$$

Theorem A.4.2. If $\mathbf{A}$ is in $\mathscr{L}(\mathscr{V};\mathscr{V})$, then

$$\mathbf{A}=A_{kj}\mathbf{i}_k\otimes\mathbf{i}_j \tag{A.4.3}$$

Proof. From (A.2.11), $\mathbf{A}\mathbf{i}_j=A_{kj}\mathbf{i}_k$. Therefore,

$$\mathbf{A}\mathbf{i}_j = A_{kq}\mathbf{i}_k\delta_{qj}$$
$$= A_{kq}\mathbf{i}_k(\mathbf{i}_q \cdot \mathbf{i}_j)$$
$$= (A_{kq}\mathbf{i}_k \otimes \mathbf{i}_q)\mathbf{i}_j$$

This result immediately yields (A.4.3).

As a special case of (A.4.3), it is possible to show that the identity $\mathbf{I}$ in $\mathscr{L}(\mathscr{V};\mathscr{V})$ has the component representation

$$\mathbf{I} = \delta_{kj}\mathbf{i}_k \otimes \mathbf{i}_j = \mathbf{i}_k \otimes \mathbf{i}_k \tag{A.4.4}$$

Theorem A.4.3. : If $\mathbf{A}$ is in $\mathscr{L}(\mathscr{V};\mathscr{V})$, then

$$\mathbf{A}^T = A_{jk}\mathbf{i}_k \otimes \mathbf{i}_j \tag{A.4.5}$$

This result follows directly from (A.4.3), (A.3.15) and (A.3.27)

Theorem A.4.4. For a symmetric endomorphism,

$$A_{kj} = A_{jk} \tag{A.4.6}$$

Theorem A.4.5. For a skew-symmetric endomorphism,

$$A_{kj} = -A_{jk} \tag{A.4.7}$$

These theorems are trivial consequences of (A.3.14) and (A.4.3). In passing, it is worthwhile to note that the component version of (A.3.18) is

$$A_{kj} = \frac{1}{2}(A_{kj} + A_{jk}) + \frac{1}{2}(A_{kj} - A_{jk}) \tag{A.4.8}$$

Theorem A.4.6. If $\mathbf{A}:\mathscr{V} \to \mathscr{V}$ and $\mathbf{B}:\mathscr{V} \to \mathscr{V}$, then

$$\mathbf{BA} = B_{qj}A_{jk}\mathbf{i}_q \otimes \mathbf{i}_k \tag{A.4.9}$$

Proof. $(\mathbf{BA})(\mathbf{v}) = \mathbf{B}(\mathbf{Av})$ for $\mathbf{v}$ in $\mathscr{V}$. From (A.4.3) it follows that

$$\mathbf{Av} = A_{jk}\upsilon_k\mathbf{i}_j$$

Therefore,

$$
\begin{aligned}
(\mathbf{BA})(\mathbf{v}) &= A_{jk}\upsilon_k\mathbf{B}(\mathbf{i}_j) = A_{jk}\upsilon_k B_{qj}\mathbf{i}_q \\
&= B_{qj}A_{jk}(\mathbf{v}\cdot\mathbf{i}_k)\mathbf{i}_q = (B_{qj}A_{jk}\mathbf{i}_q \otimes \mathbf{i}_k)\mathbf{v}
\end{aligned}
$$

The last equality yields (A.4.9)

Exercise A.4.1

Use (A.4.9) and show that

$$[\mathbf{BA}] = [\mathbf{B}][\mathbf{A}] \tag{A.4.10}$$

Exercise A.4.2

If $\mathbf{A}$ is regular show that

$$[\mathbf{A}^{-1}] = [\mathbf{A}]^{-1} \tag{A.4.11}$$

Exercise A.4.3

If $\mathbf{Q}$ is an orthogonal linear transformation in $\mathscr{L}(\mathscr{V};\mathscr{V})$ show that

$$Q_{qj}Q_{kj} = \delta_{qk} \tag{A.4.12}$$

and

$$Q_{jq}Q_{jk} = \delta_{qk} \tag{A.4.13}$$

where the Q_{qj}, $q, j = 1,\dots,n$ are the components of $\mathbf{Q}$ with respect to an orthonormal basis.

Exercise A.4.4

If $\mathbf{u}$ and $\mathbf{v}$ are in $\mathscr{V}$, show that

$$\mathbf{u}\otimes\mathbf{v} = u_j\upsilon_k\mathbf{i}_j \otimes \mathbf{i}_k \tag{A.4.14}$$

A.5. Cross Products, Determinants and the Polar Decomposition Theorem

The next concept that is needed is that of the cross product. In order to carry out this discussion, in this section the argument is restricted to the case of a vector space of dimension *three*

Definition. A *cross product* in $\mathscr{V}$ is a function, written $\mathbf{u}\times\mathbf{v}$, from $\mathscr{V}\times\mathscr{V}\to\mathscr{V}$ such that

1. $\mathbf{u} \times \mathbf{v} = -\mathbf{v} \times \mathbf{u}$ (A.5.1)

2. $\mathbf{w} \times (\mathbf{u} + \mathbf{v}) = \mathbf{w} \times \mathbf{u} + \mathbf{w} \times \mathbf{v}$ (A.5.2)

3. $\mu(\mathbf{u} \times \mathbf{v}) = (\mu \mathbf{u}) \times \mathbf{v}$ (A.5.3)

   for all $\mathbf{u}, \mathbf{v}, \mathbf{w}$ in $\mathscr{V}$ and μ in $\mathscr{R}$;

4. $\mathbf{u} \cdot (\mathbf{u} \times \mathbf{v}) = \mathbf{0}$ (A.5.4)

5. $\|\mathbf{u} \times \mathbf{v}\| = \|\mathbf{u}\| \|\mathbf{v}\| \sin \theta$ (A.5.5)

   where

$$\cos \theta = \frac{\mathbf{u} \cdot \mathbf{v}}{\|\mathbf{u}\| \|\mathbf{v}\|}$$

and $0 < \theta < 180^o$.

Geometrically, $\|\mathbf{u} \times \mathbf{v}\|$ is the area of the parallelogram with sides $\|\mathbf{u}\|$ and $\|\mathbf{v}\|$ intersecting at an angle θ.

Exercise A.5.1

Show that

$$(\mathbf{u} + \mathbf{v}) \times \mathbf{w} = \mathbf{u} \times \mathbf{w} + \mathbf{v} \times \mathbf{w} \tag{A.5.6a}$$

Exercise A.5.2

Show that

$$\mu(\mathbf{u} \times \mathbf{v}) = \mathbf{u} \times (\mu \mathbf{v}) \tag{A.5.6b}$$

Exercise A.5.3

Show that

$$\mathbf{u} \times \mathbf{u} = \mathbf{0} \tag{A.5.6c}$$

Definition. If $\mathbf{u}, \mathbf{v}$ and $\mathbf{w}$ are in $\mathscr{V}$, the *scalar triple product* is $\mathbf{u} \cdot (\mathbf{v} \times \mathbf{w})$.

Geometrically, $|\mathbf{u}\cdot(\mathbf{v}\times\mathbf{w})|$ is the volume of the parallelepiped formed by the coterminus sides $\mathbf{u},\mathbf{v}$ and $\mathbf{w}$.

Theorem A.5.1. Let $\{\mathbf{i}_1,\mathbf{i}_2,\mathbf{i}_3\}$ denote an orthonormal basis for $\mathscr{V}$. Then

$$\begin{aligned}\mathbf{i}_1\times\mathbf{i}_2 &= \pm\mathbf{i}_3\\ \mathbf{i}_3\times\mathbf{i}_1 &= \pm\mathbf{i}_2\end{aligned} \tag{A.5.7}$$

and

$$\mathbf{i}_2\times\mathbf{i}_3 = \pm\mathbf{i}_1$$

Proof. : Since $\mathbf{i}_1\times\mathbf{i}_2$ is in $\mathscr{V}$, we can express it as $\mathbf{i}_1\times\mathbf{i}_2=\mu\mathbf{i}_1+\lambda\mathbf{i}_2+\xi\mathbf{i}_3$, but $\mathbf{i}_1\cdot(\mathbf{i}_1\times\mathbf{i}_2)=0$ and $\mathbf{i}_2\cdot(\mathbf{i}_1\times\mathbf{i}_2)=0$ from (A.5.4). Therefore, $\mu=\lambda=0$ and thus $\mathbf{i}_1\times\mathbf{i}_2=\xi\mathbf{i}_3$. Since $\|\mathbf{i}_1\times\mathbf{i}_2\|=1$ from (A.5.5) we have $\xi=\pm1$. The other two results follow by an identical argument.

Exercise A.5.4

Use (A.5.10) and show that when $\mathbf{i}_1\times\mathbf{i}_2=\mathbf{i}_3$, then $\mathbf{i}_2\times\mathbf{i}_3=\mathbf{i}_1$ and $\mathbf{i}_3\times\mathbf{i}_1=\mathbf{i}_2$. Likewise when $\mathbf{i}_1\times\mathbf{i}_2=-\mathbf{i}_3$, then $\mathbf{i}_2\times\mathbf{i}_3=-\mathbf{i}_1$ and $\mathbf{i}_3\times\mathbf{i}_1=-\mathbf{i}_2$.

It follows from Theorem A.5.1 and the above exercise that $\mathscr{V}$ has two possible cross products. In the case where $\mathbf{i}_1\times\mathbf{i}_2=\mathbf{i}_3$, it is elementary to show that

$$\begin{aligned}\mathbf{u}\times\mathbf{v}=u_j\upsilon_k\mathbf{i}_j\times\mathbf{i}_k &= (u_2\upsilon_3-u_3\upsilon_2)\mathbf{i}_1+(u_3\upsilon_1-u_1\upsilon_3)\mathbf{i}_2\\ &\quad+(u_1\upsilon_2-u_2\upsilon_1)\mathbf{i}_3\end{aligned} \tag{A.5.8}$$

On the other hand, if $\mathbf{i}_1\times\mathbf{i}_2=-\mathbf{i}_3$, then

$$\mathbf{u}\times\mathbf{v}=-(u_2\upsilon_3-u_3\upsilon_2)\mathbf{i}_1-(u_3\upsilon_1-u_1\upsilon_3)\mathbf{i}_2-(u_1\upsilon_2-u_2\upsilon_1)\mathbf{i}_3 \tag{A.5.9}$$

Thus, the two possible cross products differ in sign.

Exercise A.5.5

Show that

$$\begin{aligned}\mathbf{u}\cdot(\mathbf{v}\times\mathbf{w}) &= -\mathbf{v}\cdot(\mathbf{u}\times\mathbf{w})=\mathbf{v}\cdot(\mathbf{w}\times\mathbf{u})\\ &= -\mathbf{w}\cdot(\mathbf{v}\times\mathbf{u})=\mathbf{w}\cdot(\mathbf{u}\times\mathbf{v})=-\mathbf{u}\cdot(\mathbf{w}\times\mathbf{v})\end{aligned} \tag{A.5.10}$$

Definition. A vector space $\mathscr{V}$ with the cross product (A.5.8) is said to have *positive orientation*.

Definition. A vector space $\mathscr{V}$ with the cross product (A.5.9) is said to have *negative orientation.*

Definition. The symbol ε_{ijk} is defined as follows:

$$\varepsilon_{ijk} = \begin{cases} +1 & \text{if } ijk \text{ is an even permutation of } 123 \\ -1 & \text{if } ijk \text{ is an odd permutation of } 123 \\ 0 & \text{otherwise} \end{cases}$$

The symbol ε_{ijk} is known as the *permutation symbol.*

It easily follows from this definition that

$$\begin{aligned} &\varepsilon_{123} = \varepsilon_{231} = \varepsilon_{312} = 1, \\ &\varepsilon_{213} = \varepsilon_{132} = \varepsilon_{321} = -1, \\ &\varepsilon_{112} = \varepsilon_{333} = \varepsilon_{122} = 0, \quad \text{etc} \end{aligned} \tag{A.5.11}$$

It follows from the above definition that in $\mathscr{V}$, with positive orientation,

$$\mathbf{i}_q \times \mathbf{i}_j = \varepsilon_{qjk}\mathbf{i}_k$$

and in a $\mathscr{V}$, with negative orientation,

$$\mathbf{i}_q \times \mathbf{i}_j = -\varepsilon_{qjk}\mathbf{i}_k$$

As a matter of convention, we shall always assume that $\mathscr{V}$ has *positive* orientation. Therefore,

$$\mathbf{i}_q \times \mathbf{i}_j = \varepsilon_{qjk}\mathbf{i}_k \tag{A.5.12}$$

and

$$\mathbf{i}_k \cdot (\mathbf{i}_q \times \mathbf{i}_j) = \varepsilon_{qjk} = \varepsilon_{kqj} \tag{A.5.13}$$

Exercise A.5.6

Show that

$$\varepsilon_{qjk}\varepsilon_{qst} = \delta_{js}\delta_{kt} - \delta_{jt}\delta_{ks} \tag{A.5.14}$$

Exercise A.5.7

Show that

$$\mathbf{u}\times\mathbf{v}=\varepsilon_{qjk}u_j\upsilon_k\mathbf{i}_q \tag{A.5.15}$$

and

$$\mathbf{u}\times(\mathbf{v}\times\mathbf{w})=(\mathbf{u}\cdot\mathbf{w})\mathbf{v}-(\mathbf{u}\cdot\mathbf{v})\mathbf{w} \tag{A.5.16}$$

Definition. If $\mathbf{u}$ and $\mathbf{v}$ are in $\mathscr{V}$, their *exterior product* $\mathbf{u}\wedge\mathbf{v}$ is a skew-symmetric endomorphism defined by

$$\mathbf{u}\wedge\mathbf{v}=\mathbf{u}\otimes\mathbf{v}-\mathbf{v}\otimes\mathbf{u} \tag{A.5.17}$$

If $(\mathbf{u}\wedge\mathbf{v})_{kj}$ denotes the kj of $\mathbf{u}\wedge\mathbf{v}$ with respect to an orthonormal basis, it is easily shown that

$$\mathbf{u}\times\mathbf{v}=\frac{1}{2}\varepsilon_{qjk}(\mathbf{u}\wedge\mathbf{v})_{jk}\mathbf{i}_q \tag{A.5.18}$$

Theorem A.5.2. For every set of three vectors $\{\mathbf{u},\mathbf{v},\mathbf{w}\}$ in $\mathscr{V}$, $\mathbf{u}\cdot(\mathbf{v}\times\mathbf{w})=0$ if and only if the set is linearly dependent.

Proof. First assume $\{\mathbf{u},\mathbf{v},\mathbf{w}\}$ is linearly dependent. It follows then that $\mathbf{u}=\lambda\mathbf{v}+\mu\mathbf{w}$. It easily follows from (A.5.4) and (A.5.6b) that

$$\mathbf{u}\cdot(\mathbf{v}\times\mathbf{w})=\lambda\mathbf{v}\cdot(\mathbf{v}\times\mathbf{w})+\mu\mathbf{w}\cdot(\mathbf{v}\times\mathbf{w})=0$$

Next assume that $\mathbf{u}\cdot(\mathbf{v}\times\mathbf{w})=0$. If $\mathbf{v}\times\mathbf{w}=\mathbf{0}$, then (A.5.5) shows that $\mathbf{v}$ and $\mathbf{w}$ are colinear which forces $\{\mathbf{u},\mathbf{v},\mathbf{w}\}$ to be linearly dependent. If $\mathbf{v}\times\mathbf{w}\neq\mathbf{0}$ then $\{\mathbf{v},\mathbf{w}\}$ is linearly independent. If $\mathbf{u}=\mathbf{0}$, then $\{\mathbf{u},\mathbf{v},\mathbf{w}\}$ is linearly dependent. If $\mathbf{u}\neq\mathbf{0}$, then it must be orthogonal to $\mathbf{v}\times\mathbf{w}$. Consequently, $\mathbf{u}$ lies in the subspace of $\mathscr{V}$ for which $\{\mathbf{v},\mathbf{w}\}$ is a basis. It immediately follows that $\{\mathbf{u},\mathbf{v},\mathbf{w}\}$ is linearly dependent.

In the case where $\{\mathbf{u},\mathbf{v},\mathbf{w}\}$ is linearly independent, it is possible to prove that the scalar, $\mathbf{Au}\cdot(\mathbf{Av}\times\mathbf{Aw})/\mathbf{u}\cdot(\mathbf{v}\times\mathbf{w})$, depends only on $\mathbf{A}$.

Definition. If $\mathbf{A}$ is in $\mathscr{L}(\mathscr{V};\mathscr{V})$ then the *determinant* of $\mathbf{A}$, written $\det\mathbf{A}$, is a scalar defined by

$$(\det\mathbf{A})\mathbf{u}\cdot(\mathbf{v}\times\mathbf{w})=\mathbf{Au}\cdot(\mathbf{Av}\times\mathbf{Aw}) \tag{A.5.19}$$

for all $\mathbf{u},\mathbf{v}$ and $\mathbf{w}$ in $\mathscr{V}$.

Theorem A.5.3. An endomorphism $\mathbf{A}$ in $\mathscr{L}(\mathscr{V};\mathscr{V})$ is regular if and only if $\det \mathbf{A} \neq 0$.

Proof. Without loss of generality, we can take $\mathbf{u} = \mathbf{i}_1$, $\mathbf{v} = \mathbf{i}_2$ and $\mathbf{w} = \mathbf{i}_3$ in (A.5.19). By (A.5.13), (A.5.19) can be written

$$\det \mathbf{A} = \mathbf{A}\mathbf{i}_1 \cdot (\mathbf{A}\mathbf{i}_2 \times \mathbf{A}\mathbf{i}_3) \tag{A.5.20}$$

If $\det \mathbf{A} \neq 0$, then (A.5.20) and Theorem A.5.2 show that $\{\mathbf{A}\mathbf{i}_1, \mathbf{A}\mathbf{i}_2, \mathbf{A}\mathbf{i}_3\}$ is linearly independent. If $\mathbf{b}$ is in $\mathscr{V}$, then clearly the equation $\mathbf{A}\mathbf{b} = \mathbf{0}$ can be written

$$\mathbf{A}\mathbf{b} = b_1\mathbf{A}\mathbf{i}_1 + b_2\mathbf{A}\mathbf{i}_2 + b_3\mathbf{A}\mathbf{i}_3 = \mathbf{0} \tag{A.5.21}$$

The linear independence of $\{\mathbf{A}\mathbf{i}_1, \mathbf{A}\mathbf{i}_2, \mathbf{A}\mathbf{i}_3\}$ forces $b_1 = b_2 = b_3 = 0$. Thus $\ker \mathbf{A} = \{\mathbf{0}\}$ and $\mathbf{A}$ is regular.

Next assume $\mathbf{A}$ is regular, then $\{\mathbf{A}\mathbf{i}_1, \mathbf{A}\mathbf{i}_2, \mathbf{A}\mathbf{i}_3\}$ is linearly independent and by Theorem A.5.2, $\det \mathbf{A} \neq 0$.

Theorem A.5.4. If $\mathbf{A}$ and $\mathbf{B}$ are endomorphisms of $\mathscr{V}$. then

$$\det \mathbf{AB} = \det \mathbf{A} \det \mathbf{B} \tag{A.5.22}$$

Proof.

$$\begin{aligned}
\det(\mathbf{AB})\mathbf{u} \cdot (\mathbf{v} \times \mathbf{w}) &= \mathbf{ABu} \cdot (\mathbf{ABv} \times \mathbf{ABw}) \\
&= \mathbf{A}(\mathbf{Bu}) \cdot [\mathbf{A}(\mathbf{Bv}) \times \mathbf{A}(\mathbf{Bw})] \\
&= (\det \mathbf{A})(\mathbf{Bu}) \cdot [(\mathbf{Bv}) \times (\mathbf{Bw})] \\
&= (\det \mathbf{A})(\det \mathbf{B})\mathbf{u} \cdot (\mathbf{v} \times \mathbf{w})
\end{aligned}$$

Thus,

$$\det \mathbf{AB} = \det \mathbf{A} \det \mathbf{B}$$

Corollary. If $\mathbf{A}$ is an automorphism of $\mathscr{V}$, then

$$\det \mathbf{A} = \frac{1}{\det \mathbf{A}^{-1}} \tag{A.5.23}$$

Proof. This result follows from Theorem A.5.4 and the equations

$$\mathbf{A}\mathbf{A}^{-1} = \mathbf{I}$$

and

$$\det \mathbf{I} = 1 \tag{A.5.24}$$

Theorem A.5.5. If $\mathbf{A}$ is an automorphism of $\mathscr{V}$, then

$$\mathbf{A}\mathbf{v} \times \mathbf{A}\mathbf{w} = (\det \mathbf{A})\mathbf{A}^{-1^T}(\mathbf{v} \times \mathbf{w}) \tag{A.5.25}$$

for all $\mathbf{v}, \mathbf{w}$ in $\mathscr{V}$.

Proof. From (A.3.14) and (A.5.19), it follows that

$$\mathbf{u} \cdot [\mathbf{A}^T(\mathbf{A}\mathbf{v} \times \mathbf{A}\mathbf{w})] = (\det \mathbf{A})\mathbf{u} \cdot (\mathbf{v} \times \mathbf{w})$$

Therefore,

$$\mathbf{u} \cdot \left\{[\mathbf{A}^T(\mathbf{A}\mathbf{v} \times \mathbf{A}\mathbf{w})] - (\det \mathbf{A})(\mathbf{v} \times \mathbf{w})\right\} = 0$$

which yields

$$\mathbf{A}^T(\mathbf{A}\mathbf{v} \times \mathbf{A}\mathbf{w}) = (\det \mathbf{A})(\mathbf{v} \times \mathbf{w}) \tag{A.5.26}$$

since $\mathbf{u}$ is arbitrary. Equation (A.5.25) follows immediately from (A.5.26) and $\mathbf{A}^{-1^T} = \mathbf{A}^{T^{-1}}$.

Theorem A.5.6.

$$\varepsilon_{rst} \det \mathbf{A} = \varepsilon_{qjk} A_{qr} A_{js} A_{kt} \tag{A.5.27}$$

Proof. In (A.5.19), take $\mathbf{u} = \mathbf{i}_r$, $\mathbf{v} = \mathbf{i}_s$ and $\mathbf{w} = \mathbf{i}_t$, then $\mathbf{A}\mathbf{i}_r = A_{qr}\mathbf{i}_q$, etc. Therefore, (A.5.19) can be written

$$\mathbf{i}_r \cdot (\mathbf{i}_s \times \mathbf{i}_t) \det \mathbf{A} = \mathbf{i}_q \cdot (\mathbf{i}_j \times \mathbf{i}_k) A_{qr} A_{js} A_{kt}$$

If we now use (A.5.13), we immediately obtain (A.5.27). Note that if we take $r = 1, s = 2$ and $t = 3$, then (A.5.27) yields

$$\det \mathbf{A} = \varepsilon_{qjk} A_{q1} A_{j2} A_{k3} \tag{A.5.28}$$

If we multiply (A.5.27) by ε_{rst} and use the identity,

$$\varepsilon_{rst}\varepsilon_{rst} = 3!, \tag{A.5.29}$$

it follows that

$$\det \mathbf{A} = \frac{1}{3!}\varepsilon_{qjk}\varepsilon_{rst}A_{qr}A_{js}A_{kt} \tag{A.5.30}$$

Exercise A.5.8

Use (A.5.30) and show that

$$\det \mathbf{A} = \det \mathbf{A}^T \tag{A.5.31}$$

Exercise A.5.9

Show that $\det \mathbf{Q} = \pm 1$ for all $\mathbf{Q}$ in $\mathscr{O}(\mathscr{V})$.

Exercise A.5.10

Show that

$$\det \lambda\mathbf{A} = \lambda^3 \det \mathbf{A} \qquad \text{and} \qquad \det \mathbf{A} = \det[\mathbf{A}] \tag{A.5.32}$$

Exercise A.5.11

Show that $\det \mathbf{u} \otimes \mathbf{v} = 0$

Exercise A.5.12

Define a linear transformation $\mathbf{K}_{\mathbf{A}}$ in $\mathscr{L}(\mathscr{V};\mathscr{V})$ by

$$(\mathbf{K}_{\mathbf{A}}\mathbf{u})\cdot(\mathbf{v}\times\mathbf{w}) = \mathbf{u}\cdot(\mathbf{A}\mathbf{v}\times\mathbf{A}\mathbf{w}) \tag{A.5.33}$$

for all $\mathbf{u},\mathbf{v}$ and $\mathbf{w}$ in $\mathscr{V}$. Select and orthonormal basis for $\mathscr{V}$ and show that the components of $\mathbf{K}_{\mathbf{A}}$ are related to the components of $\mathbf{A}$ by the formula

$$K_{\mathbf{A}ij} = \frac{1}{2}\varepsilon_{jpq}\varepsilon_{ist}A_{ps}A_{qt} \tag{A.5.34}$$

Show that the matrix $[\mathbf{K}_{\mathbf{A}}]$, called the *adjoint matrix* , is equal to the transposed matrix of cofactors of $[\mathbf{A}]$. In addition, show that

$$\mathbf{K}_{\mathbf{A}}\mathbf{A} = (\det \mathbf{A})\mathbf{I} \tag{A.5.35}$$

In the special case where $\mathbf{A}$ is regular, (A.5.35) yields the following formula for $\mathbf{A}^{-1}$,

$$\mathbf{A}^{-1} = \frac{\mathbf{K}_{\mathbf{A}}}{\det \mathbf{A}} \tag{A.5.36}$$

It is possible to prove that the sum $\{\mathbf{Au}\cdot(\mathbf{v}\times\mathbf{w})+\mathbf{u}\cdot(\mathbf{Av}\times\mathbf{w})+\mathbf{u}\cdot(\mathbf{v}\times\mathbf{Aw})\}/\mathbf{u}\cdot(\mathbf{v}\times\mathbf{w})$ depends only on $\mathbf{A}$.

Definition. The *trace* of an endomorphism $\mathbf{A}$ in $\mathscr{L}(\mathscr{V};\mathscr{V})$, written $\operatorname{tr}\mathbf{A}$, is a scalar defined by

$$(\operatorname{tr}\mathbf{A})\mathbf{u}\cdot(\mathbf{v}\times\mathbf{w}) = \mathbf{Au}\cdot(\mathbf{v}\times\mathbf{w})+\mathbf{u}\cdot(\mathbf{Av}\times\mathbf{w})+\mathbf{u}\cdot(\mathbf{v}\times\mathbf{Aw}) \tag{A.5.37}$$

for all $\mathbf{u},\mathbf{v}$ and $\mathbf{w}$ in $\mathscr{V}$

Theorem A.5.7. The component representation of $\operatorname{tr}\mathbf{A}$ is

$$\operatorname{tr}\mathbf{A} = A_{11} + A_{22} + A_{33} = A_{jj} \tag{A.5.38}$$

Proof. Let $\mathbf{u}=\mathbf{i}_1, \mathbf{v}=\mathbf{i}_2$, and $\mathbf{w}=\mathbf{i}_3$ in (A.5.37), then, since $\mathbf{A} = A_{kj}\mathbf{i}_k \otimes \mathbf{i}_j$ and $\mathbf{i}_1\cdot(\mathbf{i}_2\times\mathbf{i}_3)=1$,

$$\begin{aligned}(\operatorname{tr}\mathbf{A}) &= A_{r1}\mathbf{i}_r\cdot(\mathbf{i}_2\times\mathbf{i}_3)+\mathbf{i}_1\cdot(A_{r2}\mathbf{i}_r\times\mathbf{i}_3)+\mathbf{i}_1\cdot(\mathbf{i}_2\times A_{r3}\mathbf{i}_r)\\ &= A_{11}\mathbf{i}_1\cdot(\mathbf{i}_2\times\mathbf{i}_3)+\mathbf{i}_1\cdot(A_{22}\mathbf{i}_2\times\mathbf{i}_3)+\mathbf{i}_1\cdot(\mathbf{i}_2\times A_{33}\mathbf{i}_3)\\ &= A_{11}+A_{22}+A_{33}\end{aligned}$$

Show the following:

Exercise A.5.13

$$\operatorname{tr}(\lambda\mathbf{A}_1+\mu\mathbf{A}_2) = \lambda\operatorname{tr}\mathbf{A}_1+\mu\operatorname{tr}\mathbf{A}_2 \tag{A.5.39}$$

Exercise A.5.14

$$\operatorname{tr}\mathbf{AB} = \operatorname{tr}\mathbf{BA} = A_{jk}B_{kj} \tag{A.5.40}$$

Exercise A.5.15

$$\operatorname{tr}\mathbf{A} = \operatorname{tr}\mathbf{A}^T \tag{A.5.41}$$

Exercise A.5.16

$$\operatorname{tr} \mathbf{u} \otimes \mathbf{v} = \mathbf{u} \cdot \mathbf{v} \tag{A.5.42}$$

Exercise A.5.17

If $\mathbf{A}$ is in $\mathscr{L}(\mathscr{V};\mathscr{V})$ then from (A.3.18)

$$\mathbf{A} = \mathbf{B} + \mathbf{C}$$

where $\mathbf{B} = \frac{1}{2}(\mathbf{A} + \mathbf{A}^T)$ is a symmetric endomorphism and $\mathbf{C} = \frac{1}{2}(\mathbf{A} - \mathbf{A}^T)$ is a skew-symmetric endomorphism. Show that $\operatorname{tr} \mathbf{A} = \operatorname{tr} \mathbf{B}$.

Definition. If λ is in $\mathscr{R}$ and $\mathbf{A}$ is in $\mathscr{L}(\mathscr{V};\mathscr{V})$ the *characteristic polynomial* of $\mathbf{A}$ is the polynomial $f(\lambda, \mathbf{A})$ defined by

$$f(\lambda, \mathbf{A}) = \det(\mathbf{A} - \lambda \mathbf{I}) \tag{A.5.43}$$

Theorem A.5.8

$$f(\lambda, \mathbf{A}) = -\lambda^3 + I_{\mathbf{A}}\lambda^2 - II_{\mathbf{A}}\lambda + III_{\mathbf{A}} \tag{A.5.44}$$

where

$$I_{\mathbf{A}} = \operatorname{tr} \mathbf{A} \tag{A.5.45}$$

$$II_{\mathbf{A}} = \operatorname{tr} \mathbf{K}_{\mathbf{A}} \tag{A.5.46}$$

and

$$III_{\mathbf{A}} = \det \mathbf{A} \tag{A.5.47}$$

Proof. : From (A.5.19), it is true that

$$f(\lambda, \mathbf{A})\mathbf{u} \cdot (\mathbf{v} \times \mathbf{w}) = (\mathbf{A} - \lambda \mathbf{I})\mathbf{u} \cdot [(\mathbf{A} - \lambda \mathbf{I})\mathbf{v} \times (\mathbf{A} - \lambda \mathbf{I})\mathbf{w}] \tag{A.5.48}$$

By expansion of (A.5.48), it easily follows that

$$\begin{aligned} f(\lambda, \mathbf{A})\mathbf{u} \cdot (\mathbf{v} \times \mathbf{w}) = -\lambda^3 \mathbf{u} \cdot (\mathbf{v} \times \mathbf{w}) &+ \lambda^2 \left[\mathbf{A}\mathbf{u} \cdot (\mathbf{v} \times \mathbf{w}) + \mathbf{u} \cdot (\mathbf{A}\mathbf{v} \times \mathbf{w}) + \mathbf{u} \cdot (\mathbf{v} \times \mathbf{A}\mathbf{w})\right] \\ &- \lambda \left[\mathbf{A}\mathbf{u} \cdot (\mathbf{A}\mathbf{v} \times \mathbf{w}) + \mathbf{A}\mathbf{u} \cdot (\mathbf{v} \times \mathbf{A}\mathbf{w}) + \mathbf{u} \cdot (\mathbf{A}\mathbf{v} \times \mathbf{A}\mathbf{w})\right] \\ &+ \mathbf{A}\mathbf{u} \cdot (\mathbf{A}\mathbf{v} \times \mathbf{A}\mathbf{w}) \end{aligned} \tag{A.5.49}$$

From (A.5.37), the coefficient of λ^2 in (A.5.49) is $(\operatorname{tr}\mathbf{A})\mathbf{u}\cdot(\mathbf{v}\times\mathbf{w})$. Likewise, the coefficient of $-\lambda$ is $\operatorname{tr}\mathbf{K}_{\mathbf{A}}$. This fact follows from (A.5.37) and (A.5.33). Since the last term in (A.5.49) is $(\det\mathbf{A})\mathbf{u}\cdot(\mathbf{v}\times\mathbf{w})$, (A.5.44) holds. A more useful formula for $II_{\mathbf{A}}$ is

$$II_{\mathbf{A}} = \frac{1}{2}\left[(\operatorname{tr}\mathbf{A})^2 - \operatorname{tr}\mathbf{A}^2\right] \tag{A.5.50}$$

The derivation of (A.5.50) will be presented later in this section.

The quantities $I_{\mathbf{A}}, II_{\mathbf{A}}$ and $III_{\mathbf{A}}$ are the *fundamental invariants* of $\mathbf{A}$. This name arises because

$$I_{\mathbf{BAB}^{-1}} = I_{\mathbf{A}} \tag{A.5.51}$$

$$II_{\mathbf{BAB}^{-1}} = II_{\mathbf{A}} \tag{A.5.52}$$

and

$$III_{\mathbf{BAB}^{-1}} = III_{\mathbf{A}} \tag{A.5.53}$$

for all regular linear transformations $\mathbf{B}$ in $\mathscr{L}(\mathscr{V};\mathscr{V})$. The proof of (A.5.51), (A.5.52) and (A.5.53) is elementary. For example,

$$\begin{aligned}
II_{\mathbf{BAB}^{-1}} &= \frac{1}{2}\left((\operatorname{tr}\mathbf{BAB}^{-1})^2 - \operatorname{tr}(\mathbf{BAB}^{-1}\mathbf{BAB}^{-1})\right) \\
&= \frac{1}{2}\left((\operatorname{tr}\mathbf{BAB}^{-1})^2 - \operatorname{tr}(\mathbf{BA}^2\mathbf{B}^{-1})\right) \\
&= \frac{1}{2}\left((\operatorname{tr}\mathbf{B}^{-1}\mathbf{BA})^2 - \operatorname{tr}(\mathbf{B}^{-1}\mathbf{BA}^2)\right) \\
&= \frac{1}{2}\left((\operatorname{tr}\mathbf{A})^2 - \operatorname{tr}(\mathbf{A}^2)\right) = II_{\mathbf{A}}
\end{aligned}$$

If $\mathbf{A}$ is regular, it follows from $\mathbf{AA}^{-1}$ and (A.5.23) that

$$III_{\mathbf{A}^{-1}} = \frac{1}{III_{\mathbf{A}}} \tag{A.5.54}$$

Likewise, (A.5.36). (A.5.46) and (A.5.47) yield

$$I_{\mathbf{A}^{-1}} = \frac{II_{\mathbf{A}}}{III_{\mathbf{A}}} \tag{A.5.55}$$

By interchanging $\mathbf{A}$ and $\mathbf{A}^{-1}$, we see from (A.5.54) and (A.5.55) that

$$II_{\mathbf{A}^{-1}} = \frac{I_{\mathbf{A}}}{III_{\mathbf{A}}} \tag{A.5.56}$$

Theorem A.5.9. (*Cayley-Hamilton*) For every $\mathbf{A}$ in $\mathscr{L}(\mathscr{V};\mathscr{V})$,

$$-\mathbf{A}^3 + I_{\mathbf{A}}\mathbf{A}^2 - II_{\mathbf{A}}\mathbf{A} + III_{\mathbf{A}}\mathbf{I} = \mathbf{0} \tag{A.5.57}$$

Proof. From (A.5.35) and (A.5.43), it follows that

$$\mathbf{K}_{\mathbf{A}-\lambda\mathbf{I}}(\mathbf{A}-\lambda\mathbf{I}) = f(\lambda,\mathbf{A})\mathbf{I} \tag{A.5.58}$$

The definition (A.5.33) shows that $\mathbf{K}_{\mathbf{A}-\lambda\mathbf{I}}$ is a second degree polynomial in λ and, thus, we can write

$$\mathbf{K}_{\mathbf{A}-\lambda\mathbf{I}} = \lambda^2\mathbf{B}_0 - \lambda\mathbf{B}_1 + \mathbf{B}_2 \tag{A.5.59}$$

where $\mathbf{B}_0, \mathbf{B}_1$ and $\mathbf{B}_2$ are endomorphisms to be determined. If we place λ equal to zero, (A.5.59) yields

$$\mathbf{B}_2 = \mathbf{K}_{\mathbf{A}} \tag{A.5.60}$$

If (A.5.44) and (A.5.59) are substituted into (A.5.58) and the result is required to hold for all λ, it follows that

$$\mathbf{B}_2\mathbf{A} = III_{\mathbf{A}}\mathbf{I} \tag{A.5.61}$$

$$\mathbf{B}_1\mathbf{A} + \mathbf{B}_2 = II_{\mathbf{A}}\mathbf{I} \tag{A.5.62}$$

$$\mathbf{B}_0\mathbf{A} + \mathbf{B}_1 = I_{\mathbf{A}}\mathbf{I} \tag{A.5.63}$$

and

$$\mathbf{B}_0 = \mathbf{I} \tag{A.5.64}$$

With (A.5.60) and (A.5.45) it is clear that (A.5.61) is just the result (A.5.35). The result (A.5.57) follows directly from (A.5.61) through (A.5.64) by simply observing that

$$\mathbf{B}_2\mathbf{A} - (\mathbf{B}_1\mathbf{A} + \mathbf{B}_2)\mathbf{A} + (\mathbf{B}_0\mathbf{A} + \mathbf{B}_1)\mathbf{A}^2 - \mathbf{B}_0\mathbf{A}^3 = \mathbf{0}$$

identically.

Often, the Cayley-Hamilton theorem is stated by saying that an endomorphism satisfies its own characteristic equation. It is useful to note that (A.5.60), (A.5.62). (A.5.63) and (A.5.64) combine to yield the following expression for $\mathbf{K}_\mathbf{A}$:

$$\mathbf{K}_\mathbf{A} = \mathbf{A}^2 - I_\mathbf{A}\mathbf{A} + II_\mathbf{A}\mathbf{I} \tag{A.5.65}$$

Since $\operatorname{tr}\mathbf{I} = \dim \mathscr{V} = 3$, the trace of (A.5.65) yields (A.5.50)

Next we wish to briefly discuss the *eigenvalue problem* for an endomorphism $\mathbf{A}$ in $\mathscr{L}(\mathscr{V};\mathscr{V})$. In order to save space, it will be assumed that the reader is familiar with the corresponding problem in matrix algebra and, thus, proofs of theorems are not necessary.

Definition. If $\mathbf{A}$ is in $\mathscr{L}(\mathscr{V};\mathscr{V})$ an *eigenvalue* of $\mathbf{A}$ is a scalar λ such that

$$\mathbf{A}\mathbf{v} = \lambda\mathbf{v} \tag{A.5.66}$$

for some nonzero vector $\mathbf{v}$ in $\mathscr{V}$.

The vector $\mathbf{v}$ in (A.5.66) is the *eigenvector* of $\mathbf{A}$ for the eigenvalue λ. It clearly follows that in order for (A.5.66) to be satisfied for nonzero $\mathbf{v}$ it is necessary and sufficient for $\mathbf{A} - \lambda\mathbf{I}$ not to be regular. It follows then from Theorem A.5.3 that the eigenvalues of $f(\lambda, \mathbf{A}) = \det(\mathbf{A} - \lambda\mathbf{I})$ are the roots of the characteristic polynomial

$$f(\lambda, \mathbf{A}) = \det(\mathbf{A} - \lambda\mathbf{I}) = 0 \tag{A.5.67}$$

Theorem A.5.10. If $\mathbf{A}$ is a symmetric endomorphism in $\mathscr{L}(\mathscr{V};\mathscr{V})$ then the eigenvalues of $\mathbf{A}$ are real and there exists an orthonormal basis for $\mathscr{V}$ consisting of eigenvectors of $\mathbf{A}$

The proof of this theorem can be found in any linear algebra textbook.

If $\{\mathbf{n}_1, \mathbf{n}_2, \mathbf{n}_3\}$ is the orthonormal basis constructed in Theorem A.5.10, it follows from (A.4.3) that

$$\mathbf{A} = \lambda_1 \mathbf{n}_1 \otimes \mathbf{n}_1 + \lambda_2 \mathbf{n}_2 \otimes \mathbf{n}_2 + \lambda_3 \mathbf{n}_3 \otimes \mathbf{n}_3 \tag{A.5.68}$$

for a *symmetric* endomorphism in $\mathscr{L}(\mathscr{V};\mathscr{V})$. In (A.5.68)

$$\begin{aligned} \mathbf{A}\mathbf{n}_1 &= \lambda_1 \mathbf{n}_1 \\ \mathbf{A}\mathbf{n}_2 &= \lambda_2 \mathbf{n}_2 \quad \text{and} \\ \mathbf{A}\mathbf{n}_3 &= \lambda_3 \mathbf{n}_3 \end{aligned} \tag{A.5.69}$$

where λ_1, λ_2 and λ_3 are the three roots of (A.5.67). The formula (A.5.68) is known as the *spectral representation* for $\mathbf{A}$.

Exercise A.5.18

Show that

$$I_{\mathbf{A}} = \lambda_1 + \lambda_2 + \lambda_3 \tag{A.5.70}$$

Exercise A.5.19

Show that

$$II_{\mathbf{A}} = \lambda_1\lambda_3 + \lambda_2\lambda_3 + \lambda_1\lambda_2 \tag{A.5.71}$$

and

Exercise A.5.20

Show that

$$III_{\mathbf{A}} = \lambda_1\lambda_2\lambda_3 \tag{A.5.72}$$

for a symmetric endomorphism $\mathbf{A}$ in $\mathscr{L}(\mathscr{V};\mathscr{V})$.

The results (A.5.70) through (A.5.72) are actually valid for an arbitrary $\mathbf{A}$ in $\mathscr{L}(\mathscr{V};\mathscr{V})$. However, in this case the eigenvalues of $\mathbf{A}$ need not be real.

Exercise A.5.21

Show that with respect to a basis of eigenvectors

$$\mathbf{K}_{\mathbf{A}} = \lambda_2\lambda_3\mathbf{n}_1 \otimes \mathbf{n}_1 + \lambda_1\lambda_3\mathbf{n}_2 \otimes \mathbf{n}_2 + \lambda_1\lambda_2\mathbf{n}_3 \otimes \mathbf{n}_3 \tag{A.5.73}$$

for a symmetric endomorphism $\mathbf{A}$ in $\mathscr{L}(\mathscr{V};\mathscr{V})$. Note that the symmetry of $\mathbf{A}$ yielded a symmetric $\mathbf{K}_{\mathbf{A}}$

Theorem A.5.11. A symmetric endomorphism $\mathbf{A}$ in $\mathscr{L}(\mathscr{V};\mathscr{V})$ is positive definite if and only if its eigenvalues are positive.

The proof of this theorem is an immediate consequence of (A.5.68) and the definition (A.3.23). Of course corresponding results hold for negative definite, negative semidefinite and positive semidefinite symmetric linear transformations.

There are various ways to characterize when a symmetric endomorphism is positive definite. Theorem A.5.11 provides one such method. It is also true that a symmetric endomorphism is positive definite if and only if its fundamental invariants are positive. It is usually convenient to have a test for a symmetric endomorphism to be positive definite which does not require the calculation of the eigenvalues. One such test is known as the *Sylvester* criterion. It asserts that a symmetric endomorphism is positive definite if and only if it's matrx with respect to any basis has positive leading principal minors. If we adopt the orthonormal basis $\{\mathbf{i}_1, \mathbf{i}_2, \mathbf{i}_3\}$, then the matrix of the symmetric endomorphism $\mathbf{A}$ in $\mathscr{L}(\mathscr{V};\mathscr{V})$ is

$$[\mathbf{A}] = \begin{bmatrix} A_{11} & A_{12} & A_{13} \\ A_{12} & A_{22} & A_{23} \\ A_{13} & A_{23} & A_{33} \end{bmatrix} \tag{A.5.74}$$

According to the Sylvester criterion $\mathbf{A}$ is positive definite if and only if

$$\begin{aligned} A_{11} &> 0 \\ A_{11}A_{22} - A_{12}^2 &> 0 \end{aligned} \tag{A.5.75}$$

and

$$\det[\mathbf{A}] > 0$$

A proof of (A.5.75) and its generalization to higher dimensions can be found in Ref. 1.

Theorem A.5.12. (Polar Decomposition Theorem) An automorphism $\mathbf{F}$ in $\mathscr{GL}(\mathscr{V})$ has two unique multiplicative decompositions

$$\mathbf{F} = \mathbf{RU} \tag{A.5.72a}$$

and

$$\mathbf{F} = \mathbf{VR} \tag{A.5.72b}$$

where $\mathbf{R}$ is in $\mathscr{O}(\mathscr{V})$ and $\mathbf{U}$ and $\mathbf{V}$ are symmetric and positive definite.

Proof. Since $(\mathbf{F}^T\mathbf{F})^T = \mathbf{F}^T\mathbf{F}$ and $\mathbf{u}\cdot(\mathbf{F}^T\mathbf{F}\mathbf{u}) = \mathbf{F}\mathbf{u}\cdot\mathbf{F}\mathbf{u} = \|\mathbf{F}\mathbf{u}\|^2$, the linear transformation $\mathbf{F}^T\mathbf{F}$ is symmetric and positive definite. Consequently, the spectral representation (A.5.68) is valid for $\mathbf{A} = \mathbf{F}^T\mathbf{F}$. We shall write

$$\mathbf{F}^T\mathbf{F} = \lambda_1\mathbf{n}_1\otimes\mathbf{n}_1 + \lambda_2\mathbf{n}_2\otimes\mathbf{n}_2 + \lambda_3\mathbf{n}_3\otimes\mathbf{n}_3 \tag{A.5.77}$$

where $\lambda_1 > 0, \lambda_2 > 0$ and $\lambda_3 > 0$. The linear transformation $\mathbf{U}$ is defined by

$$\mathbf{U} = \lambda_1^{1/2}\mathbf{n}_1 \otimes \mathbf{n}_1 + \lambda_2^{1/2}\mathbf{n}_2 \otimes \mathbf{n}_2 + \lambda_3^{1/2}\mathbf{n}_3 \otimes \mathbf{n}_3 \tag{A.5.78}$$

It is easily established from (A.5.78) that

$$\mathbf{U}^2 = \mathbf{F}^T\mathbf{F} \tag{A.5.79}$$

It is obvious from (A.5.78) that $\mathbf{U}$ is both symmetric and positive definite. By direct calculation, it is seen that

$$\mathbf{U}^{-1} = \frac{1}{\lambda_1^{1/2}}\mathbf{n}_1 \otimes \mathbf{n}_1 + \frac{1}{\lambda_2^{1/2}}\mathbf{n}_2 \otimes \mathbf{n}_2 + \frac{1}{\lambda_3^{1/2}}\mathbf{n}_3 \otimes \mathbf{n}_3 \tag{A.5.80}$$

By definition, the regular linear transformation $\mathbf{R}$ is given by

$$\mathbf{R} = \mathbf{F}\mathbf{U}^{-1} \tag{A.5.81}$$

An elementary calculation shows that $\mathbf{R}\mathbf{R}^T = \mathbf{I}$, a result which proves that $\mathbf{R}$ is orthogonal. Equation (A.5.81) is equivalent to (A.5.72a). Clearly the above construction yields unique $\mathbf{R}$ and $\mathbf{U}$. In order to obtain (A.5.72b), we define

$$\mathbf{V} = \mathbf{R}\mathbf{U}\mathbf{R}^T \tag{A.5.82}$$

A simple computation shows that $\mathbf{V}$ is symmetric and positive definite. Given (A.5.82) and (A.5.72a), equation (A.5.72b) immediately follows. It is useful to note that (A.5.82) and (A.5.78) show that

$$\mathbf{V} = \lambda_1^{1/2}\mathbf{R}\mathbf{n}_1 \otimes \mathbf{R}\mathbf{n}_1 + \lambda_2^{1/2}\mathbf{R}\mathbf{n}_2 \otimes \mathbf{R}\mathbf{n}_2 + \lambda_3^{1/2}\mathbf{R}\mathbf{n}_3 \otimes \mathbf{R}\mathbf{n}_3 \tag{A.5.83}$$

Thus, the eigenvalues of $\mathbf{V}$ are the same as those of $\mathbf{U}$ while the eigenvectors of $\mathbf{V}$ are obtained from those of $\mathbf{U}$ by an orthogonal transformation. It follows from (A.5.72b) that

$$\mathbf{V}^2 = \mathbf{F}\mathbf{F}^T \tag{A.5.84}$$

Exercise A.5.22

Show that $\mathbf{U}$ and $\mathbf{V}$ in the polar decomposition theorem have the same fundamental invariants.

Exercise A.5.23

Apply the Polar Decomposition Theorem to the linear transformation $\mathbf{F}$ in $\mathscr{GL}(\mathscr{V})$ defined by

$$\mathbf{F}\mathbf{i}_1 = 2\sqrt{2}\mathbf{i}_1 - 2\mathbf{i}_2$$

and
$$\mathbf{F}\mathbf{i}_2 = (\frac{3}{2}\sqrt{2}+1)\mathbf{i}_1 + (3-\frac{1}{2}\sqrt{2})\mathbf{i}_2$$

$$\mathbf{F}\mathbf{i}_3 = \mathbf{i}_3$$

By calculation of the eigenvectors and eigenvalues of $\mathbf{F}^T\mathbf{F}$, show that $\mathbf{U}$ is defined by

$$\mathbf{U}\mathbf{i}_1 = (2+\sqrt{2})\mathbf{i}_1 + (2-\sqrt{2})\mathbf{i}_2$$

and
$$\mathbf{U}\mathbf{i}_2 = (2-\sqrt{2})\mathbf{i}_1 + (1+2\sqrt{2})\mathbf{i}_2$$

$$\mathbf{U}\mathbf{i}_3 = \mathbf{i}_3$$

while $\mathbf{R}$ is defined by

$$\mathbf{R}\mathbf{i}_1 = \frac{1}{2}\sqrt{2}\mathbf{i}_1 - \frac{1}{2}\sqrt{2}\mathbf{i}_2$$

$$\mathbf{R}\mathbf{i}_2 = \frac{1}{2}\sqrt{2}\mathbf{i}_1 + \frac{1}{2}\sqrt{2}\mathbf{i}_2$$

$$\mathbf{R}\mathbf{i}_3 = \mathbf{i}_3$$

It is a useful computational check in working this exercise to know that the spectral representation of $\mathbf{F}^T\mathbf{F}$ is given by

$$\mathbf{F}^T\mathbf{F} = 18\mathbf{n}_1 \otimes \mathbf{n}_1 + 9\mathbf{n}_2 \otimes \mathbf{n}_2 + \mathbf{n}_3 \otimes \mathbf{n}_3$$

where

$$\mathbf{n}_1 = \frac{1}{\sqrt{3}}\mathbf{i}_1 + \frac{\sqrt{2}}{\sqrt{3}}\mathbf{i}_2$$

and
$$\mathbf{n}_2 = -\frac{\sqrt{2}}{\sqrt{3}}\mathbf{i}_1 + \frac{1}{\sqrt{3}}\mathbf{i}_2$$

$$\mathbf{n}_3 = \mathbf{i}_3$$

Exercise A.5.24

Apply the Polar Decomposition Theorem to the linear transformation $\mathbf{F}$ in $\mathscr{GL}(\mathscr{V})$ defined by

$$\mathbf{F}\mathbf{i}_1 = \frac{246}{125}\mathbf{i}_1 + \frac{228}{125}\mathbf{i}_2$$

$$\mathbf{F}\mathbf{i}_2 = -\frac{272}{125}\mathbf{i}_1 + \frac{129}{125}\mathbf{i}_2$$

and

$$\mathbf{F}\mathbf{i}_3 = \mathbf{i}_3$$

By calculation of the eigenvectors and eigenvalues of $\mathbf{F}^T\mathbf{F}$, show that $\mathbf{U}$ is defined by

$$\mathbf{U} = 2\mathbf{n}_1 \otimes \mathbf{n}_1 + 3\mathbf{n}_2 \otimes \mathbf{n}_2 + \mathbf{n}_3 \otimes \mathbf{n}_3$$

where

$$\mathbf{n}_1 = \frac{3}{5}\mathbf{i}_1 + \frac{4}{5}\mathbf{i}_2$$

$$\mathbf{n}_2 = -\frac{4}{5}\mathbf{i}_1 + \frac{3}{5}\mathbf{i}_2$$

and

$$\mathbf{n}_3 = \mathbf{i}_3$$

Exercise A.5.25

Apply the Polar Decomposition Theorem to the linear transformation $\mathbf{F}$ in $\mathscr{GL}(\mathscr{V})$ defined by

$$\mathbf{F}\mathbf{i}_1 = 2.38\mathbf{i}_1 + 1.02\mathbf{i}_2 - .74\mathbf{i}_3$$
$$\mathbf{F}\mathbf{i}_2 = -1.34\mathbf{i}_1 + 1.61\mathbf{i}_2 + 1.18\mathbf{i}_3$$

and

$$\mathbf{F}\mathbf{i}_3 = 2\mathbf{i}_1 - 1.5\mathbf{i}_2 + 4.33\mathbf{i}_3$$

By calculation of the eigenvectors and eigenvalues of $\mathbf{F}^T\mathbf{F}$, show that $\mathbf{U}$ is defined by

$$\mathbf{U}\mathbf{i}_1 = 2.65\mathbf{i}_1 - .48\mathbf{i}_2$$
$$\mathbf{U}\mathbf{i}_2 = -.48\mathbf{i}_1 + 2.35\mathbf{i}_2$$

and

$$\mathbf{U}\mathbf{i}_3 = 4.99\mathbf{i}_3$$

while $\mathbf{R}$ is defined by

$$\mathbf{Ri}_1 = .8253\mathbf{i}_1 + .529\mathbf{i}_2 - .1953\mathbf{i}_3$$
$$\mathbf{Ri}_2 = -.40\mathbf{i}_1 + .7928\mathbf{i}_2 + .4596\mathbf{i}_3$$

$$\mathbf{Ri}_3 = .4003\mathbf{i}_1 - .3005\mathbf{i}_2 + .8657\mathbf{i}_3$$

It is a useful computational check in working this exercise to know that the spectral representation of $\mathbf{F}^T\mathbf{F}$ is given by

$$\mathbf{F}^T\mathbf{F} = 3.9865\mathbf{n}_1 \otimes \mathbf{n}_1 + 9.0459\mathbf{n}_2 \otimes \mathbf{n}_2 + 24.9989\mathbf{n}_3 \otimes \mathbf{n}_3$$

where

$$\mathbf{n}_1 = .5953\mathbf{i}_1 + .8034\mathbf{i}_2$$
$$\mathbf{n}_2 = -.8034\mathbf{i}_1 + .5953\mathbf{i}_2$$

and

$$\mathbf{n}_3 = \mathbf{i}_3$$

Exercise A.5.26

Prove that a symmetric linear transformation is positive definite if and only if its fundamental invariants are positive.

A.6. Multilinear Functionals and Tensor Algebra

In this section, the concepts of multilinear functionals and tensor algebra will be briefly discussed. We allow the vector space $\mathscr{V}$ to have any finite dimension.

Definition. A *linear functional* is a function $f : \mathscr{V} \to \mathscr{R}$ such that

$$f(\lambda\mathbf{u} + \mu\mathbf{v}) = \lambda f(\mathbf{u}) + \mu f(\mathbf{v}) \tag{A.6.1}$$

for all $\mathbf{u}, \mathbf{v}$ in $\mathscr{V}$ and λ, μ in $\mathscr{R}$.

Example. If $\mathbf{u}$ is a *fixed vector* in $\mathscr{V}$, then $f : \mathscr{V} \to \mathscr{R}$ defined by

$$f(\mathbf{v}) = \mathbf{u} \cdot \mathbf{v} \tag{A.6.2}$$

is a linear functional.

Definition. A *bilinear functional* is a function $\mathscr{V} \times \mathscr{V} \to \mathscr{R}$ such that

$$\mathbf{A}(\lambda \mathbf{u}_1 + \mu \mathbf{u}_2, \mathbf{v}) = \lambda \mathbf{A}(\mathbf{u}_1, \mathbf{v}) + \mu \mathbf{A}(\mathbf{u}_2, \mathbf{v}) \tag{A.6.3}$$

and

$$\mathbf{A}(\mathbf{u}, \lambda \mathbf{v}_1 + \mu \mathbf{v}_2) = \lambda \mathbf{A}(\mathbf{u}, \mathbf{v}_1) + \mu \mathbf{A}(\mathbf{u}, \mathbf{v}_2) \tag{A.6.4}$$

for all $\mathbf{v}_1, \mathbf{v}_2, \mathbf{v}, \mathbf{u}_1, \mathbf{u}_2, \mathbf{u}$ in $\mathscr{V}$ and λ, μ in $\mathscr{R}$.

Examples.

(1) The function defined by

$$\mathbf{A}(\mathbf{u}, \mathbf{v}) = \mathbf{u} \cdot \mathbf{v} \tag{A.6.5}$$

for all $\mathbf{u}, \mathbf{v}$ in $\mathscr{V}$ is a bilinear functional.

(2) More generally, if $\mathbf{A}$ is in $\mathscr{L}(\mathscr{V};\mathscr{V})$ then the function defined by

$$\mathbf{A}(\mathbf{u}, \mathbf{v}) = \mathbf{u} \cdot (\mathbf{A}\mathbf{v}) \tag{A.6.6}$$

is a bilinear functional.

Definition. If p is a positive integer, a *p-linear functional* is a function with p arguments

$$\underbrace{\mathscr{V} \times \mathscr{V} \times \mathscr{V} \times \cdots \times \mathscr{V}}_{p \text{ times}} \to \mathscr{R}$$

that is linear, in the sense of (A.6.1) in each of its arguments.

Definition. A *tensor of order* p $(p \geq 1)$ on $\mathscr{V}$ is a p-linear functional.

The set of tensors of order p on $\mathscr{V}$ is denoted by $\mathscr{T}_p(\mathscr{V})$. Addition and scalar multiplication in this set is defined by

$$\begin{gathered}(\mathbf{A} + \mathbf{B})(\mathbf{v}_1, ..., \mathbf{v}_p) = \mathbf{A}(\mathbf{v}_1, ..., \mathbf{v}_p) + \mathbf{B}(\mathbf{v}_1, ..., \mathbf{v}_p) \\ \text{and} \\ (\lambda \mathbf{A})(\mathbf{v}_1, ..., \mathbf{v}_p) = \lambda(\mathbf{A}(\mathbf{v}_1, ..., \mathbf{v}_p))\end{gathered} \tag{A.6.7}$$

for all $\mathbf{v}_1, ..., \mathbf{v}_p$ in $\mathscr{V}$ and λ in $\mathscr{R}$.

With these definitions, $\mathscr{T}_p(\mathscr{V})$ becomes a vector space. The zero vector in $\mathscr{T}_p(\mathscr{V})$ is the p-linear functional that maps all sets of p vectors in $\mathscr{V}$ into zero.

Examples

(1) If $\dim\mathscr{V}=3$, the functional $\mathbf{D}$

$$\mathbf{D}(\mathbf{u},\mathbf{v},\mathbf{w})=\mathbf{u}\cdot(\mathbf{v}\times\mathbf{w}) \tag{A.6.8}$$

is a tensor of order three. The tensor $\mathbf{D}$ clearly has additional properties not shared by all elements of $\mathscr{T}_p(\mathscr{V})$. They are

$$\mathbf{D}(\mathbf{v}_i,\mathbf{v}_j,\mathbf{v}_k)=\varepsilon_{ijk}\mathbf{D}(\mathbf{v}_1,\mathbf{v}_2,\mathbf{v}_3) \tag{A.6.9}$$

Equations (A.6.9) follows easily from (A.6.8). The tensor $\mathbf{D}$ with the properties (A.6.9), is a *completely skew-symmetric tensor* of order three.

(2) The function $\mathbf{u}$ defined by

$$\mathbf{u}(\mathbf{v})=\mathbf{u}\cdot\mathbf{v} \tag{A.6.10}$$

where $\mathbf{u}$ is in $\mathscr{V}$ is a tensor of order one. It can be shown that the tensor $\mathbf{u}$ is uniquely determined by the vector $\mathbf{u}$ and conversely. Thus $\mathscr{V}$ and $\mathscr{T}_1(\mathscr{V})$ are isomorphic vector spaces. It is customary to suppress this isomorphism and, thus, regard elements of these two spaces as the same. This procedure will be followed here.

(3) The function $\mathbf{A}$ defined by

$$\mathbf{A}(\mathbf{u},\mathbf{v})=\mathbf{u}\cdot(\mathbf{A}\mathbf{v}) \tag{A.6.11}$$

is a tensor of order 2. It can also be shown here that the second order tensor $\mathbf{A}$ is uniquely determined by the linear transformation $\mathbf{A}$ and conversely. Thus, $\mathscr{L}(\mathscr{V};\mathscr{V})$ and $\mathscr{T}_2(\mathscr{V})$ are isomorphic vector spaces. This isomorphism shall be suppressed as was the one in Example (2).

A.7. Euclidean Point Spaces, Coordinate Systems

In vector analysis, the idea of a rectangular Cartesian coordinate system is accepted and used. In this section, it will be shown how this idea can be introduced in a somewhat formal fashion. The first concept we shall introduce is that of a *Euclidean point space*.

Definition. A set $\mathscr{E}$ is a *Euclidean point space* if there exists a function $f:\mathscr{E}\times\mathscr{E}\rightarrow\mathscr{V}$ such that

1. $f(\mathbf{x},\mathbf{y})=f(\mathbf{x},\mathbf{z})+f(\mathbf{z},\mathbf{y})$ for all $\mathbf{x},\mathbf{y},\mathbf{z}$ in $\mathscr{E}$.

2. For every $\mathbf{x}$ in $\mathscr{E}$ and $\mathbf{v}$ in $\mathscr{V}$, there exists a unique element $\mathbf{y}$ in $\mathscr{E}$ such that $f(\mathbf{x},\mathbf{y}) = \mathbf{v}$.

The elements of $\mathscr{E}$ are called *points*. The vector space $\mathscr{V}$ is called the *translation space*. We say that $f(\mathbf{x},\mathbf{y})$ is the vector determined by the *end point* $\mathbf{x}$ and *initial point* $\mathbf{y}$. Condition (2) above is equivalent to requiring the function $f_{\mathbf{x}}$ defined by $f_{\mathbf{x}}(\mathbf{y}) = f(\mathbf{x},\mathbf{y})$ to be one to one. The dimension of $\mathscr{E}$, written $\dim\mathscr{E}$, is defined to be n, the dimension of $\mathscr{V}$.

A Euclidean point space is not a vector space. However, a vector space with inner product is a Euclidean point space with $f(\mathbf{v}_1,\mathbf{v}_2) = \mathbf{v}_1 - \mathbf{v}_2$. For an arbitrary point space $\mathscr{E}$, the function f is called the *point difference*, and it is customary to use the suggestive notation

$$f(\mathbf{x},\mathbf{y}) = \mathbf{x} - \mathbf{y} \tag{A.7.1}$$

In this notation (1) and (2) above take the forms

$$\mathbf{x} - \mathbf{y} = \mathbf{x} - \mathbf{z} + \mathbf{z} - \mathbf{y} \tag{A.7.2}$$

and

$$\mathbf{x} - \mathbf{y} = \mathbf{v} \tag{A.7.3}$$

Theorem A.7.1. In a Euclidean point space $\mathscr{E}$

1. $\mathbf{x} - \mathbf{x} = \mathbf{0}$,
2. $\mathbf{x} - \mathbf{y} = -(\mathbf{y} - \mathbf{x})$,
3. if $\mathbf{x} - \mathbf{y} = \mathbf{x}' - \mathbf{y}'$, then $\mathbf{x} - \mathbf{x}' = \mathbf{y} - \mathbf{y}'$.

Proof. For (1) take $\mathbf{x} = \mathbf{y} = \mathbf{z}$ in (A.7.2) then

$$\mathbf{x} - \mathbf{x} = \mathbf{x} - \mathbf{x} + \mathbf{x} - \mathbf{x}$$

which implies $\mathbf{x} - \mathbf{x} = \mathbf{0}$. To obtain (2) take $\mathbf{y} = \mathbf{x}$ in (A.7.2) and use (1). For (3) observe that

$$\mathbf{x} - \mathbf{y}' = \mathbf{x} - \mathbf{y} + \mathbf{y} - \mathbf{y}' = \mathbf{x} - \mathbf{x}' + \mathbf{x}' - \mathbf{y}'$$

from (A.7.2). However we are given $\mathbf{x} - \mathbf{y} = \mathbf{x}' - \mathbf{y}'$ which implies (3).

The equation

$$\mathbf{x} - \mathbf{y} = \mathbf{v} \tag{A.7.4}$$

has the property that given $\mathbf{v}$ and $\mathbf{x}$, $\mathbf{y}$ is uniquely determined or given $\mathbf{v}$ and $\mathbf{y}$, $\mathbf{x}$ is uniquely determined. For this reason, it is customary to write

$$\mathbf{x} = \mathbf{y} + \mathbf{v} \tag{A.7.5}$$

for the point $\mathbf{x}$ uniquely determined by $\mathbf{y}$ in $\mathscr{E}$ and $\mathbf{v}$ in $\mathscr{V}$. The distance from $\mathbf{x}$ to $\mathbf{y}$ written $d(\mathbf{x},\mathbf{y})$, is defined by

$$d(\mathbf{x},\mathbf{y}) = \left[(\mathbf{x}-\mathbf{y})\cdot(\mathbf{x}-\mathbf{y})\right]^{1/2} \tag{A.7.6}$$

Definition. A *rectangular Cartesian coordinate system* consists of a fixed element of $\mathscr{E}$, denoted by $\mathbf{0}$, called the *origin,* and an orthonormal basis $\{\mathbf{i}_1,\mathbf{i}_2,\ldots,\mathbf{i}_n\}$ for $\mathscr{V}$.

For every point $\mathbf{x}$ in $\mathscr{E}$, the vector $\mathbf{x}-\mathbf{0}$ is in $\mathscr{V}$ and can be represented by

$$\mathbf{x} - \mathbf{0} = x_j \mathbf{i}_j \tag{A.7.7}$$

Definition. The n-tuple of scalars $(x_1, x_2, \ldots, x_n)$ are called the *coordinates* of $\mathbf{x}$ relative to the rectangular Cartesian coordinate system.

The origin has the coordinates $(0,0,0,\ldots,0)$. In a three-dimensional case, $\mathscr{E}$ is represented by Fig. A.7.1. The reader is cautioned that we have introduced three different zeros. The zero scalar, the zero vector and the origin. The context will always indicate which of these zeros is being used.

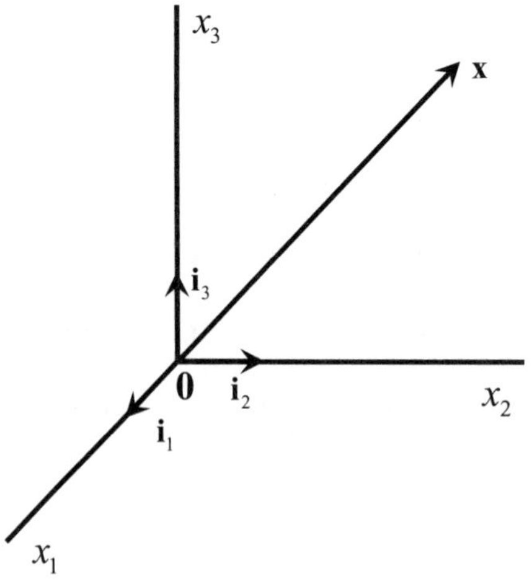

Figure A.7.1

In Chapters III and IV we had occasion to make use of a certain type of mapping of $\mathscr{E}$ onto $\mathscr{E}$. This mapping was of the form

$$\mathbf{x}^* = \mathbf{c} + \mathbf{Q}(\mathbf{x} - \mathbf{0}) \tag{A.7.8}$$

where $\mathbf{c}$ and $\mathbf{0}$ are fixed elements of $\mathscr{E}$ and $\mathbf{Q}$ is in $\mathscr{O}(\mathscr{V})$. Such mappings are called *rigid affine mappings*.

Exercise A.7.1

Show that the mapping (A.7.8) has the following properties:

1. $\mathbf{x}_1 - \mathbf{y}_1 = \mathbf{x}_2 - \mathbf{y}_2$ implies $\mathbf{x}_1^* - \mathbf{y}_1^* = \mathbf{x}_2^* - \mathbf{y}_2^*$,
2. $\left\|\mathbf{x}^* - \mathbf{y}^*\right\| = \left\|\mathbf{x} - \mathbf{y}\right\|$.

Equation (A.7.8) can be viewed geometrically as follows:

For simplicity, we take $n = 2$ and then $\mathscr{E}$ can be represented by the plane of the page. The element $\mathbf{0}$ is shown below. Equation (A.7.8) then yields the geometric construction shown in the following figure:

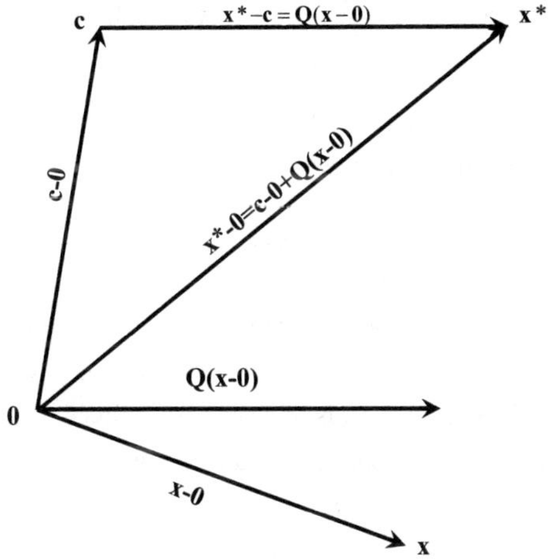

Figure A.7.2

It is important *not* to view the mapping (A.7.8) as a change of coordinates. The idea of a rectangular Cartesian coordinate system was introduced earlier in this section and these ideas were not needed in order to discuss a rigid affine mapping.

A.8. Vector Analysis

In this section we shall record certain elementary results from vector analysis. The reader is assumed to be familiar with elementary definitions from analysis such as an *open set* and a *continuous function*.

A function $f:\mathscr{U}\rightarrow\mathscr{E}'$, where $\mathscr{U}$ is an open subset of $\mathscr{E}$ and $\mathscr{E}'$ is a Euclidean point space or a vector space, is *differentiable* at $\mathbf{x}$ in $\mathscr{U}$ if there exists a linear transformation $\mathbf{A}_{\mathbf{x}}$ in $\mathscr{L}(\mathscr{V};\mathscr{V}')$ such that

$$f(\mathbf{x}+\mathbf{v})=f(\mathbf{x})+\mathbf{A}_{\mathbf{x}}\mathbf{v}+o(\mathbf{x},\|\mathbf{v}\|) \tag{A.8.1}$$

where

$$\lim_{\|\mathbf{v}\|\to 0}\frac{o(\mathbf{x},\|\mathbf{v}\|)}{\|\mathbf{v}\|}=\mathbf{0} \tag{A.8.2}$$

In this definition, $\mathscr{V}$ denotes the translation space of $\mathscr{E}$ and $\mathscr{V}'$ denotes the translation space of $\mathscr{E}'$ if $\mathscr{E}'$ is a point space and denotes $\mathscr{E}'$ if $\mathscr{E}'$ is a vector space. It is possible to prove that $\mathbf{A}_{\mathbf{x}}$ is unique.

If f is differentiable at *every* point of $\mathscr{U}$, then we can define a function $\operatorname{grad} f:\mathscr{U}\rightarrow\mathscr{L}(\mathscr{V};\mathscr{V}')$, called the *gradient* of f, by

$$\operatorname{grad} f(\mathbf{x})=\mathbf{A}_{\mathbf{x}} \tag{A.8.3}$$

for all $\mathbf{x}$ in $\mathscr{U}$. If $\operatorname{grad} f$ is continuous on $\mathscr{U}$, then f is said to be of class C^1. If $\operatorname{grad} f$ exists and is itself of class C^1, then f is of class C^2. More generally f is of class C^r if it is of class C^{r-1} and its (r-1)st gradient, written $\operatorname{grad}^{r-1} f$, is of class C^1. A function f is of class C^0 if it is merely continuous on $\mathscr{U}$. If f is a C^r, one to one, function with a C^r inverse f^{-1} defined on $f(\mathscr{U})$, then f is called a C^r*-diffeomorphism*.

If f is differentiable at $\mathbf{x}$ then we can show from (A.8.1) and (A.8.2) that

$$\mathbf{A}_{\mathbf{x}}\mathbf{v} = \frac{d}{d\tau} f(\mathbf{x}+\tau\mathbf{v})\Big|_{\tau=0} \tag{A.8.4}$$

for all $\mathbf{v}$ in $\mathscr{V}$. To obtain (A.8.4), replace $\mathbf{v}$ by $\tau\mathbf{v}$, $\tau > 0$, in (A.8.1) and write the result as

$$\mathbf{A}_{\mathbf{x}}\mathbf{v} = \frac{f(\mathbf{x}+\tau\mathbf{v}) - f(\mathbf{x})}{\tau} - \frac{o(\mathbf{x},|\tau|\|\mathbf{v}\|)}{\tau} \tag{A.8.5}$$

By (A.8.2) the limit of the last term is zero as $\tau \to 0$ and (A.8.4) is obtained. Equation (A.8.4) holds for all $\mathbf{v}$ in $\mathscr{V}$ because we can always choose τ in (A.8.5) small enough to insure that $\mathbf{x}+\tau\mathbf{v}$ is in $\mathscr{U}$, the domain of f. If f is differentiable at every $\mathbf{x}$ in $\mathscr{U}$, then (A.8.4) can be written

$$\left[\operatorname{grad} f(\mathbf{x})\right]\mathbf{v} = \frac{d}{d\tau} f(\mathbf{x}+\tau\mathbf{v})\Big|_{\tau=0} \tag{A.8.6}$$

A function $f:\mathscr{U}\to\mathscr{R}$, where $\mathscr{U}$ is an open subset of $\mathscr{E}$, is called a *scalar field.* Similarly, $f:\mathscr{U}\to\mathscr{V}$ is a vector field, and $f:\mathscr{U}\to\mathscr{T}_p(\mathscr{V})$ is a *tensor field* of order p.

If $f:\mathscr{U}\to\mathscr{T}_p(\mathscr{V})$ is a tensor field, we can construct a function $\hat{f}$ defined on an open subset of $\mathscr{R}^n$ with values in $\mathscr{T}_p(\mathscr{V})$ by

$$\hat{f}(x_1,\dots,x_n) = f(\mathbf{x}) \tag{A.8.7}$$

where $x_1,\dots,x_n$ are the coordinates of $\mathbf{x}$ as defined by (A.7.7). In order to simplify the notation in the following, we shall often not distinguish the function f in (A.8.7) and its *coordinate function* $\hat{f}$.

If $\varphi:\mathscr{U}\to\mathscr{R}$ denotes a differentiable scalar field defined on an open subset $\mathscr{U}$ of $\mathscr{E}$, then

$$\operatorname{grad}\varphi(\mathbf{x}) = \frac{\partial\varphi}{\partial x_k}\mathbf{i}_k \tag{A.8.8}$$

with respect to a rectangular cartesian coordinate system. To establish (A.8.8), we use (A.8.6) to write

$$\operatorname{grad}\varphi(\mathbf{x})\cdot\mathbf{v} = \frac{d}{d\tau}\varphi(\mathbf{x}+\tau\mathbf{v})\Big|_{\tau=0} \tag{A.8.9}$$

By the definition (A.8.7)

$$\varphi(\mathbf{x}+\tau\mathbf{v}) = \varphi(x_1+\tau\upsilon_1, x_2+\tau\upsilon_2, \ldots, x_n+\tau\upsilon_n) \tag{A.8.10}$$

Therefore,

$$\left.\frac{d}{d\tau}\varphi(\mathbf{x}+\tau\mathbf{v})\right|_{\tau=0} = \frac{\partial\varphi(x_1, x_2, \ldots, x_n)}{\partial x_k}\upsilon_k \tag{A.8.11}$$

Since $\mathbf{v}$ in $\mathscr{V}$ is arbitrary, (A.8.11) and (A.8.9) combine to yield (A.8.8).

Exercise A.8.1

Show that

$$\operatorname{grad}(\varphi+\psi) = \operatorname{grad}\varphi + \operatorname{grad}\psi \tag{A.8.12}$$

and

$$\operatorname{grad}(\varphi\psi) = \varphi\operatorname{grad}\psi + \psi\operatorname{grad}\varphi \tag{A.8.13}$$

If $\mathbf{g}:\mathscr{U}\to\mathscr{V}$ denotes a differentiable vector field defined on an open subset $\mathscr{U}$ of $\mathscr{E}$, then

$$\operatorname{grad}\mathbf{g} = \frac{\partial g_j}{\partial x_k}\mathbf{i}_j\otimes\mathbf{i}_k \tag{A.8.14}$$

with respect to a rectangular cartesian coordinate system. To establish this result, we again use (A.8.6) to write

$$\left[\operatorname{grad}\mathbf{g}(\mathbf{x})\right]\mathbf{v} = \left.\frac{d}{d\tau}\mathbf{g}(\mathbf{x}+\tau\mathbf{v})\right|_{\tau=0} \tag{A.8.15}$$

By the same argument which produced (A.8.11)

$$\left.\frac{d}{d\tau}\mathbf{g}(\mathbf{x}+\tau\mathbf{v})\right|_{\tau=0} = \frac{\partial\mathbf{g}(x_1, x_2, \ldots, x_n)}{\partial x_k}\upsilon_k \tag{A.8.16}$$

Since $\mathbf{g} = g_j\mathbf{i}_j$, and the $\mathbf{i}_j$ are constant vector fields, (A.8.16) can be replaced by

$$\left.\frac{d}{d\tau}\mathbf{g}(\mathbf{x}+\tau\mathbf{v})\right|_{\tau=0}=\frac{\partial g_j(x_1,x_2,\ldots,x_n)}{\partial x_k}\upsilon_k\mathbf{i}_j \tag{A.8.17}$$

If (A.8.17) is substituted into (A.8.15), the result can be written

$$\left[\operatorname{grad}\mathbf{g}(\mathbf{x})-\frac{\partial g_j(x_1,x_2,\ldots,x_n)}{\partial x_k}\mathbf{i}_j\otimes\mathbf{i}_k\right]\mathbf{v}=\mathbf{0} \tag{A.8.18}$$

Since $\mathbf{v}$ is arbitrary, (A.8.18) yields (A.8.14).

Definition. The *divergence* of $\mathbf{g}:\mathscr{U}\to\mathscr{V}$ is a scalar field, written $\operatorname{div}\mathbf{g}$, defined by

$$\operatorname{div}\mathbf{g}=\operatorname{tr}(\operatorname{grad}\mathbf{g}) \tag{A.8.19}$$

From (A.5.38) and (A.8.14), the component representation for $\operatorname{div}\mathbf{g}$ is

$$\operatorname{div}\mathbf{g}=\frac{\partial g_k}{\partial x_k} \tag{A.8.20}$$

Exercise A.8.2

Show that

$$\operatorname{div}(\varphi\mathbf{g})=\varphi\operatorname{div}\mathbf{g}+\mathbf{g}\cdot\operatorname{grad}\varphi \tag{A.8.21}$$

By the same type of argument which produced (A.8.14), the gradient of a field $\mathbf{A}:\mathscr{U}\to\mathscr{L}(\mathscr{V};\mathscr{V})$ has the component representation

$$\operatorname{grad}\mathbf{A}=\frac{\partial A_{jq}}{\partial x_k}\mathbf{i}_j\otimes\mathbf{i}_q\otimes\mathbf{i}_k \tag{A.8.22}$$

where $\mathbf{i}_j\otimes\mathbf{i}_q\otimes\mathbf{i}_k$, for each j,q and k, is a third order tensor defined by

$$(\mathbf{i}_j\otimes\mathbf{i}_q\otimes\mathbf{i}_k)(\mathbf{u},\mathbf{v},\mathbf{w})=(\mathbf{i}_j\cdot\mathbf{u})(\mathbf{i}_q\cdot\mathbf{v})(\mathbf{i}_k\cdot\mathbf{w}) \tag{A.8.23}$$

for all $\mathbf{u},\mathbf{v},\mathbf{w}$ in $\mathscr{V}$. The *divergence* of the field $\mathbf{A}(\mathbf{x})$ is

$$\operatorname{div}\mathbf{A}=\frac{\partial A_{jk}}{\partial x_k}\mathbf{i}_j \tag{A.8.24}$$

The *Laplacian* of a vector field $\mathbf{g}(\mathbf{x})$ is a vector field $\Delta\mathbf{g}$ defined by

$$\Delta\mathbf{g} = \operatorname{div}(\operatorname{grad}\mathbf{g}) = \frac{\partial^2 g_k}{\partial x_j \partial x_j}\mathbf{i}_k \tag{A.8.25}$$

The Laplacian of a scalar field $\varphi(\mathbf{x})$ is a scalar field $\Delta\varphi$ defined by

$$\Delta\varphi = \operatorname{div}(\operatorname{grad}\varphi) = \frac{\partial^2 \varphi}{\partial x_j \partial x_j} \tag{A.8.26}$$

For a vector field $\mathbf{g}(\mathbf{x})$, the *divergence theorem* is the statement that

$$\oint_{\partial\mathscr{P}} \mathbf{g}(\mathbf{x}) \cdot d\mathbf{s} = \int_{\mathscr{P}} \operatorname{div}\mathbf{g}(\mathbf{x}) dv \tag{A.8.27}$$

where $\mathscr{P}$ denotes a region of $\mathscr{E}$ with closed surface $\partial\mathscr{P}$, dv is the element of volume, and $d\mathbf{s}$ is the outward drawn vector element of area. Equation (A.8.27) is a special case of a more general result, called *Gauss' theorem*,

$$\oint_{\partial\mathscr{P}} \mathbf{A}(\mathbf{x}) \otimes d\mathbf{s} = \int_{\mathscr{P}} \operatorname{grad}\mathbf{A}(\mathbf{x}) dv \tag{A.8.28}$$

where $\mathbf{A}(\mathbf{x})$ is an arbitrary tensor field and the tensor product $\mathbf{A}(\mathbf{x}) \otimes d\mathbf{s}$ is defined in a manner similar to that used in (A.8.23). A special case of (A.8.28) we shall need is

$$\oint_{\partial\mathscr{P}} \mathbf{A}(\mathbf{x}) d\mathbf{s} = \int_{\mathscr{P}} \operatorname{div}\mathbf{A}(\mathbf{x}) dv \tag{A.8.29}$$

for a linear transformation $\mathbf{A}(\mathbf{x})$.

Exercises A.8.3

Take $\mathbf{g}(\mathbf{x}) = \mathbf{A}(\mathbf{x})^T \mathbf{u}(\mathbf{x})$ in (A.8.27) and require $\mathbf{u}$ to be an arbitrary constant vector field. Derive equation (A.8.29) from the result of this substitution. This exercise can be generalized to a derivation of (A.8.28) from (A.8.27).

Exercise A.8.4

There are applications in mechanics for which (A.8.28) is not valid because $\mathbf{A}(\mathbf{x})$ is not differentiable throughout $\mathscr{P}$. In the case where $\mathbf{A}$ is differentiable everywhere in $\partial\mathscr{P}$ except on a surface Σ, where $\mathbf{A}$ suffers a discontinuity, show that

$$\oint_{\partial\mathscr{P}} \mathbf{A}(\mathbf{x}) \otimes d\mathbf{s} = \int_{\mathscr{P}} \operatorname{div}\mathbf{A}(\mathbf{x}) dv - \int_{\Sigma} [\mathbf{A}] \otimes \mathbf{n} d\sigma \tag{A.8.30}$$

where $[\mathbf{A}]$ denotes the *jump* of $\mathbf{A}$ at Σ defined below. The geometric arrangement appropriate for (A.8.30) is shown in the Fig. A.8.1. The region $\mathscr{P}$ consists of the two parts $\mathscr{P}^-$ and $\mathscr{P}^+$ with surfaces $\partial\mathscr{P}^-$ and $\partial\mathscr{P}^+$ respectively. The surface $\partial\mathscr{P}$ is the union of $\partial\mathscr{P}^-$ and $\partial\mathscr{P}^+$. The unit vector $\mathbf{n}$ is normal to the surface Σ and is directed into $\mathscr{P}^+$. The quantity $\mathbf{n}d\sigma$ is the vector element of area of Σ.

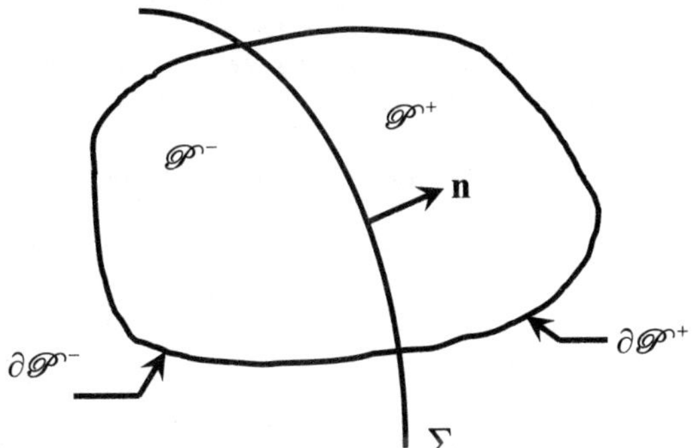

Figure A.8.1

The jump $[\mathbf{A}]$ is defined by

$$[\mathbf{A}] = \mathbf{A}^- - \mathbf{A}^+ \tag{A.8.31}$$

where $\mathbf{A}^+$ is the limiting value of $\mathbf{A}$ as a point on Σ is approached from $\mathscr{P}^+$ while $\mathbf{A}^-$ is the limiting value of $\mathbf{A}$ as the same point is approached from $\mathscr{P}^-$.

It should be evident that (A.8.30) is the three dimensional generalization of (1.1.30). Important special cases of (A.8.30) are

$$\oint_{\partial\mathscr{P}} \mathbf{A}(\mathbf{x})d\mathbf{s} = \int_{\mathscr{P}} \operatorname{div} \mathbf{A}(\mathbf{x})dv - \int_{\Sigma} [\mathbf{A}]\mathbf{n}d\sigma \tag{A.8.32}$$

and

$$\oint_{\partial\mathscr{P}} \mathbf{g}(\mathbf{x})\cdot d\mathbf{s} = \int_{\mathscr{P}} \operatorname{div} \mathbf{g}(\mathbf{x})dv - \int_{\Sigma} [\mathbf{g}]\cdot \mathbf{n}d\sigma \tag{A.8.33}$$

where $\mathbf{A}(\mathbf{x})$ is a linear transformation and $\mathbf{g}(\mathbf{x})$ is a vector.

References

1. BELLMAN, RICHARD, *Introduction to Matrix Analysis*, McGraw-Hill, New York (1960).

Bibliography

1. BOWEN, RAY M., and C.- C. WANG, *Introduction to Vectors and Tensors*, Vol. I and II, Plenum Press, New York (1976).
2. NICKERSON, H. K., D. C. SPENCER and N. E. STEENROD, *Advanced Calculus*, Van Nostrand, Princeton, New Jersey (1959).
3. HALMOS, P. R., *Finite Dimensional Vector Spaces*, 2nd ed., Van Nostrand, Princeton, New Jersey (1958).
4. GREUB, W. H., *Linear Algebra*, 3rd ed., Springer-Verlag, New York (1967).
5. MOSTOW, G. D., J. H. SAMPSON, and J. P. MEYER, *Fundamental Structures of Algebra*, McGraw-Hill, New York (1963).
6. SHEPARD, G. C., *Vector Spaces of Finite Dimension*, Interscience, New York (1966).

Appendix B

Representation Theorems

In Chapter 4 it was necessary to utilize three representation theorems involving certain isotropic functions and isotropic tensors. In this appendix, short proofs will be given of these representations.

In Sections 4.4 and 4.10 it was asserted that if a scalar valued function u of a symmetric linear transformation $\mathbf{B}$ obeys

$$u(\mathbf{B}) = u(\mathbf{QBQ}^T) \tag{B.1}$$

for all orthogonal $\mathbf{Q}$ in $\mathcal{O}(\mathscr{V})^+$, then there exists a function u^* of the fundamental invariants $I_{\mathbf{B}}, II_{\mathbf{B}}$ and $III_{\mathbf{B}}$ such that

$$u(\mathbf{B}) = u^*(I_{\mathbf{B}}, II_{\mathbf{B}}, III_{\mathbf{B}}) \tag{B.2}$$

In order to establish this result, first recall from (A.5.50) through (A.5.52) that $\mathbf{B}$ and $\mathbf{QBQ}^T$ have the same fundamental invariants for all $\mathbf{Q}$ in $\mathcal{O}(\mathscr{V})^+$

Conversely, if a symmetric linear transformation $\mathbf{B}$ and a symmetric linear transformation $\mathbf{A}$ satisfy

$$I_{\mathbf{A}} = I_{\mathbf{B}} \tag{B.3}$$

$$II_{\mathbf{A}} = II_{\mathbf{B}} \tag{B.4}$$

and

$$III_{\mathbf{A}} = III_{\mathbf{B}} \tag{B.5}$$

then (A.5.43) and (A.5.66) show that $\mathbf{A}$ and $\mathbf{B}$ have the same eigenvalues λ_1, λ_2 and λ_3. If $\{\mathbf{e}_1, \mathbf{e}_2, \mathbf{e}_3\}$ is the basis of orthonormal eigenvectors of $\mathbf{B}$ and $\{\mathbf{f}_1, \mathbf{f}_2, \mathbf{f}_3\}$ is the basis of orthonormal eigenvectors of $\mathbf{A}$, then (A.5.65) shows that

$$\mathbf{Be}_j = \lambda_j \mathbf{e}_j \tag{B.6}$$

and

and

$$\mathbf{A}\mathbf{f}_j = \lambda_j \mathbf{f}_j \tag{B.7}$$

where there is no sum on the index j. Because $\{\mathbf{e}_1, \mathbf{e}_2, \mathbf{e}_3\}$ and $\{\mathbf{f}_1, \mathbf{f}_2, \mathbf{f}_3\}$ are orthonormal sets, we can define a $\mathbf{Q}$ in $\mathscr{O}(\mathscr{V})^+$ by

$$\mathbf{Q}\mathbf{e}_j = \mathbf{f}_j \tag{B.8}$$

for $j = 1,2,3$. Therefore, from (B.6), (B.7) and (B.8)

$$\mathbf{Q}\mathbf{B}\mathbf{e}_j = \lambda \mathbf{Q}\mathbf{e}_j = \lambda_j \mathbf{f}_j = \mathbf{A}\mathbf{f}_j = \mathbf{A}\mathbf{Q}\mathbf{e}_j \tag{B.9}$$

for $j = 1,2,3$. Therefore, $\mathbf{A}$ and $\mathbf{B}$ must be related by

$$\mathbf{A} = \mathbf{Q}\mathbf{B}\mathbf{Q}^T \tag{B.10}$$

What we have shown is that (B.3) through (B.5) hold if and only if there exists a $\mathbf{Q}$ in $\mathscr{O}(\mathscr{V})^+$ such that (B.10) holds. Therefore, the only way (B.1) can be satisfied is for (B.2) to hold.

In Section 4.11 it was asserted that if the linear transformation $\mathbf{K}$ obeyed

$$\mathbf{Q}\mathbf{K} = \mathbf{K}\mathbf{Q} \tag{B.11}$$

for all $\mathbf{Q}$ in $\mathscr{O}(\mathscr{V})^+$, then $\mathbf{K}$ must be given by

$$\mathbf{K} = \kappa \mathbf{I} \tag{B.12}$$

Equation (B.12) is clearly sufficient for (B.11) to be satisfied for all $\mathbf{Q}$ in $\mathscr{O}(\mathscr{V})^+$. Conversely, if in (B.11) we consider a smooth family of proper orthogonal linear transformations $\mathbf{Q}(\tau)$ such that $\mathbf{Q}(0) = \mathbf{I}$, it follows by differentiation of (B.11) that

$$\dot{\mathbf{Q}}(0)\mathbf{K} = \mathbf{K}\dot{\mathbf{Q}}(0) \tag{B.13}$$

Because $\mathbf{Q}(\tau)$ is orthogonal for each τ and because $\mathbf{Q}(0) = \mathbf{I}$, it follows that $\dot{\mathbf{Q}}(0)$ is skew symmetric [See equation (4.9.22) of Section 4.9]. Thus, (B.13) implies that

$$\mathbf{Z}\mathbf{K} = \mathbf{K}\mathbf{Z} \tag{B.14}$$

for all skew-symmetric linear transformations $\mathbf{Z}$. Because the matrix of $\mathbf{Z}$ must be of the form

$$[\mathbf{Z}] = \begin{bmatrix} 0 & Z_{12} & Z_{13} \\ -Z_{12} & 0 & Z_{23} \\ -Z_{13} & -Z_{23} & 0 \end{bmatrix}$$

we can easily express (B.14) in components and conclude that

$$K_{11} = K_{22} = K_{33} \tag{B.15}$$

$$K_{12} = K_{21} = 0 \tag{B.16}$$

$$K_{23} = K_{32} = 0 \tag{B.17}$$

and

$$K_{13} = K_{31} = 0 \tag{B.18}$$

If we *define*

$$\kappa = K_{11} \tag{B.19}$$

then (B.15) through (B.19) yield (B.12) as a necessary condition. As observed above, (B.12) is sufficient to insure that (B.11) is satisfied for all $\mathbf{Q}$ in $\mathscr{O}(\mathscr{V})^{+}$.

It also was asserted in Section 4.11 that if $\mathbf{\Lambda}$ is a fourth-order tensor that obeys

$$\mathbf{Q}(\mathbf{\Lambda}[\mathbf{D}])\mathbf{Q}^{T} = \mathbf{\Lambda}[\mathbf{Q}\mathbf{D}\mathbf{Q}^{T}] \tag{B.20}$$

for all $\mathbf{Q}$ in $\mathscr{O}(\mathscr{V})^{+}$ and for all symmetric linear transformations $\mathbf{D}$, then

$$\mathbf{\Lambda}[\mathbf{D}] = \lambda(\operatorname{tr}\mathbf{D}) + 2\mu\mathbf{D} \tag{B.21}$$

It is an elementary calculation to establish that (B.21) is sufficient for (B.20) to hold. In order to prove necessity, we shall first prove that (B.20) forces $\mathbf{D}$ and $\mathbf{\Lambda}[\mathbf{D}]$ to have the same eigenvectors. If $\mathbf{n}_1, \mathbf{n}_2$ and $\mathbf{n}_3$ are eigenvectors of $\mathbf{D}$, then

$$\mathbf{D}\mathbf{n}_1 = d_1\mathbf{n}_1 \tag{B.22}$$

$$\mathbf{D}\mathbf{n}_2 = d_2\mathbf{n}_2 \tag{B.23}$$

and

$$\mathbf{D}\mathbf{n}_3 = d_3\mathbf{n}_3 \tag{B.24}$$

By Theorem (A.5.10), the eigenvectors form an orthonormal basis of $\mathscr{V}$ and the eigenvalues d_1, d_2 and d_3 are real. Since $\mathbf{Q}$ in (B.20) is an arbitrary element of $\mathscr{O}(\mathscr{V})^+$, we select one defined by

$$\mathbf{Q}\mathbf{n}_1 = \mathbf{n}_1 \tag{B.25}$$

$$\mathbf{Q}\mathbf{n}_2 = -\mathbf{n}_2 \tag{B.26}$$

and

$$\mathbf{Q}\mathbf{n}_3 = -\mathbf{n}_3 \tag{B.27}$$

A straight forward calculation based upon (B.22) through (B.27) yields

$$\mathbf{Q}\mathbf{D}\mathbf{Q}^T\mathbf{n}_1 = \mathbf{D}\mathbf{n}_1 \tag{B.28}$$

$$\mathbf{Q}\mathbf{D}\mathbf{Q}^T\mathbf{n}_2 = \mathbf{D}\mathbf{n}_2 \tag{B.29}$$

and

$$\mathbf{Q}\mathbf{D}\mathbf{Q}^T\mathbf{n}_3 = \mathbf{D}\mathbf{n}_3 \tag{B.30}$$

Since $\{\mathbf{n}_1, \mathbf{n}_2, \mathbf{n}_3\}$ is a basis for $\mathscr{V}$, (B.28) through (B.30) yield

$$\mathbf{Q}\mathbf{D}\mathbf{Q}^T = \mathbf{D} \tag{B.31}$$

Utilizing (B.31), we see that (B.20) reduces to

$$\mathbf{Q}(\mathbf{\Lambda}[\mathbf{D}])\mathbf{Q}^T = \mathbf{\Lambda}[\mathbf{D}] \tag{B.32}$$

We need to prove that $\mathbf{\Lambda}[\mathbf{D}]\mathbf{n}_j$ is parallel to $\mathbf{n}_j$ for $j = 1,2,3$. This result follows if we can prove that $\mathbf{\Lambda}[\mathbf{D}]\mathbf{n}_j$ is orthogonal to $\mathbf{n}_k$ and $\mathbf{n}_q$ $j \neq k \neq q$. For example,

$$\begin{aligned}
\mathbf{n}_2 \cdot (\mathbf{\Lambda}[\mathbf{D}]\mathbf{n}_1) &= \mathbf{n}_2 \cdot (\mathbf{Q}\mathbf{\Lambda}[\mathbf{D}]\mathbf{Q}^T\mathbf{n}_1) \\
&= \mathbf{Q}^T\mathbf{n}_2 \cdot (\mathbf{\Lambda}[\mathbf{D}]\mathbf{Q}^T\mathbf{n}_1) \\
&= -\mathbf{n}_2 \cdot (\mathbf{\Lambda}[\mathbf{D}]\mathbf{n}_1)
\end{aligned}$$

where (B.25), (B.26) and (B.32) have been used. This result yields

$$\mathbf{n}_2 \cdot (\mathbf{\Lambda}[\mathbf{D}]\mathbf{n}_1) = 0 \tag{B.33}$$

Similar calculations yield

$$\mathbf{n}_j \cdot (\mathbf{\Lambda}[\mathbf{D}]\mathbf{n}_k) = 0 \tag{B.34}$$

for $j \neq k$, which proves that $\mathbf{D}$ and $\mathbf{\Lambda}[\mathbf{D}]$ have the same eigenvectors.

If we write $\mathbf{N}_j = \mathbf{n}_j \otimes \mathbf{n}_j$, for $j = 1,2,3$, then the result just established shows that

$$\mathbf{\Lambda}[\mathbf{N}_j] = \Lambda_{kj}\mathbf{N}_k \tag{B.35}$$

where the Λ_{kj} are the components of the forth order tensor $\mathbf{\Lambda}$ with respect to the basis of eigenvectors. An elementary manipulation of (B.35) shows that

$$\Lambda_{kj} = \text{tr}(\mathbf{N}_k(\mathbf{\Lambda}[\mathbf{N}_j])) \tag{B.36}$$

With (B.20), we can rewrite (B.36) as

$$\Lambda_{kj} = \text{tr}(\mathbf{Q}\mathbf{N}_k\mathbf{Q}^T(\mathbf{\Lambda}[\mathbf{Q}\mathbf{N}_j\mathbf{Q}^T])) \tag{B.37}$$

where $\mathbf{Q}$ is an arbitrary element of $\mathcal{O}(\mathscr{V})^+$. If we take $\mathbf{Q}$ to be defined by $\mathbf{Q}\mathbf{n}_1 = \mathbf{n}_2$, $\mathbf{Q}\mathbf{n}_2 = \mathbf{n}_3$ and $\mathbf{Q}\mathbf{n}_3 = \mathbf{n}_1$, then an elementary calculation shows that $\mathbf{Q}\mathbf{N}_1\mathbf{Q}^T = \mathbf{N}_2$, $\mathbf{Q}\mathbf{N}_2\mathbf{Q}^T = \mathbf{N}_3$ and $\mathbf{Q}\mathbf{N}_3\mathbf{Q}^T = \mathbf{N}_1$. When these facts are used in (B.37) one finds that

$$\Lambda_{11} = \Lambda_{22} = \Lambda_{33} \tag{B.38}$$

and

$$\Lambda_{12} = \Lambda_{21} = \Lambda_{23} = \Lambda_{32} = \Lambda_{13} = \Lambda_{31} \tag{B.39}$$

Without loss of generality, we can write

$$\lambda = \Lambda_{12} \tag{B.40}$$

and

$$2\mu = \Lambda_{11} - \Lambda_{12} = \Lambda_{11} - \lambda \tag{B.41}$$

Thus, (B.35) reduces to

$$\mathbf{\Lambda}[\mathbf{N}_j] = \lambda\mathbf{I} + 2\mu\mathbf{N}_j \tag{B.42}$$

after (B.38) through (B.41) are used. By the spectral representation of $\mathbf{D}$, [see (A.5.67)] we can write

$$\mathbf{D} = d_1\mathbf{N}_1 + d_2\mathbf{N}_2 + d_3\mathbf{N}_3 \tag{B.43}$$

Therefore,

$$\mathbf{\Lambda}[\mathbf{D}] = d_1\mathbf{\Lambda}[\mathbf{N}_1] + d_2\mathbf{\Lambda}[\mathbf{N}_2] + d_3\mathbf{\Lambda}[\mathbf{N}_3] \tag{B.44}$$

Given (B.42), (B.44) can be shown to reduce to the desired result (B.21).

Bibliography

1. TRUESDELL, C., and W. NOLL, *The Non-Linear Field Theories of Mechanics*, Handbuch der Physik, Vol. III/3 (ed. S. Flugge), Springer-Verlag (1965).
2. JEFFREYS, H., *Cartesian Tensors*, Cambridge University Press, Cambridge (1931).
3. THOMAS, T. Y., *Concepts from Tensor Analysis and Differential Geometry*, Academic Press, New York (1961).

Index

A CATALOG OF SELECTED

DOVER BOOKS

IN SCIENCE AND MATHEMATICS

Engineering

DE RE METALLICA, Georgius Agricola. The famous Hoover translation of greatest treatise on technological chemistry, engineering, geology, mining of early modern times (1556). All 289 original woodcuts. 638pp. 6¾ x 11. 0-486-60006-8

FUNDAMENTALS OF ASTRODYNAMICS, Roger Bate et al. Modern approach developed by U.S. Air Force Academy. Designed as a first course. Problems, exercises. Numerous illustrations. 455pp. 5⅜ x 8½. 0-486-60061-0

DYNAMICS OF FLUIDS IN POROUS MEDIA, Jacob Bear. For advanced students of ground water hydrology, soil mechanics and physics, drainage and irrigation engineering and more. 335 illustrations. Exercises, with answers. 784pp. 6⅛ x 9¼. 0-486-65675-6

THEORY OF VISCOELASTICITY (SECOND EDITION), Richard M. Christensen. Complete consistent description of the linear theory of the viscoelastic behavior of materials. Problem-solving techniques discussed. 1982 edition. 29 figures. xiv+364pp. 6⅛ x 9¼. 0-486-42880-X

MECHANICS, J. P. Den Hartog. A classic introductory text or refresher. Hundreds of applications and design problems illuminate fundamentals of trusses, loaded beams and cables, etc. 334 answered problems. 462pp. 5⅜ x 8½. 0-486-60754-2

MECHANICAL VIBRATIONS, J. P. Den Hartog. Classic textbook offers lucid explanations and illustrative models, applying theories of vibrations to a variety of practical industrial engineering problems. Numerous figures. 233 problems, solutions. Appendix. Index. Preface. 436pp. 5⅜ x 8½. 0-486-64785-4

STRENGTH OF MATERIALS, J. P. Den Hartog. Full, clear treatment of basic material (tension, torsion, bending, etc.) plus advanced material on engineering methods, applications. 350 answered problems. 323pp. 5⅜ x 8½. 0-486-60755-0

A HISTORY OF MECHANICS, René Dugas. Monumental study of mechanical principles from antiquity to quantum mechanics. Contributions of ancient Greeks, Galileo, Leonardo, Kepler, Lagrange, many others. 671pp. 5⅜ x 8½. 0-486-65632-2

STABILITY THEORY AND ITS APPLICATIONS TO STRUCTURAL MECHANICS, Clive L. Dym. Self-contained text focuses on Koiter postbuckling analyses, with mathematical notions of stability of motion. Basing minimum energy principles for static stability upon dynamic concepts of stability of motion, it develops asymptotic buckling and postbuckling analyses from potential energy considerations, with applications to columns, plates, and arches. 1974 ed. 208pp. 5⅜ x 8½. 0-486-42541-X

BASIC ELECTRICITY, U.S. Bureau of Naval Personnel. Originally a training course; best nontechnical coverage. Topics include batteries, circuits, conductors, AC and DC, inductance and capacitance, generators, motors, transformers, amplifiers, etc. Many questions with answers. 349 illustrations. 1969 edition. 448pp. 6½ x 9¼. 0-486-20973-3

ROCKETS, Robert Goddard. Two of the most significant publications in the history of rocketry and jet propulsion: "A Method of Reaching Extreme Altitudes" (1919) and "Liquid Propellant Rocket Development" (1936). 128pp. $5^3/_8$ x $8^1/_2$. 0-486-42537-1

STATISTICAL MECHANICS: PRINCIPLES AND APPLICATIONS, Terrell L. Hill. Standard text covers fundamentals of statistical mechanics, applications to fluctuation theory, imperfect gases, distribution functions, more. 448pp. $5^3/_8$ x $8^1/_2$. 0-486-65390-0

ENGINEERING AND TECHNOLOGY 1650–1750: ILLUSTRATIONS AND TEXTS FROM ORIGINAL SOURCES, Martin Jensen. Highly readable text with more than 200 contemporary drawings and detailed engravings of engineering projects dealing with surveying, leveling, materials, hand tools, lifting equipment, transport and erection, piling, bailing, water supply, hydraulic engineering, and more. Among the specific projects outlined-transporting a 50-ton stone to the Louvre, erecting an obelisk, building timber locks, and dredging canals. 207pp. $8^3/_8$ x $11^1/_4$. 0-486-42232-1

THE VARIATIONAL PRINCIPLES OF MECHANICS, Cornelius Lanczos. Graduate level coverage of calculus of variations, equations of motion, relativistic mechanics, more. First inexpensive paperbound edition of classic treatise. Index. Bibliography. 418pp. $5^3/_8$ x $8^1/_2$. 0-486-65067-7

PROTECTION OF ELECTRONIC CIRCUITS FROM OVERVOLTAGES, Ronald B. Standler. Five-part treatment presents practical rules and strategies for circuits designed to protect electronic systems from damage by transient overvoltages. 1989 ed. xxiv+434pp. $6^1/_8$ x $9^1/_4$. 0-486-42552-5

ROTARY WING AERODYNAMICS, W. Z. Stepniewski. Clear, concise text covers aerodynamic phenomena of the rotor and offers guidelines for helicopter performance evaluation. Originally prepared for NASA. 537 figures. 640pp. $6^1/_8$ x $9^1/_4$. 0-486-64647-5

INTRODUCTION TO SPACE DYNAMICS, William Tyrrell Thomson. Comprehensive, classic introduction to space-flight engineering for advanced undergraduate and graduate students. Includes vector algebra, kinematics, transformation of coordinates. Bibliography. Index. 352pp. $5^3/_8$ x $8^1/_2$. 0-486-65113-4

HISTORY OF STRENGTH OF MATERIALS, Stephen P. Timoshenko. Excellent historical survey of the strength of materials with many references to the theories of elasticity and structure. 245 figures. 452pp. $5^3/_8$ x $8^1/_2$. 0-486-61187-6

ANALYTICAL FRACTURE MECHANICS, David J. Unger. Self-contained text supplements standard fracture mechanics texts by focusing on analytical methods for determining crack-tip stress and strain fields. 336pp. $6^1/_8$ x $9^1/_4$. 0-486-41737-9

STATISTICAL MECHANICS OF ELASTICITY, J. H. Weiner. Advanced, self-contained treatment illustrates general principles and elastic behavior of solids. Part 1, based on classical mechanics, studies thermoelastic behavior of crystalline and polymeric solids. Part 2, based on quantum mechanics, focuses on interatomic force laws, behavior of solids, and thermally activated processes. For students of physics and chemistry and for polymer physicists. 1983 ed. 96 figures. 496pp. $5^3/_8$ x $8^1/_2$. 0-486-42260-7

Mathematics

FUNCTIONAL ANALYSIS (Second Corrected Edition), George Bachman and Lawrence Narici. Excellent treatment of subject geared toward students with background in linear algebra, advanced calculus, physics and engineering. Text covers introduction to inner-product spaces, normed, metric spaces, and topological spaces; complete orthonormal sets, the Hahn-Banach Theorem and its consequences, and many other related subjects. 1966 ed. 544pp. $6^1/_8$ x $9^1/_4$. 0-486-40251-7

DIFFERENTIAL MANIFOLDS, Antoni A. Kosinski. Introductory text for advanced undergraduates and graduate students presents systematic study of the topological structure of smooth manifolds, starting with elements of theory and concluding with method of surgery. 1993 edition. 288pp. $5^3/_8$ x $8^1/_2$. 0-486-46244-7

VECTOR AND TENSOR ANALYSIS WITH APPLICATIONS, A. I. Borisenko and I. E. Tarapov. Concise introduction. Worked-out problems, solutions, exercises. 257pp. $5^5/_8$ x $8^1/_4$. 0-486-63833-2

AN INTRODUCTION TO ORDINARY DIFFERENTIAL EQUATIONS, Earl A. Coddington. A thorough and systematic first course in elementary differential equations for undergraduates in mathematics and science, with many exercises and problems (with answers). Index. 304pp. $5^3/_8$ x $8^1/_2$. 0-486-65942-9

FOURIER SERIES AND ORTHOGONAL FUNCTIONS, Harry F. Davis. An incisive text combining theory and practical example to introduce Fourier series, orthogonal functions and applications of the Fourier method to boundary-value problems. 570 exercises. Answers and notes. 416pp. $5^3/_8$ x $8^1/_2$. 0-486-65973-9

COMPUTABILITY AND UNSOLVABILITY, Martin Davis. Classic graduate-level introduction to theory of computability, usually referred to as theory of recurrent functions. New preface and appendix. 288pp. $5^3/_8$ x $8^1/_2$. 0-486-61471-9

AN INTRODUCTION TO MATHEMATICAL ANALYSIS, Robert A. Rankin. Dealing chiefly with functions of a single real variable, this text by a distinguished educator introduces limits, continuity, differentiability, integration, convergence of infinite series, double series, and infinite products. 1963 edition. 624pp. $5^3/_8$ x $8^1/_2$. 0-486-46251-X

METHODS OF NUMERICAL INTEGRATION (SECOND EDITION), Philip J. Davis and Philip Rabinowitz. Requiring only a background in calculus, this text covers approximate integration over finite and infinite intervals, error analysis, approximate integration in two or more dimensions, and automatic integration. 1984 edition. 624pp. $5^3/_8$ x $8^1/_2$. 0-486-45339-1

INTRODUCTION TO LINEAR ALGEBRA AND DIFFERENTIAL EQUATIONS, John W. Dettman. Excellent text covers complex numbers, determinants, orthonormal bases, Laplace transforms, much more. Exercises with solutions. Undergraduate level. 416pp. $5^3/_8$ x $8^1/_2$. 0-486-65191-6

RIEMANN'S ZETA FUNCTION, H. M. Edwards. Superb, high-level study of landmark 1859 publication entitled "On the Number of Primes Less Than a Given Magnitude" traces developments in mathematical theory that it inspired. xiv+315pp. $5^3/_8$ x $8^1/_2$. 0-486-41740-9

CALCULUS OF VARIATIONS WITH APPLICATIONS, George M. Ewing. Applications-oriented introduction to variational theory develops insight and promotes understanding of specialized books, research papers. Suitable for advanced undergraduate/graduate students as primary, supplementary text. 352pp. $5^{3}/_{8}$ x $8^{1}/_{2}$.
0-486-64856-7

MATHEMATICIAN'S DELIGHT, W. W. Sawyer. "Recommended with confidence" by *The Times Literary Supplement,* this lively survey was written by a renowned teacher. It starts with arithmetic and algebra, gradually proceeding to trigonometry and calculus. 1943 edition. 240pp. $5^{3}/_{8}$ x $8^{1}/_{2}$. 0-486-46240-4

ADVANCED EUCLIDEAN GEOMETRY, Roger A. Johnson. This classic text explores the geometry of the triangle and the circle, concentrating on extensions of Euclidean theory, and examining in detail many relatively recent theorems. 1929 edition. 336pp. $5^{3}/_{8}$ x $8^{1}/_{2}$. 0-486-46237-4

COUNTEREXAMPLES IN ANALYSIS, Bernard R. Gelbaum and John M. H. Olmsted. These counterexamples deal mostly with the part of analysis known as "real variables." The first half covers the real number system, and the second half encompasses higher dimensions. 1962 edition. xxiv+198pp. $5^{3}/_{8}$ x $8^{1}/_{2}$. 0-486-42875-3

CATASTROPHE THEORY FOR SCIENTISTS AND ENGINEERS, Robert Gilmore. Advanced-level treatment describes mathematics of theory grounded in the work of Poincaré, R. Thom, other mathematicians. Also important applications to problems in mathematics, physics, chemistry and engineering. 1981 edition. References. 28 tables. 397 black-and-white illustrations. xvii + 666pp. $6^{1}/_{8}$ x $9^{1}/_{4}$. 0-486-67539-4

COMPLEX VARIABLES: Second Edition, Robert B. Ash and W. P. Novinger. Suitable for advanced undergraduates and graduate students, this newly revised treatment covers Cauchy theorem and its applications, analytic functions, and the prime number theorem. Numerous problems and solutions. 2004 edition. 224pp. $6^{1}/_{2}$ x $9^{1}/_{4}$. 0-486-46250-1

NUMERICAL METHODS FOR SCIENTISTS AND ENGINEERS, Richard Hamming. Classic text stresses frequency approach in coverage of algorithms, polynomial approximation, Fourier approximation, exponential approximation, other topics. Revised and enlarged 2nd edition. 721pp. $5^{3}/_{8}$ x $8^{1}/_{2}$. 0-486-65241-6

INTRODUCTION TO NUMERICAL ANALYSIS (2nd Edition), F. B. Hildebrand. Classic, fundamental treatment covers computation, approximation, interpolation, numerical differentiation and integration, other topics. 150 new problems. 669pp. $5^{3}/_{8}$ x $8^{1}/_{2}$.
0-486-65363-3

MARKOV PROCESSES AND POTENTIAL THEORY, Robert M. Blumental and Ronald K. Getoor. This graduate-level text explores the relationship between Markov processes and potential theory in terms of excessive functions, multiplicative functionals and subprocesses, additive functionals and their potentials, and dual processes. 1968 edition. 320pp. $5^{3}/_{8}$ x $8^{1}/_{2}$. 0-486-46263-3

ABSTRACT SETS AND FINITE ORDINALS: An Introduction to the Study of Set Theory, G. B. Keene. This text unites logical and philosophical aspects of set theory in a manner intelligible to mathematicians without training in formal logic and to logicians without a mathematical background. 1961 edition. 112pp. $5^{3}/_{8}$ x $8^{1}/_{2}$. 0-486-46249-8

Physics

OPTICAL RESONANCE AND TWO-LEVEL ATOMS, L. Allen and J. H. Eberly. Clear, comprehensive introduction to basic principles behind all quantum optical resonance phenomena. 53 illustrations. Preface. Index. 256pp. $5^3/_8$ x $8^1/_2$. 0-486-65533-4

QUANTUM THEORY, David Bohm. This advanced undergraduate-level text presents the quantum theory in terms of qualitative and imaginative concepts, followed by specific applications worked out in mathematical detail. Preface. Index. 655pp. $5^3/_8$ x $8^1/_2$. 0-486-65969-0

ATOMIC PHYSICS (8th EDITION), Max Born. Nobel laureate's lucid treatment of kinetic theory of gases, elementary particles, nuclear atom, wave-corpuscles, atomic structure and spectral lines, much more. Over 40 appendices, bibliography. 495pp. $5^3/_8$ x $8^1/_2$. 0-486-65984-4

A SOPHISTICATE'S PRIMER OF RELATIVITY, P. W. Bridgman. Geared toward readers already acquainted with special relativity, this book transcends the view of theory as a working tool to answer natural questions: What is a frame of reference? What is a "law of nature"? What is the role of the "observer"? Extensive treatment, written in terms accessible to those without a scientific background. 1983 ed. xlviii+172pp. $5^3/_8$ x $8^1/_2$. 0-486-42549-5

AN INTRODUCTION TO HAMILTONIAN OPTICS, H. A. Buchdahl. Detailed account of the Hamiltonian treatment of aberration theory in geometrical optics. Many classes of optical systems defined in terms of the symmetries they possess. Problems with detailed solutions. 1970 edition. xv + 360pp. $5^3/_8$ x $8^1/_2$. 0-486-67597-1

PRIMER OF QUANTUM MECHANICS, Marvin Chester. Introductory text examines the classical quantum bead on a track: its state and representations; operator eigenvalues; harmonic oscillator and bound bead in a symmetric force field; and bead in a spherical shell. Other topics include spin, matrices, and the structure of quantum mechanics; the simplest atom; indistinguishable particles; and stationary-state perturbation theory. 1992 ed. xiv+314pp. $6^1/_8$ x $9^1/_4$. 0-486-42878-8

LECTURES ON QUANTUM MECHANICS, Paul A. M. Dirac. Four concise, brilliant lectures on mathematical methods in quantum mechanics from Nobel Prize-winning quantum pioneer build on idea of visualizing quantum theory through the use of classical mechanics. 96pp. $5^3/_8$ x $8^1/_2$. 0-486-41713-1

THIRTY YEARS THAT SHOOK PHYSICS: THE STORY OF QUANTUM THEORY, George Gamow. Lucid, accessible introduction to influential theory of energy and matter. Careful explanations of Dirac's anti-particles, Bohr's model of the atom, much more. 12 plates. Numerous drawings. 240pp. $5^3/_8$ x $8^1/_2$. 0-486-24895-X

ELECTRONIC STRUCTURE AND THE PROPERTIES OF SOLIDS: THE PHYSICS OF THE CHEMICAL BOND, Walter A. Harrison. Innovative text offers basic understanding of the electronic structure of covalent and ionic solids, simple metals, transition metals and their compounds. Problems. 1980 edition. 582pp. $6^1/_8$ x $9^1/_4$. 0-486-66021-4

HYDRODYNAMIC AND HYDROMAGNETIC STABILITY, S. Chandrasekhar. Lucid examination of the Rayleigh-Benard problem; clear coverage of the theory of instabilities causing convection. 704pp. $5^5/_8$ x $8^1/_4$. 0-486-64071-X

INVESTIGATIONS ON THE THEORY OF THE BROWNIAN MOVEMENT, Albert Einstein. Five papers (1905–8) investigating dynamics of Brownian motion and evolving elementary theory. Notes by R. Fürth. 122pp. $5^3/_8$ x $8^1/_2$. 0-486-60304-0

THE PHYSICS OF WAVES, William C. Elmore and Mark A. Heald. Unique overview of classical wave theory. Acoustics, optics, electromagnetic radiation, more. Ideal as classroom text or for self-study. Problems. 477pp. $5^3/_8$ x $8^1/_2$. 0-486-64926-1

GRAVITY, George Gamow. Distinguished physicist and teacher takes reader-friendly look at three scientists whose work unlocked many of the mysteries behind the laws of physics: Galileo, Newton, and Einstein. Most of the book focuses on Newton's ideas, with a concluding chapter on post-Einsteinian speculations concerning the relationship between gravity and other physical phenomena. 160pp. $5^3/_8$ x $8^1/_2$. 0-486-42563-0

PHYSICAL PRINCIPLES OF THE QUANTUM THEORY, Werner Heisenberg. Nobel Laureate discusses quantum theory, uncertainty, wave mechanics, work of Dirac, Schroedinger, Compton, Wilson, Einstein, etc. 184pp. $5^3/_8$ x $8^1/_2$. 0-486-60113-7

ATOMIC SPECTRA AND ATOMIC STRUCTURE, Gerhard Herzberg. One of best introductions; especially for specialist in other fields. Treatment is physical rather than mathematical. 80 illustrations. 257pp. $5^3/_8$ x $8^1/_2$. 0-486-60115-3

AN INTRODUCTION TO STATISTICAL THERMODYNAMICS, Terrell L. Hill. Excellent basic text offers wide-ranging coverage of quantum statistical mechanics, systems of interacting molecules, quantum statistics, more. 523pp. $5^3/_8$ x $8^1/_2$. 0-486-65242-4

THEORETICAL PHYSICS, Georg Joos, with Ira M. Freeman. Classic overview covers essential math, mechanics, electromagnetic theory, thermodynamics, quantum mechanics, nuclear physics, other topics. First paperback edition. xxiii + 885pp. $5^3/_8$ x $8^1/_2$. 0-486-65227-0

PROBLEMS AND SOLUTIONS IN QUANTUM CHEMISTRY AND PHYSICS, Charles S. Johnson, Jr. and Lee G. Pedersen. Unusually varied problems, detailed solutions in coverage of quantum mechanics, wave mechanics, angular momentum, molecular spectroscopy, more. 280 problems plus 139 supplementary exercises. 430pp. $6^1/_2$ x $9^1/_4$. 0-486-65236-X

THEORETICAL SOLID STATE PHYSICS, Vol. 1: Perfect Lattices in Equilibrium; Vol. II: Non-Equilibrium and Disorder, William Jones and Norman H. March. Monumental reference work covers fundamental theory of equilibrium properties of perfect crystalline solids, non-equilibrium properties, defects and disordered systems. Appendices. Problems. Preface. Diagrams. Index. Bibliography. Total of 1,301pp. $5^3/_8$ x $8^1/_2$. Two volumes. Vol. I: 0-486-65015-4 Vol. II: 0-486-65016-2

WHAT IS RELATIVITY? L. D. Landau and G. B. Rumer. Written by a Nobel Prize physicist and his distinguished colleague, this compelling book explains the special theory of relativity to readers with no scientific background, using such familiar objects as trains, rulers, and clocks. 1960 ed. vi+72pp. $5^3/_8$ x $8^1/_2$. 0-486-42806-0